JUN 1 1 2018

EAST MEADOW PUBLIC LIBRARY
3 1299 00980 2738

P9-CRA-961

East Meadow Public Library
1886 Front Street, East Meadow, NY 11554
(516) 794-2570
www.eastmeadow.info

YOUNG WASHINGTON

ALSO BY PETER STARK

Astoria

At the Mercy of the River

Last Breath

The Last Empty Places

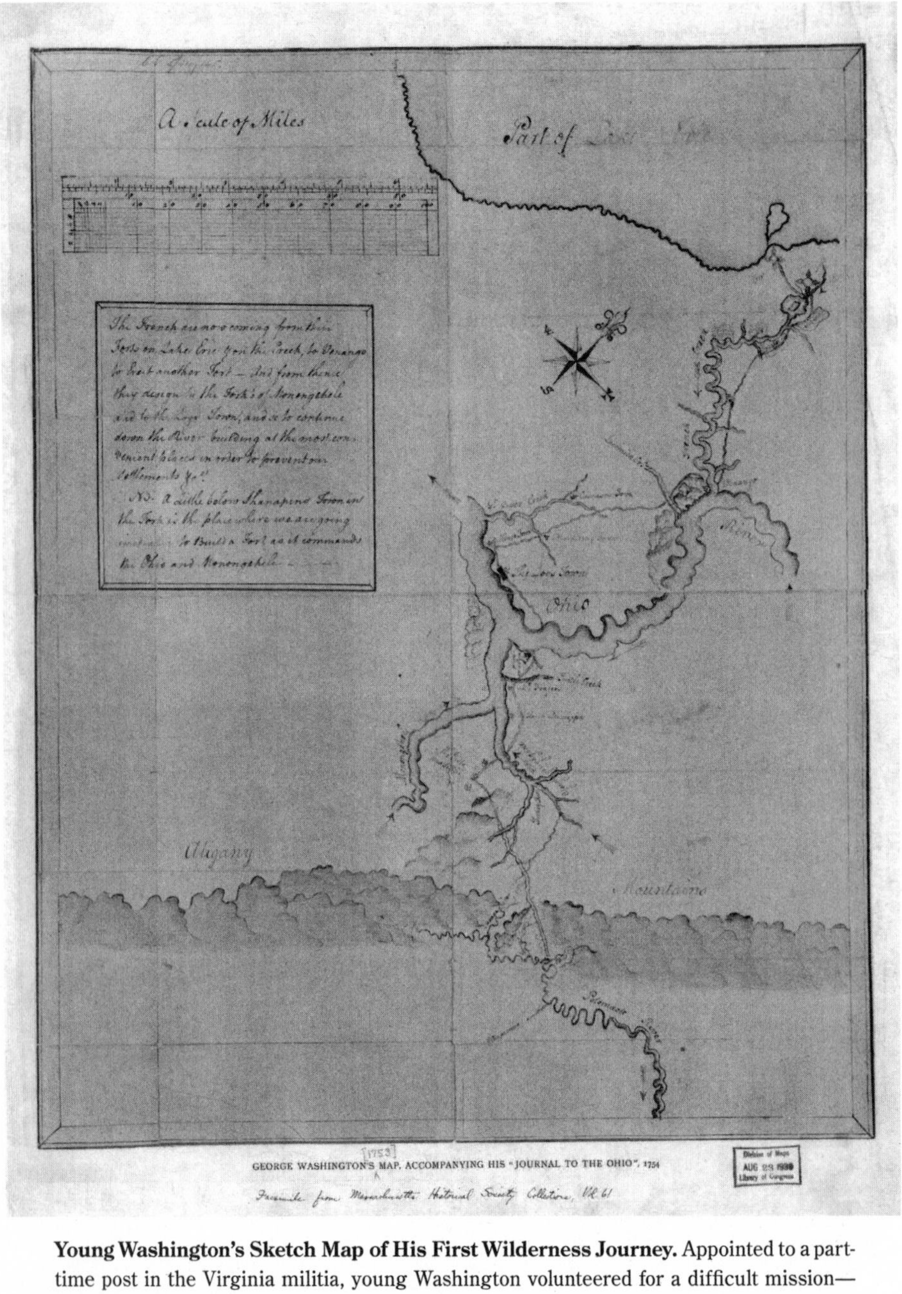

GEORGE WASHINGTON'S MAP, ACCOMPANYING HIS "JOURNAL TO THE OHIO", 1754

Division of Maps
AUG 23 1930
Library of Congress

Young Washington's Sketch Map of His First Wilderness Journey. Appointed to a part-time post in the Virginia militia, young Washington volunteered for a difficult mission—deliver an urgent message from Virginia's British governor to the French commandant deep in the Ohio wilderness: *Stay out!*

YOUNG WASHINGTON

HOW WILDERNESS AND WAR FORGED AMERICA'S FOUNDING FATHER

PETER STARK

An Imprint of HarperCollins*Publishers*

YOUNG WASHINGTON. Copyright © 2018 by Peter Stark. All rights reserved. Printed in the United States of America. No part of this book may be used or reproduced in any manner whatsoever without written permission except in the case of brief quotations embodied in critical articles and reviews. For information, address HarperCollins Publishers, 195 Broadway, New York, NY 10007.

HarperCollins books may be purchased for educational, business, or sales promotional use. For information, please e-mail the Special Markets Department at SPsales@harpercollins.com.

FIRST EDITION

Designed by Suet Yee Chong
Maps designed by Kevin W. McCann, Cartographics, LLC

Library of Congress Cataloging-in-Publication Data

Names: Stark, Peter, 1954– author.
Title: Young Washington: How wilderness and war forged America's founding father / Peter Stark.
Description: New York: Ecco, 2018. | Includes bibliographical references and index.
Identifiers: LCCN 2017052000 (print) | LCCN 2017052689 (ebook) | ISBN 9780062416087 () | ISBN 9780062416063 (hardcover) | ISBN 9780062416070 (paperback) | ISBN 9780062845993 (large print) | ISBN 9780062847454 (digital audio)
Subjects: LCSH: Washington, George, 1732–1799—Childhood and youth. | United States—History—French and Indian War, 1754–1763—Biography. | Presidents—United States—Biography.
Classification: LCC E312.2 (ebook) | LCC E312.2 .S79 2018 (print) | DDC 973.4/1092 [B]—dc23
LC record available at https://lccn.loc.gov/2017052000

---

18 19 20 21 22 LSC 10 9 8 7 6 5 4 3 2 1

For Amy

CONTENTS

MAPS

CAST OF CHARACTERS

General Edward Braddock—Member of the Coldstream Guards who rose up through the ranks of the British Royal Army to head the 1755 British expedition against the French and Indians in the Ohio wilderness.

Claude-Pierre Pécaudy de Contrecoeur—French colonial commanding officer of Fort Duquesne at the Forks of the Ohio.

Martha Dandridge Custis—Wife, then widow, of wealthy Virginia landowner Daniel Parke Custis and mother of their two young children. Upon his death, she inherited vast holdings of her husband's land.

Captain John Dagworthy—Officer from Maryland who held a commission from the regular British Royal Army, thus technically ranking him above Colonel George Washington, who had a colonial officer's commission from the governor of Virginia.

Governor Robert Dinwiddie—Scottish merchant, colonial Virginia's acting governor, and young Washington's commanding officer.

Colonel William Fairfax—Cousin to Lord Fairfax, Colonel Fairfax was the patriarch of the part of the family that had settled in Virginia and built Belvoir Manor not far from Lawrence Washington's plantation known as Mount Vernon.

George William Fairfax—Son of Colonel Fairfax and nephew of Lord Fairfax, George William was a polished young "gentleman" who had attended school in England, returned to Virginia, and took young George Washington on a surveying party into frontier lands. Shortly after his return from that surveying expedition, George William married Sally Cary, with whom young George Washington became enamored.

Sarah "Sally" Cary Fairfax—One of four daughters of the aristocratic Cary family, Sally married the young gentleman George William Fairfax.

Thomas Fairfax, Sixth Lord Fairfax of Cameron—The inheritor or "proprietor" of five million acres of land in the Virginia region known as the Northern Neck, Thomas, Sixth Lord Fairfax of Cameron, left his home of Leeds Castle in Kent, England, and moved to a hunting lodge in Virginia's Shenandoah Valley.

General John Forbes—After earlier attempts to drive the French from the Ohio wilderness had failed, Forbes was the commanding officer sent by British secretary of state William Pitt to lead an army and accomplish the task.

Lieutenant Colonel Thomas Gage—Second son of a British viscount and officer in the Royal Army with previous experience in European wars, Gage commanded the advance party of General Braddock's troops.

Christopher Gist—Frontiersman, explorer, settler, and wilderness guide who accompanied Washington on his forays into the Ohio wilderness.

Half King—Leader of the Mingoes, or Iroquois living in the Ohio wilderness. Known to the British as "Half King," because he had to clear his decisions with Iroquois superiors, his Indian name was Tanaghrisson.

Ensign Joseph Coulon de Villiers de Jumonville—French colonial officer leading what was said to be a diplomatic party from Fort Duquesne through the Ohio wilderness to approach the Virginia troops who were under Washington's command.

Marquis Duquesne—Governor-general of New France, overseeing French territory in North America, including Canada, Acadia, and Louisiana.

Captain Robert Orme—Officer in the British Royal Army and Braddock's principal aide-de-camp. He kept a journal of Braddock's march into the Ohio wilderness.

Christian Frederick Post—Prussian cabinet maker and devotee of the

Moravian Church who came to the frontier of British North America to bring Christianity to the Indians. Married to an Indian woman, he was fluent in Indian languages.

James Smith—Young frontier settler in Pennsylvania who joined the woodsmen chopping a road for Braddock's army and was captured by the Indians.

Sir John St. Clair—Scottish gentry who became a career officer in the British Royal Army, served in European campaigns, and as assistant quartermaster general oversaw the difficult logistics of the Braddock campaign.

Adam Stephen—Scottish doctor who had served aboard a Royal Navy ship, emigrated to Virginia, and served in the Virginia Regiment alongside Washington.

Captain Louis Coulon de Villiers—French colonial officer and older brother of Ensign Jumonville. Villiers led the French and Indian attack against Washington and his troops at Fort Necessity.

Augustine "Gus" Washington—Father of George Washington, who died when George was eleven, leaving his wife, Mary Ball Washington, a widow with several young children.

John Augustine "Jack" Washington—George Washington's younger brother (like George, a son of Gus and his second wife, Mary Ball Washington) who oversaw some of his properties and with whom George corresponded.

Lawrence Washington—Older half-brother of George (Lawrence was eldest son of Gus and his first wife, Jane Butler) and a model and hero to his younger half-brother. Lawrence married one of the Fairfax daughters.

Mary Ball Washington—Mother of George Washington and Gus's second wife.

PART ONE

CHAPTER ONE

THE TIRED, SODDEN TRAVELERS RODE INTO THE OLD Indian town of Venango at the mouth of Rivière Le Boeuf. Twenty-one-year-old George Washington had never traveled this deep into the wilds. He had now ridden nearly five hundred miles from the elegant plantations of coastal Virginia, over the crest of the Appalachians, far into the Ohio Valley wilderness, a great forest in America's interior approximately the size of Spain. Smoke plumed from bark-covered longhouses and hung low in the damp, rainy air. It was early December 1753. A deep chill wrapped the forest clearing. Children and scrawny dogs ran on muddy paths. The arrival of the little caravan—the tall young stranger who carried himself with a proud bearing, the shaggy frontiersman guide, the French-speaking interpreter, the hired men handling the packhorses, the three Indian chiefs—brought out onlookers as word spread among the longhouses. The youthful white planter from the distant Atlantic coast of British America self-consciously rode through the native village, anxious to carry out this crucial step in his first mission for Virginia's Governor Dinwiddie—to deliver a message to the French commandant somewhere in the Ohio wilderness.

A stouter log cabin squatted amid the bark longhouses, a geometric cube intruding on an organic world. The French flag draped limply from a pole. The cabin had served as the trading post of British subject John Fraser until the French had recently evicted him. This, too, informed Washington's mission—to learn why the French had so boldly removed

British colonist and trader Fraser from the Indian village of Venango and occupied his post.

Washington headed directly for the former trading post, now under the French flag, accompanied by his frontiersman guide, Christopher Gist, and his interpreter for French, Jacob van Braam. Three French military officers received him with formal introductions and polite respect. Captain Phillipe Thomas Joncaire stepped forward as the ranking officer. Son of a French trader and Seneca mother, forty-six-year-old Captain Joncaire had lived in the interior wilds since the age of eleven. Known to the Indians as Nitachinon, he had great influence among the tribes of the region, moving easily between the two worlds and working diplomatically to ally them to the French. Amid the fluid French graciousness, Washington abruptly asked who was the commander of French forces in the Ohio Valley.

Captain Joncaire politely replied that while he had command of the Ohio, another officer, Saint-Pierre, ranked higher. Washington should deliver to Saint-Pierre the letter from the governor of Virginia that he carried in his satchel. He could be found at the headwaters of Rivière Le Boeuf—River of the Buffalo—where the French had recently erected a fortification.

How far away?

Several days' ride, he was told.

The veteran Captain Joncaire had more or less brushed Washington off. Anticipating a formal encounter after his long journey into the Ohio wilderness, Washington may have felt belittled by Joncaire's dismissal. He surely felt intimidated by the wilderness veteran and military officer twenty-five years his senior, and perhaps angered that he wasn't being taken seriously enough. George Washington at twenty-one was a very different Washington from the one we know and hold sacred, different from the stately commander of the Continental Army, the selfless first president, the unblemished father of our country gazing off into posterity. This is not the Washington possessed of nearly superhuman virtue, who, given the chance to consolidate power and rule indefinitely over the just-born nation, willingly stepped down and returned to a quiet life on his Virginia plantation. Rather, this is the young Washington. But not the Washington of the cherry-tree bedtime story. This young Washington is ambitious, temperamental, vain, thin-skinned, petulant, awkward, demanding, stubborn, annoying, hasty, passionate. This Washington has not yet learned to cultivate his image or contain his emotions. Here, instead, is a raw young man struggling toward maturity and in love with a close friend's wife. This is the Washington of emotional neediness, personal ambition, and mistakes—many mistakes.

Everything about Washington's life played out on a grander scale than most people's, including his maturing during his younger years. Most young people make mistakes. Many learn from them. The difference with Washington is that the mistakes he made occurred in an arena that quickly expanded from local, to regional, and finally to global, with far-reaching historical consequences. While the older, mature Washington would lead the Continental Army during the Revolutionary War, this young, raw Washington personally bears responsibility for inadvertently striking the spark that lit the tinder that exploded into the French and Indian War. This young Washington was accused of being a war criminal, an assassin, a murderer, an incompetent leader, negligent, and an international embarrassment. The war that he touched off would last seven years and spread around the world—the first truly global war. Unfamiliar to many Americans, this war in the late 1750s and early 1760s won much of the North American continent for the British. At the same time, it unleashed the forces that twenty years later would mushroom into the American Revolution against those same British. These pivotal early years of Washington's life give us a picture much at odds with his popular image. This war and his personal passage through the wilderness laid the groundwork for the great leader that Washington would one day become.

For over two centuries, adventurous young men had struck out for the wilderness of the New World. They sought their fortunes, they tested themselves. The passage through the wilderness would become a ritual of life in the New World. Some young men set off and never returned. Others struck on great discoveries—gold, fur, ancient civilizations, vast lands. Still others, taken captive, returned with stories to tell about what lay out there. In five crucial years during his early twenties, George Washington traversed between the "civilization" of the Atlantic coast—with its gracious plantations, linen tablecloths, and attentive servants and slaves—and the wilderness of the North American interior that lay over the mountains, a few hundred miles away. These vastly different worlds would mold Washington as he underwent the transition from adolescence to adulthood and moved from self-centered youth to empathetic adult, from ambitious individual to selfless leader.

Washington's passage parallels what in mythology Joseph Campbell calls the "hero's journey." As the hero travels from the known to the unknown world, often a wilderness, he encounters supernatural powers and mythical beasts and undergoes a series of trials. He returns, ultimately, the "master of two worlds." Through this five-year journey from the civilized coast through wilderness and war and the long series of struggles and mistakes that accompanied them, Washington eventually became a master of two worlds. As master of two worlds, and finally as master of himself, George Washington would go on to accomplish extraordinary things.

On this, his first mission as a part-time junior officer in the Virginia militia, eager to make a name and propel himself into the Virginia aristocracy, young Washington felt a surge of anxiety. How much longer to his destination? Governor Dinwiddie had urged him to go with all possible haste, as every delay could be costly. Now Captain Joncaire was telling him he faced several days' additional ride to reach the French commandant who resided upriver at the fort of the buffalo—French Fort Le Boeuf.

Business completed, mission diverted, Captain Joncaire invited Washington, Gist, and van Braam to dine with him and his fellow French officers. The travelers received impeccable hospitality. One can imagine the crude log cabin with a fire blazing in the hearth, candles burning on the table's rough boards. Outside, the light faded quickly in early December and damp twilight dimmed to blackness. But the warmth and guttering light within revealed a table laden with roasted venison, perhaps beaver tail, elk, and other meaty delicacies of the forest, brought in by Indian hunters. The French officers offered the luxury of bread baked with wheat flour, transported by canoe down the Rivière Le Boeuf from French outposts on the Great Lakes, supplied from Montreal. After a four-week journey from the coast, the dinner came as a sumptuous banquet for Washington and his companions.

And the wine flowed, abundantly. The French officers drank, copiously. Washington held back, tracked the conversation, perhaps tried to ask some questions.

On what basis could they claim possession of the Ohio? It was well known that the region fell within the capacious western borders of the British province of Virginia.

France, they told him, claimed the Ohio on the basis of the explorations of La Salle.

Washington did not understand. Amid the wine and the candlelight, they explained that La Salle had explored the great Ohio River valley many decades earlier—at least sixty years ago—and claimed it for France. They had used it for many decades as a route for their fur trade, and now they would occupy it, manning a series of forts under construction on the Great Lakes and along the Ohio and tributaries. The French had heard that settler families from British coastal colonies such as Virginia and Pennsylvania had crossed the Appalachians to clear homesteads in the Ohio Valley. British fur traders already operated in the region. The French forts would halt this British encroachment.

These were strong words. The French had determined to take the Ohio—land that Virginia claimed as its own. With the wine giving "License to their tongues," as Washington, earnestly trying to gather intelligence, put it in his journal, the French officers laid out their strategic thinking. "[T]hey were sensible the *English* could raise two Men for their one; yet they knew [that the English] Motions were too slow and dilatory to prevent any Undertaking of theirs."

The wine flowed. The light flickered on the log walls of the British trader Fraser's former outpost, now occupied by the French. The officers

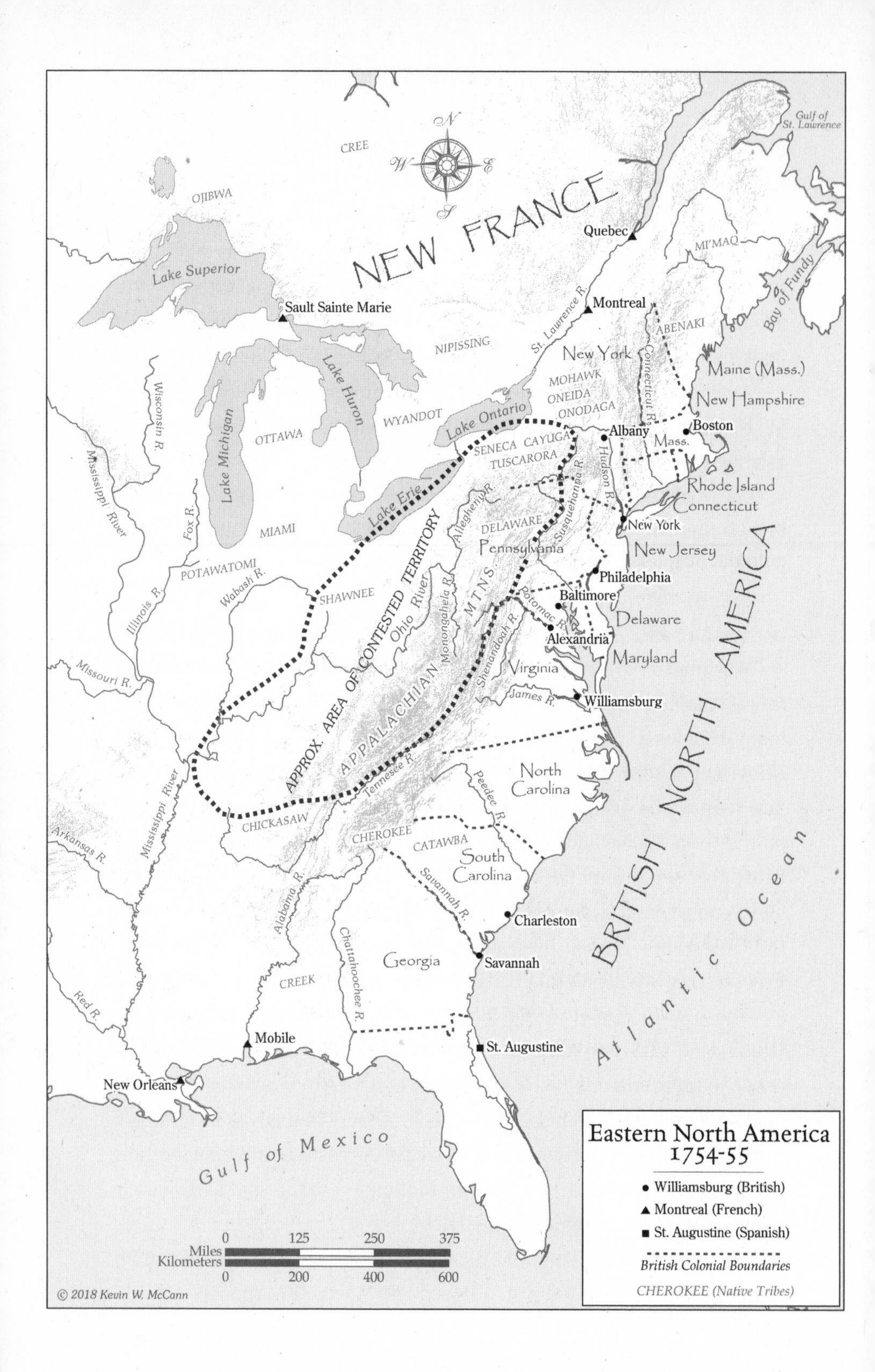

Eastern North America
1754-55
Williamsburg (British)
Montreal (French)
St. Augustine (Spanish)
British Colonial Boundaries
CHEROKEE (Native Tribes)
NEW FRANCE
BRITISH NORTH AMERICA
APPROX. AREA OF CONTESTED TERRITORY
APPALACHIAN MTNS.
Atlantic Ocean
Gulf of Mexico
Gulf of St. Lawrence
Bay of Fundy
Lake Superior
Lake Michigan
Lake Huron
Lake Ontario
Lake Erie
Quebec
Montreal
Sault Sainte Marie
Albany
Boston
New York
Philadelphia
Baltimore
Alexandria
Williamsburg
Charleston
Savannah
St. Augustine
Mobile
New Orleans
New York
Maine (Mass.)
New Hampshire
Mass.
Rhode Island
Connecticut
New Jersey
Pennsylvania
Delaware
Maryland
Virginia
North Carolina
South Carolina
Georgia
CREE
OJIBWA
MI'MAQ
ABENAKI
NIPISSING
MOHAWK
ONEIDA
ONODAGA
SENECA
CAYUGA
TUSCARORA
WYANDOT
OTTAWA
MIAMI
POTAWATOMI
SHAWNEE
DELAWARE
CHICKASAW
CHEROKEE
CATAWBA
CREEK
St. Lawrence R.
Connecticut R.
Hudson R.
Susquehanna R.
Allegheny R.
Ohio River
Monongahela R.
Potomac R.
Shenandoah R.
James R.
Tennesee R.
Peedee R.
Savannah R.
Chattahoochee R.
Alabama R.
Mississippi River
Wisconsin R.
Fox R.
Illinois R.
Wabash R.
Missouri R.
Arkansas R.
Red R.
Miles
Kilometers
0
125
250
375
0
200
400
600
© 2018 Kevin W. McCann

drank. "They told me," Washington recorded, "That it was their absolute Design to take Possession of the *Ohio,* and by G— they would do it."

HERE, in this wilderness outpost on the Ohio headwaters, over a dinner of wild game, with a determined twenty-one-year-old in attendance, two world empires bumped against each other like great weighty vessels on a rolling sea. For decades France and Britain had regarded each other as bitter rivals. One an island nation, the other locked in a continental mass, they jealously faced each other across a narrow channel of water, each looking for advantage. They fed their growth by acquiring colonies around the globe, their fleets of ships briskly hauling traded goods back and forth between mother country and far-flung colonial outposts, with every shipload—furs, sugar, rum—boosting the royal treasuries.

France had landed her first North American settlers in 1604 at today's Nova Scotia. By sailing ship and birch-bark canoe, she had worked her way up the Saint Lawrence River into the continent's interior, pursuing the trade in furs and befriending the Indian nations among whom the French colonists lived and intermarried. The English established their first colonies to the south, on the Atlantic coast, in 1607 at Jamestown in today's Virginia and then at Plymouth in today's Massachusetts. They cleared land and planted crops, gradually pushing aside and often warring with the native inhabitants. At what became Virginia, the earliest colonists had, in effect, planted two flags in the ground along the coast, separated by two hundred miles, and said, *We own all the land from here westward to the island of California.* Initially believed to be an island, the Pacific coast had been explored in the 1500s by Spanish ships and by Sir Francis Drake. The early British colonists had little idea how far the continent actually extended, or what lay on the far side, and they initially ventured no farther inland from the flat green Atlantic Seaboard than the edge of the Appalachian Mountains.

But now, in the mid-1700s, nearly a century and a half after planting those first flags, after clearing forests and building farms and tobacco plantations manned by African slaves, the British colonists had begun to outgrow that narrow coastal plain. First the settlers bumped up against the Appalachian Mountains. Then the very bravest and most intrepid, crossing on foot and horseback, began to spill over, just a few of them, like trader John Fraser, into the vast wilderness of the Ohio Valley, which British Virginia claimed on the basis of those two flags. The thousand-mile Ohio River, like a great vein draining the eastern continent, rises near Lake Erie. It flows west and south, fed by countless smaller rivers and streams, until it spreads to nearly a mile in width and pours into the Mississippi River near today's Saint Louis. The French had used the Ohio—the "good river" in the Iroquoian languages, or the "beautiful river"—for decades not only as a trading route but as a link between their Canadian colony to the north and their French colony far down at the mouth of the Mississippi, New Orleans. Plied by voyageur canoes and bateaux—a kind of flat-bottomed cargo canoe—the Ohio River was, as historian Francis Parkman put it, a French line of communication between the snows of Canada and the canebrakes of Louisiana.

For centuries the two powers, France and Britain, had vied for dominance in Europe, most recently in the War of the Austrian Succession. This had lasted nearly a decade and ended only five years earlier, in 1748. At issue, at least to start, was who would inherit the realm of Charles VI of the Habsburgs. France and her allies wanted the realm broken up. Britain and her allies, worried about French dominance on the continent, wanted it to remain intact. Such was the infinitely complex balance of power in Europe, where alliances and treaties blossomed and faded like flowers in spring rains. While the Austrian succession issue was settled in 1748, and a treaty signed, the tensions and rivalry between France and Britain still simmered just beneath the calm surface.

And now the great empires bumped and jostled softly during this polite, wine-soaked evening in a log cabin in the Indian town of Venango. One can picture Washington at the table, quiet, socially awkward, over six feet tall with broad shoulders, a fine rider and good athlete, but lacking a formal education. He knows the French and British have been enemies and are now at peace. He knows that his leader, Governor Dinwiddie, bewigged and jowly, back in the Virginia capital at Williamsburg, has land investments in the Ohio Valley. He knows that he must move quickly, at the governor's orders, to deliver the message to the French commandant. But young Washington cannot conceive of the weight of empires pressing down inside the warm cabin on this cold, damp December evening in the Ohio wilderness.

THE NEXT DAY IT RAINED——HARD. A frustrated Washington, holed up in his tent, with rain pattering on the woven fabric, couldn't travel. Even one day's delay wore on his patience. He had arrived in the wrong place. His mission had been shunted in a new direction by Captain Joncaire. Now rain fell with no end in sight and he would need several days more to deliver the governor's urgent letter.

Other tensions besides Washington's haste hung over the dripping Indian town. Both the French and the British were attempting to win the Indians to their side, working within an intertribal web of relationships as complex as the alliance-bound kingdoms of Europe. Whichever nation, France or Britain, earned the allegiance of the Ohio tribes and the nearby Iroquois would possess a great advantage in laying claim to the vast wilderness. But the tribes did not speak with a single voice. Of the several different tribes at Venango, the Delaware Indians there "lived under the French colors," as frontier guide Gist put it in his journal. The three chiefs traveling with Washington's party—the imposing Iroquois chief known as the Half King, Jesakake, and White Thunder—were partial to the British. They tried to convince the Delaware Indians living under the French flag at Venango to switch allegiances and pledge friendship to the British, too.

They refused to switch.

Washington, meanwhile, struggled to keep "his" Indians—the three chiefs, or sachems—away from Captain Joncaire. He worried that Joncaire would use his clever arts of persuasion to win them to the French side. The first night at dinner, Washington had kept them from coming to the French officers' quarters, out of sight in the Indian town. The next day, when Joncaire realized that the Half King and other sachems had arrived in Venango with Washington's party the day before, he asked Washington why he had not mentioned their presence.

Young Washington told a diplomatic lie. He had heard the captain speak disparagingly of Indians, he said to Joncaire. Having heard the captain speak thus, Washington thought that the captain would not want to see them.

"I excused it in the best Manner I was capable," Washington reported.

Their presence revealed, Captain Joncaire employed all his skill to win over the three chiefs, giving them presents and pouring generous servings of brandy until they had become thoroughly drunk.

By the next day, however, the Half King had sobered up. In a formal speech, he told Captain Joncaire and the French to leave the Indians' land in the Ohio and go back to Montreal.

IT RAINED AND IT RAINED AND IT RAINED. Heavy gray skies hung low over the wilderness. They rode their horses through dark forest groves and misty open meadows. Roughly sixty miles of travel lay between them and Fort Le Boeuf, at the head of Rivière Le Boeuf, where they hoped to find the French commandant. The little caravan traced the winding course of the small river northward. Horses' hooves sank in the marshy mires. They splashed through low spots. The feeder streams ran full from the rains, the high waters submerging the tall grasses that grew along the riverbank. The grass clumps bent underwater, like long beards blowing in the wind, rippling and wavering with the eddies of the stream flow.

Making camp, they unrolled thick furs or scratchy blankets or bearskins in the tents while rain pattered on the canvas. A fire leaped and hissed outside. They huddled round it, drying themselves, eating their supper. Salt pork, hardtack—a large, tough cracker made from flour—dried meat, Indian cornmeal mush, and, on days when the Indian hunters met success, fresh roasted venison or bear.

It was the Congo of its day, the Heart of Darkness. Heading into this vast wilderness of forest and swamp, river and mountain, peopled by strange tribes whose alliances you did not know, you felt at times that you were silently watched from the trees. Odd, hybrid characters of indeterminate race and origin appeared at distant villages and trading outposts—characters like Andrew Montour, who costumed himself in a mix of haute couture and animism, wearing a European coat of fine sky-blue fabric, a red damask waistcoat, breeches, and stockings but circling his face with a broad ring of bear fat and paint and braiding his ears with brass wire, as one traveler put it, “like a handle on a basket.” Those who ventured here did so at their own risk, leaving behind the familiar safety of the lands over the Appalachian Mountains, the civilized havens, the provincial capitals of Philadelphia and Williamsburg along the Atlantic coast. There was much to fear, and much to gain—fortunes to be made in fur, in land—although to do so one risked losing more than one understood. A certain callousness attended life here. The blood feuds raged as passionately and the sense of honor burned as brightly as in the lands on the far side of the mountains, but bound by rules you didn’t know.

This was the land into which young Washington journeyed.

CHAPTER TWO

THINGS COULD HAVE TURNED OUT VERY DIFFERENTLY. If Mary Ball Washington had given permission for her eldest son, fourteen-year-old George, to go to sea, he might never have set foot in the Ohio wilderness. She almost relented. Her husband, Augustine "Gus" Washington, had died when George was eleven, leaving her at home with five young children. Not a wealthy family, the Washingtons occupied a lower tier in the hierarchy of Virginia planters. George lacked a formal education, unlike his older half-brothers, whom Gus had sent to England for rigorous schooling and who, upon Gus's death, inherited the bulk of his estate. Without a large plantation to inherit or a career to pursue, how would George make his way in the world?

When George turned fourteen, his oldest half-brother from his father's first marriage, twenty-eight-year-old Lawrence, took the adolescent under his wing and dispensed some brotherly advice. Lawrence had served as a Virginia colonial officer accompanying the British Royal Navy in 1741 when it laid siege to Spain's gold-shipping port of Cartagena, in South America, to avenge insults to British shipping.* A massive fleet of nearly two hundred ships and twenty-five thousand men had arrived under Admiral Edward Vernon, but the Spanish defenders repelled Ad-

* The conflict got its quirky name, The War of Jenkins' Ear, after a Spanish naval commander boarded a British ship off Florida, accused the captain of smuggling in Spanish waters, raised his sword, and sliced off the British captain's left ear, which the captain later supposedly showed to Parliament.

miral Vernon's land forces at the fortress walls, and the arrival of the rainy season hatched countless mosquitos in swamps and puddles, injecting the British troops with yellow fever virus, effectively finishing them off. Safely aboard the admiral's great eighty-gun flagship, the HMS *Princess Caroline,* Captain Lawrence Washington of Virginia was spared mosquito-borne disease and the sharpened bayonets of Spanish countercharges, writing home with a mix of bravado and grimness, "War is horrid in fact but much more so in imagination. We there have learned to live on ordinary diet; to watch much and disregard the noise or shot of cannon."

George was deeply impressed on Lawrence's return by his erect military bearing, his uniform, and his stories of Spanish castles and British warships. After their father's death, Lawrence showed a special liking for his younger sibling and urged George to consider a career in the Royal Navy or aboard a commercial ship. In 1746 the adolescent George readily accepted this advice from the brother he called his "best friend." In order to sign on, however, he would need Mary Washington's blessing. For several months Mary remained undecided. "[S]he offers several trifling objections," one family friend reported back to Lawrence, "such as fond and unthinking mothers naturally suggest. . . ."

Mary finally wrote to her brother in London, Joseph Ball, asking for his advice. Joseph recoiled at the idea of his young nephew going to sea as a common sailor on a commercial ship where the officers would "cut him and staple him and use him like a Negro, or rather, like a dog." Nor did Joseph think George could secure a good position in the Royal Navy due to the stiff competition from well-born youths in Britain. George could make more money than a ship captain, wrote Joseph to his sister, if he had three or four hundred acres of Virginia tobacco land and three or four slaves—unless he tried to live too much like a "gentleman" and spent extravagantly on luxury goods shipped from London, sinking himself in debt.

That settled it for Mary. She vetoed the seafaring idea. But George's prospects as a planter did not look great, either. The modest piece of cultivated land, Ferry Farm, that he had inherited from his father and where he and his mother now lived, on the Rappahannock River near Fredericksburg, fell well short of the plantation acreage his London uncle had recommended for a decent living. He needed a skill that would supplement its slender output. His late father, Gus, had done some surveying. It was at this point, in his mid-teens, that George began to look at the land itself as a profession—how to measure land, how to value land, how to invest in land.

* * *

Scholars have identified twenty-one leading families in Virginia's early aristocracy. The Washingtons were not among them. Bearing names like Byrd, Carter, Lee, and Randolph, the founders of these family lines had sailed from England to Virginia in the mid-1600s, a few decades after the Jamestown landings in 1607. While the earliest Virginia colonists had struggled for survival—many perishing and others resorting to cannibalism during what became known as the Starving Time—by the mid-1600s, Virginians, at least those with means, had found ways to prosper and to govern themselves.

In 1610, a new arrival to Jamestown, John Rolfe, brought with him seeds of a sweet strain of tobacco and experimented with its cultivation. For millennia, Native Americans of both North and South America had raised tobacco and held it sacred, using it for rituals, social occasions, and medicine. A member of the nightshade family, its leaves contain a small percentage of an alkaloid, nicotine, that acts as a stimulant in the human body by triggering the release of adrenaline and dopamine. "[T]hey suck, absorb, or receive that smoke inside with the breath," wrote an early Spanish chronicler of Caribbean Indians, "by which they become benumbed and almost drunk. . . ." Introduced to Europe, tobacco became a profitable and highly regulated trade for the Spanish Crown, which prescribed the death penalty for anyone caught smuggling the precious seeds. Rolfe somehow acquired and brought to Jamestown the seeds of a variety of tobacco, *Nicotiana tabacum,* that grew in tropical climates controlled by the Spanish and that tasted sweet and mellow compared with the harsh strain grown by North American Indians.

Within two years, the enterprising Rolfe (soon to marry a Powhatan Indian girl named Pocahontas) had learned how to cultivate the sweet tobacco—Orinoco, as he named it. Rolfe's first export shipment to England, four barrels of the dried leaves, launched what would become a huge and highly profitable enterprise for the struggling colony—Virginia's equivalent of Spain's gold-filled rooms of the Incan king and silver-veined Andean mountains.

Virginia tobacco, however, was not easy to grow. It exhausted the soil after a few years, and demanded tremendous labor. This meant that Virginia tobacco planters constantly needed new forest lands to clear and, through negotiations or violence or disease, constantly pushed the Powhatan farther into the interior. To work the fields, the planters imported indentured servants from England, Germany, and other countries, paying their ship passage across the Atlantic in exchange for several years of free labor before releasing them, perhaps to own land of their own.

As tobacco cultivation spread from Jamestown along the many fingers of Virginia's coastal rivers, the planters' hunger for cheap labor proved insatiable. A new source of human energy first arrived in 1619 when two British privateers raided a Portuguese ship bound for Mexico that carried in her hold hundreds of captured black Angolans. The privateers landed near Jamestown and exchanged some of their human cargo for provisions. Some of the earliest African laborers in Virginia may not have been slaves but indentured servants who eventually won their freedom. By the mid-1600s, however, slavery had become institutionalized in Virginia, and by century's end slave traders shipped large numbers of black Africans to the colony. The entire tobacco enterprise rested on this cheap, abundant labor. It demanded countless hands to clear the forest, bury the seeds, transplant thousands of seedlings, weed the rows, trim the tops, prune the tendrils, slice the stems, cure the leaves, and pack them in tuns—thousand-pound barrels—for shipping to Europe.

Certain families rose to prominence as tobacco cultivation took hold in Virginia. The family founders mostly arrived from England in the mid-1600s, young men seeking adventure and opportunity in the New World. Some were merchants or merchants' sons, others were second or third sons of landed families whose eldest brothers, through the ancient practice of primogeniture, had inherited the bulk of the family estate. For a small "summe of money," wrote the colony's governor, Sir William Berkeley, in 1662, a younger son of good English upbringing could "erect a flourishing family" in Virginia.

Once in Virginia these young men procured land for tobacco and other crops and earned money however they could—building gristmills and sawmills, trading in furs, exporting wheat to the West Indies, importing sugar and rum. A surprising number practiced law while also trading and planting. Working connections to the governor or back in England, they won coveted appointments to public offices that guaranteed an income, such as auditors and collectors. And they acquired more land—great acreages of land, pushing ever deeper up river estuaries that riddled the forests and marshes of Virginia's low-lying coastal plain. They called it the Tidewater—the fecund region where the briny sea rising at high tide creeps up the freshwater rivers.

But within decades, Virginia society lost its early fluidity and, as if mimicking English society, settled into strata. By some counts, Virginia society had eight strata, with an aristocracy entrenched on top, a slave class anchoring the bottom, and small farmers, merchants, sailors, frontier settlers, servants, and convicts sandwiched in between. The acreages expanded and families consolidated as the 1600s turned to the 1700s, with the leading families each owning an average of nearly twelve thousand acres. Families intermarried, weaving together their lands and fortunes, so that sorting out relations, it was said, resembled trying to separate a box of fishhooks. Fewer indentured servants arrived from Europe, meaning fewer possible future landowners or tradesmen who might be upwardly mobile, and far more enslaved Africans, who were not. "Virginia society," notes historian Daniel Boorstin, "was beginning to be frozen."

Voting rights steadily eroded until the right to elect members of the Virginia House of Burgesses resided only in landowning men, as in the English system. Along Tidewater rivers rose "great houses" fitted with pavilions, porches, arcades, and marble-floored central halls where dances could be held. Surrounding these great houses like discreet villages were kitchens, dairies, smokehouses, workshops, dovecotes, stables, barns, henhouses, gardens, lawns, and lanes leading down to private river wharves where slaves loaded tobacco tuns onto ships and carried from their holds agricultural implements and crates of English porcelain destined for the main house. The big plantations had to house and feed hundreds of slaves, the field hands as well as cooks, maids, butlers, coachmen, blacksmiths, carpenters, wheelwrights, and gardeners, and white staff such as overseers, managers, secretaries, and tutors and governesses for the family's children. By the early 1700s, a century after the Jamestown landings and the Starving Time, a few elite families with their great estates dominated Virginia society—aspiring, in many ways, to the life of the British aristocracy and the English country gentleman.

* * *

The Washingtons had occupied a spot on the second tier of society for generations, first in England and then in Virginia. They had worked hard, cultivated connections, seized opportune moments, and, often, married well—betrothing wealthy widows whose lands and assets and status boosted their own. Nevertheless, despite their efforts, with remarkable consistency they had not achieved top rank in the English aristocracy.

One of the Washington family's more illustrious members worked in the early 1500s as the sheep estate manager for a knight of King Henry VIII. Apparently, while traveling on estate business to sell its wool, this ancestral Lawrence Washington met and then married the widow of a prosperous Northampton wool merchant, took over the merchant's business, and became the town's mayor. After his first wife died childless, Lawrence married another wealthy widow, this one holding leases on lands owned by Catholic monasteries. When Henry VIII fought with the Catholic Church over annulment of his marriage to Catharine of Aragon and dissolved England's Catholic monasteries, Lawrence Washington purchased from the Crown the former monastery lands leased by his wife and built his own estate, Sulgrave Manor, thus acquiring the status of "gentleman." The Washington family procreated so abundantly—Lawrence had eleven children, and his eldest son had fifteen with two wives—that, even in this era of primogeniture, the wealth dispersed. Within a century Sulgrave Manor had left the family's hands.

The Washingtons, however, had achieved a certain status. A great-grandson of "Lawrence the Builder" became the first Washington known to attend Oxford, although his fraught experience there would prove fateful for the family line. Completing his master's in 1626, the great-grandson, also named Lawrence, won an appointment as lector and then proctor at Oxford's Brasenose College. Loyal to the university administration, King Charles I, and the Anglican Church, he helped Oxford's tough-minded chancellor purge from its ranks "heretical" Puritans—individuals who wanted to radically reform the official Church of England for being too close to the Roman Catholic Church.* It was during this time that one party of impassioned religious separatists fled to Holland, then jammed aboard a small ship named the *Mayflower* and landed at Cape Cod, calling themselves Pilgrims and founding the Plymouth Colony six hundred miles north of Jamestown.

Lawrence Washington, however, was still back in England and allied with the Puritans' persecutors. He married in 1633—his wife-to-be may have been pregnant at the time—left Oxford, and was appointed rector of a wealthy Essex parish. The two sides of the English Civil War soon squared off, Parliamentarians against Royalists. England's Puritans, having been persecuted by the Crown, sided with the Parliamentarians. Washington was accused by the Parliamentarians of being a "Malignant Royalist." He ranked high on the hit list when war openly broke out in 1642 and Parliament began purging prominent Anglican clergy. "[He is] a common frequenter of alehouses" went the charge against Washington,

* At the time, England's most outspoken Puritans faced brutal punishments, such as William Prynne (whose last name Nathaniel Hawthorne may have borrowed for the outcast Hester Prynne in *The Scarlet Letter*). A strict Puritan disciplinarian who condemned toasting as sinful, opposed celebrating Christmas, and published a thousand-page screed arguing that stage plays, and actresses especially, were immoral (this despite the queen's love of theater), he was sentenced in 1634 by Archbishop of Canterbury William Laud to be "cropped"—ears sliced off while held fast in a pillory. Archbishop Laud, when chancellor at Oxford a few years earlier, had been Proctor Lawrence Washington's patron. When an unrepentant Prynne continued to write pamphlets from prison, Archbishop Laud ordered the stumps of his ears cropped and his cheeks branded with the letters *S.L.*, for "Seditious Libeler." Released in 1640 as tensions mounted toward the civil war, Prynne vowed revenge. Five years later his tormentor, Archbishop Laud, was beheaded.

"not onely himselfe sitting dayly tipling there, but also encouraging others in that beastly vice. . . ."

Dismissed, the penniless Washington, who as a fifth son had inherited little, struggled to support his six children. The children, too, now bore the taint of having been Royalists. This was not a good label to carry, as the Parliamentarian forces defeated the king's Royalist armies and forced the ultimate Royalist, King Charles, to lay his head on the executioner's block. His son and heir to the throne, Charles II, fled to Europe, the old order was overturned, and the Commonwealth and Oliver Cromwell rose to power.

The ousted Washington's oldest son, John, realized that a better future might lie across the Atlantic than amid these upheavals. He signed on in his early twenties as "second man"—meaning he would help sail the ship and share the profits—on the ketch *Sea Horse of London* for a trading voyage to the colony of Virginia. Relatives may have pointed the way for the young man, for distant Washington cousins had thrived in Virginia. The *Sea Horse* sailed up the colony's coastal rivers and called at plantation wharves where she traded her cargo of English household goods for tobacco destined for Europe's fashionable users. Her holds full of cured leaf, the *Sea Horse* was sailing out of the Potomac River in early winter 1657 when she "run upon a Middland ground or shole" and was pounded by a winter storm whose waves crusted her with ice and sank her, soaking her precious cargo. There followed an effort to float her again and months of acrimony and legal action between the *Sea Horse*'s captain and young John Washington over who owed what for the loss.

One of the magistrates in the case, a wealthy planter named Nathaniel Pope, apparently took a liking to the young man and invited Washington to the Pope plantation. Here Pope introduced him to his twenty-year-old daughter, Anne, and also offered to pay off Washington's *Sea Horse* debts in beaver skins, then worth 8 shillings a pound. In another serendipitous conjunction of land and marriage for a Washington male, within eighteen months of his arrival in Virginia, John Washington had married Anne Pope. The pleased father-in-law awarded his daughter seven hundred acres of land and lent John Washington £80. Setting up as a tobacco planter,

John soon gained more land through the "headrights" policy, which was intended to boost the underpopulated colony's workforce. For each indentured servant that a planter brought across the Atlantic from Europe, he would receive fifty acres of land. John Washington, in partnership with his brother-in-law, imported at least sixty-three indentured servants.

By the time of his death at age forty-six in the late 1670s, only two decades after his inauspicious arrival aboard the *Sea Horse* as the son of a disgraced minister, the industrious John Washington held eighty-five hundred acres of land—although, unlike the larger planters, much of his lay in scattered parcels, including a large wilderness tract on the Potomac known as Little Hunting Creek. His wife, Anne Pope Washington, had died in her early thirties after ten years of marriage, having borne at least five children, three of whom survived. John had remarried, lost his second wife, too, and married a third time. He had climbed to local prominence by holding a variety of offices—coroner, county judge, member of Virginia's House of Burgesses—and had won the title of colonel in the Virginia militia, which was called out to punish Indians for attacks on settlers. "John the Sailor" had established a life cycle for the coming generations of Washington males in British America—born in America, they were sent to England for a proper education, returned to Virginia colony, inherited or acquired land, planted tobacco, married a planter's daughter, raised a family, held public office, and died young. George Washington, born in 1732, belonged to the fourth generation of American Washingtons.

* * *

He had begun the previous summer to experiment with his father's surveying instruments. Starting with simple techniques and keeping careful, neat notes, George documented his first survey in August 1747—just weeks after his London uncle had discouraged a sailor's life. It happened that George took up his father's surveyor's compass at a time when a great deal of Virginia land needed surveying.

That same summer, a fifty-four-year-old bachelor, Lord Fairfax, left his home of Leeds Castle and sailed across the Atlantic to disembark at his cousin's elegant plantation on the Potomac River. Named Belvoir

Manor, this estate was just up the Potomac riverbank from the house and large tract Lawrence Washington had inherited from his father, Gus—a tract formerly known as Little Hunting Creek. Lawrence, in honor of his former commander in the Caribbean, Admiral Vernon, had renamed it Mount Vernon. Both Belvoir and Mount Vernon lay in Virginia's Northern Neck—a large peninsula of land, or "neck," separating the Potomac and Rappahannock rivers. Lord Fairfax as a young man in England had inherited a kind of New World fiefdom known as the Northern Neck Proprietary, essentially making him the landlord of five million acres of Virginia land and possibly much more.

The equivalent of a tract ninety miles by ninety miles, these lands had been given a century earlier by King Charles II to reward several of his faithful retainers during his years of exile in Europe, when he finally returned to England and was restored to the throne in 1660. No one paid much attention to the sprawling Northern Neck parcels given to his retainers. Over decades, the parcels consolidated into one ownership and eventually passed in the early 1700s to a single heir—Thomas, Sixth Lord Fairfax of Cameron. Acting through Virginia agents while residing at Leeds Castle in Kent, he sold off pieces of these lands to Virginia planters. The proprietary's western boundary, however, had remained in dispute between Lord Fairfax and the Virginia colony. Finally, in 1745, the Crown ruled in Lord Fairfax's favor, giving him an enormous further expanse that reached into the Appalachian Mountain valleys. Soon thereafter, he packed up from Leeds Castle and sailed to America to take the measure of his lands and reap their annual rents. He arrived one day in 1747 at the Belvoir estate of his cousin, Colonel William Fairfax, who had been acting as his land agent.

It was not a difficult thing for Lawrence Washington to properly introduce his younger brother to Lord Fairfax soon after his arrival from England. The Fairfaxes were neighbors, after all, and also his in-laws. Lawrence had married Anne Fairfax (nicknamed "Nancy"), daughter of Colonel William Fairfax and his late first wife, Sarah.* It was a social step

---

* Nancy was only fourteen when she married Lawrence Washington. She had told her parents that a trusted family friend, the Reverend Charles Green of Truro Par-

up for a Washington to marry into a family of titled British aristocracy, and, like a number of his ancestors, he married into large landholdings—Nancy Fairfax would soon come into four thousand acres of her own.

As Lawrence and sixteen-year-old George trotted up to the entrance portico of the Fairfax's two-story, nine-room brick manor house on the bluff overlooking the Potomac, one imagines black servants grabbing the bridles of their horses. Perhaps a carriage brought Nancy from Mount Vernon. George rested in the saddle with natural grace and fluid ease, more at home atop a horse than in society. When he swung his long leg over the saddle and stepped down onto the crunching path, his stretched frame revealed that he had inherited his father's height, and his torso would soon bulk out with his father's strength. Through Lawrence's marriage to Nancy, George knew her father, Colonel William Fairfax; her stepmother, Deborah; and her siblings, especially her gentlemanly younger brother, George William, who was seven years George Washington's senior and whom he admired. But he had never met a true British lord. It may have intimidated him to see Lord Fairfax at the Belvoir estate, perhaps standing on the front portico with Nancy's family or waiting in the manor's spacious front hall.

In recent years, the teenage George had practiced manners along with his penmanship, carefully copying out in an exercise book the one hundred and ten entries of the seventeenth-century self-help manual *The Rules of Civility and Decent Behaviour in Company and Conversation.* "[B]edew no mans face with your Spittle, by approaching too near him when you Speak," the rules cautioned. "Strive not with your Superiors in argument, but always Submit your Judgment to others with Modesty. . . ."

George deferred modestly to his lordship. It may have surprised him to see Lord Fairfax so casually—even, shockingly, informally—dressed. Among other eccentricities, Lord Fairfax vastly preferred the company of

ish, had recently sexually molested her, or attempted to. These accusations were aired in a very public fashion when Lawrence Washington accused Green of this abuse and the matter went to trial. [See "The Scandal That Rocked Old Virginny," *Washington Post,* December 19, 1990.]

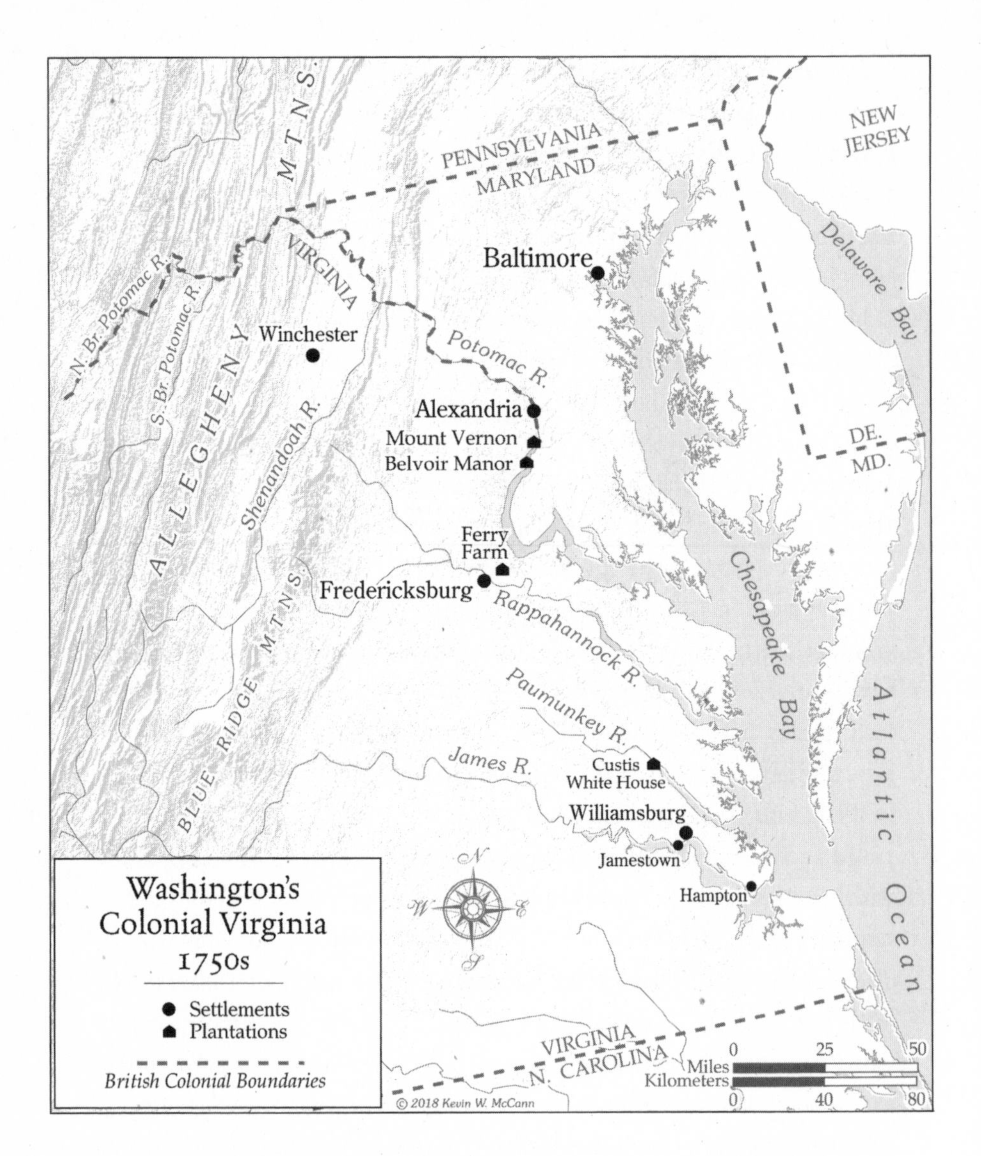

men to women (it was rumored that he had never recovered from a broken-off engagement), eventually retreating to a remote hunting lodge in the Shenandoah Valley where he became known for his hospitality to guests. He loved hunting, and possessed a literary flair that he had employed at Oxford during his twenties to write for *The Spectator,* Joseph Addison and Richard Steele's witty coffeehouse publication. At Belvoir Manor—over dinner, perhaps, or riding on the estate—the men must have talked about hunting, military matters, literature, and most definitely about land. The Privy Council's recent decision had officially awarded to Lord Fairfax an enormous expanse of land that extended from the flat green coastal plain known as the Tidewater region into the ascending ridges and valleys of the mountains. Much of this western sector of his lands had not been surveyed. With the idea of selling more parcels, his lordship wished to send a survey party to take measure of it.

Would George like to join the surveyors?

* * *

In early March 1748 they headed their horses north and west away from the lush Tidewater plantations fringing the estuaries of the Potomac and eased across the rolling, rising ground of the Piedmont toward the hazy barrier in the distance—the Blue Ridge, the first low, forested rise of the Appalachian Mountains. They climbed its flank and crested the ridge through a "gap"—a natural notch—on an ancient Indian trail to emerge on the far flank. Their view swept a broad valley below, the Shenandoah Valley, a fertile mat of forest, marsh, and meadow dotted with a few settlers' cabins. Twenty miles away across the valley lifted the next hazy ridge of the Appalachians, and beyond that rose another ridge, as if a great, stationary set of ocean waves barred the continent's interior from the Atlantic coast.

It was in the Shenandoah Valley and beyond that they would survey. Led by veteran surveyor James Genn, the expedition numbered among its seven members various assistants and "chainmen." The Fairfax family's interests were represented by polished, British-educated, twenty-three-year-old George William. George William and sixteen-year-old George Washington constituted the two Tidewater gentlemen, setting them apart from the expedition's scruffier members. The journey may have been something of a bachelor's last hurrah for George William, for in a few months he would marry the charming Sarah "Sally" Cary, who would play a major role in young Washington's life.

She, too, belonged to a prominent Virginia family. Sally's ancestors had been successful merchants in Bristol, one of the first English ports engaging in the trans-Atlantic trade, and had arrived in Virginia in the mid-1600s, soon establishing themselves in what some early Virginia historians called "the ruling oligarchy." The family seat, on the lower James River near Hampton Roads, lay at the entryway for ships bound to upriver ports such as Jamestown. With several generations of Carys serving as naval officials, the family's brick mansion received a worldly parade of ship's captains, wealthy passengers, and other travelers. The Carys also maintained a townhouse in Williamsburg that they used when the House of Burgesses was in session.

It was here, at the Governor's Palace, that Colonel Wilson Cary's four teenage daughters were introduced at formal receptions and quickly attracted wealthy young Virginian suitors from the nearby College of William and Mary. Sally, the eldest Cary daughter, was said to be, in the words of a Cary family memoirist, "the cleverest and far the most fascinating," having read widely under her father's tutelage, possessing elegant manners; a bright, lively, almost teasing personality; and slender, dark-eyed good looks. At age seventeen, she and George William Fairfax met at the Governor's Ball in 1747. Spending time with the couple, Washington, a year or two younger than Sally, would soon become deeply enamored of her.

Beyond the Blue Ridge, riding beside the soon-to-be-married George William, young Washington first tasted a near-wilderness very different from the manicured plantations and well-staffed orderliness of the Tidewater region. At one of their first stops in the Shenandoah, at a remote "ordinary," or traveler's inn, George—"not being so good a Woodsman as the rest of my Company," as he admitted in his brief diary of the journey—was shocked to discover upon carefully undressing and crawling into his bed that, instead of a feather mattress and clean white sheets, it consisted of a straw pallet and ragged blanket ridden with bedbugs. Once the candle had been taken away from the room, he dressed again and lay down on the floor fully clothed as his companions knew to do, making a promise to himself to sleep thereafter in the open air in front of a campfire. The dining arrangements took him aback at another stop: with neither tablecloth nor tableware, diners ate in medieval style with fingers and a sharp knife.

Away from the few small settlements and inns, the outdoor life had its own compensations. There was an excitement to it, an adrenaline edge, a raw sense of having to fend for yourself. You could take satisfaction in simple things: learning how to build a fire in deep woods, how to cook simple food in a single pot or with a handheld spit over an open flame, how to erect a tent in cold rain, how to attain a body position best suited to sleep on hard ground, wrapped in bearskin or blanket. Unlike the Tidewater region, with its clean sheets and bountiful tables provided by liveried servants and field slaves, one constantly had to tend to one's own most basic human needs—food, shelter, and warmth—while traversing landscapes that could transform from benign and embracing to hostile and forbidding, depending on terrain and weather, from clambering over rotted logs and crawling through underbrush in blustering sleet to strolling through grassy meadows under fleecy white clouds drifting past in a blue sky. Social status did not matter. You had to figure it out, work together, acquire confidence and self-sufficiency, sharply engage in the *here* and the *now*. You had to pay attention in the wilds to stay alive.

He recorded the moments in his diary:

> *Saterday April 2d Last Night was a blowing & Rainy night Our Straw catch d a Fire yt. we were laying upon & was luckily Preserv d by one of our Mens awaking. . . .*
>
> *Sunday 3 d Last Night . . . we had our Tent carried Quite of with ye Wind and was obliged to Lie ye Latter part of ye night without covering. . . .*
>
> *Wednesday 6th Last Night was so Intolerably smoky that we were obliged all hands to leave ye Tent to ye Mercy of ye Wind and Fire. . . .*

During this 1748 Fairfax surveying expedition at age sixteen, young Washington also first encountered the strangeness of the people who traveled this fringe where the coastal plain crumpled into the corrugations of the mountains. They did not fit into the orderly, British-based social matrix he knew. As an aspiring Tidewater gentleman who did not know quite what to make of them, he reflexively diminished these people on the fringe. For a time, the Fairfax party camped at the crossing of two major Indian trails where a British-born trader named Cresap, also known as "Big Spoon," had erected a post and permanently hung a big iron pot over a fire to welcome Indians. Here the Fairfax survey party encountered thirty warriors returning from a long-distance raid against an enemy tribe to the south. The Indians danced that night to the rhythmic shakes of a gourd rattle and the drumbeat of a deerskin stretched over a pot of water. As Washington recorded in his diary:

> *They clear a Large Circle & make a Great Fire in y. middle then seats themselves around it, [The] Speaker makes a grand Speech telling them in what Manner they are to Daunce after he has finish d y. best Dauncer Jumps up as one awaked out of a Sleep & Runs & Jumps about y. Ring in a most comicle Manner he is followed by y. Rest. . . .*

Literally and figuratively, the Indians spoke a language young Washington did not understand, nor did he seem to want to learn. Likewise, the frontier settlers on their isolated homesteads represented aliens without a place in his Tidewater social matrix. As he described them in his diary:

> *a great Company of People Men Women and Children . . . attended us through the Woods as we went showing there Antick tricks I really think they seemed to be as Ignorant a Set of People as the Indians they would never speak English but when spoken to they speak all Dutch. . . .*

But he did learn how to measure and evaluate land. Over the next four years of his late teens, he took what he learned on the Fairfax expedition and parlayed it into a profession. He surveyed nearly two hundred plots of land, encompassing over sixty thousand acres, often working for Lord

Fairfax in the Shenandoah and other mountain valleys where he had traveled on that first trip. He learned well the different species of trees—red oak, white oak, box oak, Spanish oak, chestnut oak, hickory, pine—and identified them in his surveys to mark boundaries. Becoming a certified county surveyor and making good money for a teenager, he began to purchase his own lots on the frontier fringe as his forebears had before him and his older brothers were doing now, making his first acquisition, in the autumn of 1750 at the age of eighteen, along the Shenandoah Valley's Bullskin Creek.

Compared with George, however, Lawrence and another half-brother, Augustine, or Austin, speculated in western lands much farther afield and on a much greater scale. While George surveyed in the Shenandoah and other near valleys, they looked beyond the Appalachian Mountains to that vast wilderness called the Ohio. The older half-brothers, along with Tidewater planter friends and other investors who included Thomas "Big Spoon" Cresap, founded a land speculation company to profit from the vast region, naming it the Ohio Company of Virginia. Using connections with a rich London merchant and an English duke, in 1749 they received from King George II a grant of two hundred thousand acres near the Forks of the Ohio—a strategic spot over the mountains and about two hundred miles inland where two rivers, the Allegheny and the Monongahela, join to form the Ohio River. If the Ohio Company developed the land for the British Crown by settling it with one hundred families and building a fort, King George would grant three hundred thousand acres more. The problem, however, was that both the French in Canada and land speculators in Pennsylvania were eyeing the same spot, not to mention the Indian tribes for whom the whole Ohio wilderness served as a hunting ground.

Then Lawrence fell ill. He fought an unrelenting cough. Doctors were consulted, treatments sought. George accompanied his older half-brother and role model from Mount Vernon on a journey over the Blue Ridge to a warm springs whose waters were said to restore health. When that did not help, Lawrence aimed for the island of Barbados in the Caribbean Sea, where the warm, dry tropical winter was said to be good for the lungs. At Mount Vernon, Lawrence's wife, twenty-two-year-old Nancy, had recently given birth to a daughter after three of their earlier children had died in infancy. It was feared that neither mother nor child could survive the tropical climate, so George, now nineteen, went as Lawrence's travel companion instead.

Sailing in late September of 1751 they made landfall six weeks later at the Barbados port of Bridgetown in Carlisle Bay.* With its narrow, winding streets, medieval feel, and massive stone forts amid waving palms, this was a central economic hub of the British Empire, the island's great wealth rooted in the sugar cane growing tall in the tropical sun. They rented rooms in a British officer's two-story stone house on a hilltop, with a big portico, sea breezes through the windows, and a spectacular view over the bay. George, in his own words, was "perfectly ravished" by the island's beauty, sugar plantations, and lush tropical fruits. They engaged in a busy social life with the island's elite, including dinners, the theater, and fireworks displays, although Lawrence remained a shut-in during the day to avoid the hot sun, and George contracted—and survived—the acute fevers of a case of smallpox. This was a blessing in disguise, for it immunized him for life.

They had hardly stayed six weeks in the British colony when Lawrence, feeling oppressed by the heat, opted for the cooler climate of Bermuda. He put his younger brother on a ship back to Virginia, and George, fighting seasickness on the gray winter ocean voyage home, tried to process the brief, spectacularly colorful visit. A few months later Lawrence himself arrived back in Virginia so frail that he immediately wrote out his will. Among other things, it contained the fateful stipulation that George should inherit Mount Vernon if Lawrence's infant daughter, Sarah, died without children and his wife, Nancy, was deceased also. Lawrence died a few weeks later during the summer of 1752, in his midthirties, with his family around him at Mount Vernon, due to what was then called "consumption" but is now known as tuberculosis.

---

* This was George Washington's first and last journey overseas. [See www.mountvernon.org/george-washington/washingtons-youth/journey-to-barbados.]

With Lawrence's death, George's life began an inexorable transformation. He was now liberated from his enthrallment to his role model and childhood hero while also weighted with a new sense of gravitas—even urgency—with the feeling that his own adulthood had commenced. Time seemed short. His brother had died young, and no one in his Washington male line in Virginia had made it out of his forties. Lacking a formal British education, young Washington had strived for self-improvement during his teen years—practicing mathematics, copying out rules of civility, learning to survey—and had displayed conscientiousness, focus, a knack for quick learning. Now, with the death of Lawrence, a powerful dynamic upwelled from somewhere deep to the placid surface of his life—a relentless sense of ambition.

Where it originated is a matter of speculation. The centuries-long legacy of his ancestry back to John the Builder, the Reverend Lawrence, and John the Sailor may have delineated one current. Washington would later claim to know little about his ancestors in England, but he knew stories of his Virginia forebears and their enterprise and success. He was also acutely aware that the Washingtons had never quite arrived in the colony's upper aristocracy, despite their diligent efforts. He had looked to the adventurous Lawrence and the elegant young gentleman George William Fairfax for models of what he wished to become. Other family currents may have powered his ambition, such as a father who died when George was very young and may have left behind lofty expectations for his son—whether overtly stated, or implied, or existing only in George's mind. His father's untimely death also left behind his widowed mother, whose own emotional needs may have contributed to his ambition. She never seemed satisfied with what her oldest son had done, or done for her.

Some historians point to George's six weeks in Barbados as a pivotal moment in his life. Here, rather than a nineteen-year-old surveyor from a modest Northern Neck planter family, he was accepted as a gentleman equal in an elite British society, where he dined, to his amazement, in the glittering company of commodores, admirals, judges, wealthy merchants, and the governor of Tortola. The table conversation ranged far beyond that in his familiar corner of Virginia, exposing him to the cosmopolitan centers and farthest reaches of the British Empire and the greater political forces at play in the world. Seeing this, he may have aimed his ambition much higher than he had before.

When he arrived back in Virginia and stopped in the colony's small capital of Williamsburg, he sarcastically referred to it as "the great metropolis." He had delivered some letters from Barbados to the Governor's Palace. Governor Dinwiddie himself had invited him to dine. This was a new experience, too. The governor had asked about his brother's health, knowing Lawrence from public service. In addition to being a planter and member of the House of Burgesses, Lawrence had held the office of adjutant for the Virginia milita—the administrative officer in charge of the colony's citizen forces that the governor could call up in time of need. This office now took on a great deal of appeal for George, despite his total lack of military experience and his youth. He had witnessed among the British elite of Barbados and its resplendent commodores and generals some of the glamour and reward of the life of a royal officer, commanding the astounding power of empire that resided in its coastal defense's massive stone walls and heavy cannon. As Lawrence's health had faded, it was clear his adjutancy would fall vacant. Governor Dinwiddie decided to split it into four smaller adjutancies, each carrying the title of major. Not long after their dinner together in Williamsburg, George petitioned Governor Dinwiddie to award him the district adjutancy for his home region, Virginia's Northern Neck.

"If I could have the Honour of obtaining that [adjutancy]," wrote George, with almost indecipherable rhetorical curlicues, ". . . [I] should take the greatest pleasure in punctually obeying from time, to time, your Honours commands. . . . I am sensible my best endeavors will not be wanting, and doubt not, but by a constant application to fit myself for the Office, could I presume Your Honour had not in view a more deserving Person. . . ."

He soon won the appointment. What he would find in his first mission for Governor Dinwiddie, however, did not at all resemble the gold-embroidered glamour and palm-studded splendor that he had seen among the Barbadian elite.

CHAPTER THREE

A COLD DECEMBER RAIN FELL INCESSANTLY ON WASHINGton's little party as their horses slogged through the miry marshes and islands of trees along Rivière Le Boeuf. In a wax-sealed envelope wrapped in a waterproof satchel lay the urgent letter. They had been traveling for over a month from the Tidewater region into the Ohio wilderness and had, in effect, ridden off the edge of the map. They had now reached the last leg of their journey—leaving the old Indian town of Venango, they would follow the Rivière Le Boeuf upstream sixty or more miles north to its source near Lake Erie. Here Washington hoped to find the French commandant and deliver Governor Dinwiddie's letter.

Late on the second day, the horses stopped. A large feeder stream, Sugar Creek, poured its swirling waters into Rivière Le Boeuf, blocking their progress along the riverbank. Engorged with rain, the creek had climbed its banks, breaking away chunks of debris. A rider stepped his horse down the bank of Sugar Creek and splashed into the water up to the horse's belly to test its depth, then reined it back toward shore. Sugar Creek ran too deep for the horses to wade. The animals would have to swim. But they could not easily swim encumbered by the heavy baggage—tents, bedrolls, provisions—tied atop them or along their flanks.

Trees had fallen across the creek, their exposed root balls undermined by the current. The party consulted. They would unload the baggage and camp here for the night. In the morning, they would haul the

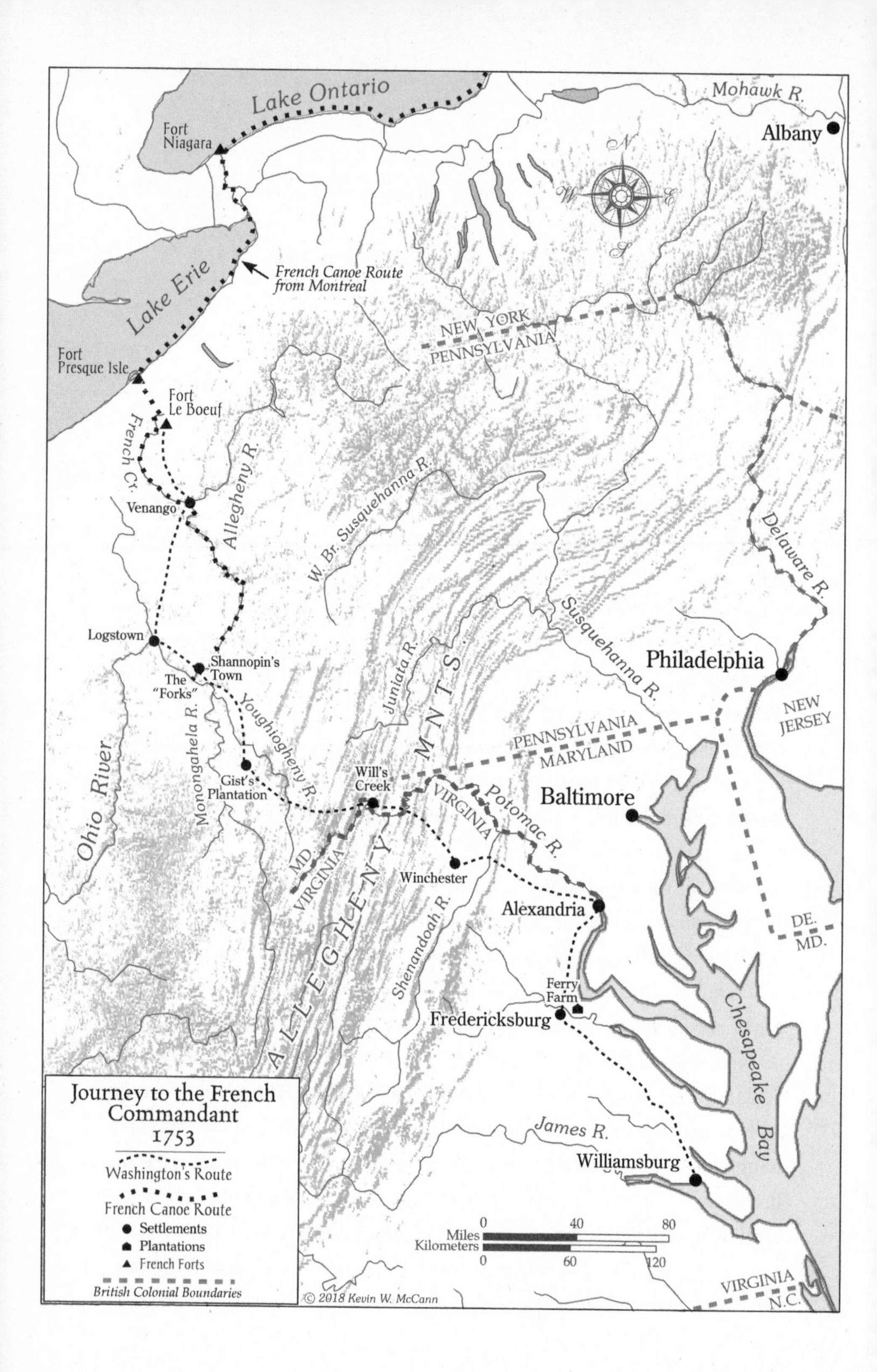
Lake Ontario
Mohawk R.
Albany
Fort Niagara
N
W
E
S
Lake Erie
French Canoe Route from Montreal
NEW YORK
PENNSYLVANIA
Fort Presque Isle
Fort Le Boeuf
French Cr.
Allegheny R.
W. Br. Susquehanna R.
Delaware R.
Venango
Logstown
Shannopin's Town
The "Forks"
Susquehanna R.
Philadelphia
Juniata R.
NEW JERSEY
PENNSYLVANIA
MARYLAND
Youghiogheny R.
Monongahela R.
Ohio River
Gist's Plantation
Will's Creek
VIRGINIA
Potomac R.
Baltimore
MD.
VIRGINIA
Winchester
Alexandria
ALLEGHENY MNTS.
Shenandoah R.
DE.
MD.
Ferry Farm
Fredericksburg
Chesapeake Bay
James R.
Williamsburg
Journey to the French Commandant 1753
Washington's Route
French Canoe Route
Settlements
Plantations
French Forts
British Colonial Boundaries
Miles
Kilometers
0
40
80
0
60
120
VIRGINIA
N.C.
© 2018 Kevin W. McCann

baggage piece by piece across the fallen logs over Sugar Creek, swim the horses, and, once beyond the feeder creek, resume their journey up the bank of Rivière Le Boeuf.

The servitors loosened cinches, sliding bags and tents and saddles down the horses' wet flanks and dropping them into heaps on the ground. They hobbled the horses, tying their legs close together to prevent them from wandering far. The booted men swished through the tall grasses, feeling for the flattest, driest patches of ground, to pitch the tents. They snapped off dead branches from the surrounding groves of trees—oak, beech, sugar maple, ash—and from fallen logs, then crouched on the ground and swept away the wet leaves. Opening a tinderbox, one man removed fire-making materials and struck flint to steel, aiming the blow so the spark dropped onto a bit of charred cloth. The spark ignited a tiny orange circle in the charred cloth. Setting the glowing cloth onto a nest of tinder—dandelion fluff, dried grass, or wood slivers—he bent over on hands and knees, head near the ground, and blew gently until the tinder caught flame, then laid on tiny dry twigs and thick branches that popped and jumped until the flames climbed upward from the riverbank like a living thing, and an expanding sphere of warmth and light pushed back the chill, soggy dusk.

Soon they gathered around the fire, sitting or standing, drying their sodden clothes, roasting pieces of venison on spits. A pot of Indian corn mush simmered. Led by Major Washington, the party had grown since leaving Venango, now numbering more than a dozen—wilderness guide Christopher Gist, French interpreter Jacob van Braam and Indian interpreter John Davison, and the four servitors who handled horses and baggage for Washington's party; a French officer and three French soldiers provided by Captain Joncaire as an escort for Washington and his men on this last leg of their journey; plus the three Indian sachems and their hunter.

Sitting or standing in the fire's warm circle with Indian chiefs and a French officer who knew the way, Washington must have asked: Is there a quicker route than through these marshes to the river's headwaters?

On this, his first mission, Washington took Governor Dinwiddie's charge with the utmost gravity—deliver the letter to the French commandant, wherever he was, with "all convenient and possible dispatch." But delay upon delay had bogged his progress—delays to wait for Indian chiefs, delays from rain, and now delays due to rain-swollen streams like Sugar Creek. Governor Dinwiddie had received alarming reports that the French in Canada were moving soldiers toward the Ohio Valley and perhaps even building forts there, this on lands claimed by His Majesty, King George II. Governor Dinwiddie, as well as Washington's own family, had a personal financial stake in the young major's mission: the two hundred thousand acres of Ohio Valley wilderness lands granted by King George II to Lawrence Washington's Ohio Company of Virginia, where another half-brother, Austin, was also a shareholder. With an eye to gaining maximum political leverage in an era without conflict-of-interest rules, the Ohio Company partnership had cleverly given a 5 percent interest to Virginia's recently appointed acting governor, Robert Dinwiddie. Much hung on the timely delivery of the letter. If the French moved in before Washington had delivered it, the Ohio Company's investment, Washington's family's interest, and the governor's stake—and the great valley itself—could be lost.

Now rain-swollen Sugar Creek had stopped the party, at least for the night. How many days more until they reached the headwaters of the Rivière Le Boeuf?

One imagines the three Indian chiefs around the fire gazing at Washington solemnly, topknots protruding from bald heads, outer ear-rims sliced so they dangled like earrings. In traditional Native American fashion, one may have picked up a stick and sketched on the earth a mental map of the Rivière Le Boeuf and their route along it. In his surveyor's fashion, Washington had been sketching his own paper map of his route through the largely uncharted region to take back to his patron, Governor Dinwiddie. Noting Washington's impatience, the Half King may have spoken, or perhaps the frontiersman Gist or the wilderness-savvy French officer, La Force, who was said to be fluent in fourteen Indian languages as well as French and English. Or perhaps it was left unsaid: The natural world keeps its own rhythms. Events here do not conform to a schedule, by hour or day or date, no matter how great the urgency that one imports from the coast. It will defy your efforts to impose your will.

And there were other, more delicate matters that young Major Washington may or may not have understood. How did these three parties around the campfire that night—Iroquois, French, and British—dance around the fact that each claimed the Ohio Valley as theirs?

The Iroquois claimed it because they had vanquished the tribes who lived there. The French claimed it because their explorers had canoed it decades before other Europeans and then used it for years as a fur-trading route. The British claimed it at the time of the Jamestown landings when they staked for the Crown a theoretical claim extending to the west coast.

The dynamics and diplomacy among Indian tribes were as complex as the dynastic rivalries and ever-shifting alliances in the kingdoms and principalities of Europe—between the Habsburgs, the Stuarts, the Bourbons, and all the rest. The first British sailing ships that anchored along the Eastern Seaboard had discovered countless Native American villages dotting the shores, inhabited by dozens of different Indian tribes. Some had bonded into greater alliances. In what's now Maine, for instance, an estimated twenty-five thousand people of at least twenty tribes joined together in the Abenaki Confederacy, while in today's Virginia, the Powhatan Confederacy embraced about two hundred villages. As the Europeans spread inland from their tiny coastal toeholds, disputes flared over land use and trading rights. Diseases like smallpox imported from Europe devastated native villages. Violence erupted—the Powhatan Wars, the Pequot War, King Philip's War. Peace treaties were drawn and alliances forged between whites and natives, shifting as the relative strength of tribes and colonies waxed and waned. But as a ceaseless procession of white sails lifted above the blue horizon, more settlers relentlessly demanded more land. Some British colonists seized Indian lands by divine proclamation or by war, others bought it through treaty or purchase, until British colonists had pushed Native Americans from the fertile coast toward the interior mountains. Tribes had always fought each other in small-scale wars for honor and revenge.

Now, however, tribes vied against each other for trading opportunities with the colonists, exchanging the abundant wild animal furs of the American forests for valuable European manufactured goods such as knives, beads, and guns.

In the mid-1600s in the eastern American woodlands, a series of conflicts erupted over trading rights with the Europeans and other tribal matters that came to be known as "The Beaver Wars." The Iroquois League emerged from the Beaver Wars as a ruling power, similar to a European kingdom that vanquished rival kingdoms and made them vassal states. A union of five tribes (later six), the Iroquois League occupied a swath just south of the Great Lakes. This created a buffer zone between the two rival European empires that were colonizing North America—the French in Canada and the British along the Atlantic coast.

The party clustered in the fire's warm circle at Sugar Creek mirrored the region's power dynamics: eight British colonials, four French soldiers, and, in between the rival empires, three Indian chiefs—the Half King, White Thunder, and Jesakake, with their young hunter.

Did young Major Washington grasp the Half King's difficult, ambiguous position, literally teetering between the two empires? The Iroquois League had long maintained its neutrality between the French in Canada and the British on the east coast, at times deftly playing the two powers against each other to its own advantage. The league's power reached across a large region of eastern North America in the late 1600s and early 1700s. It had pushed rival tribes out of the lower Great Lakes area and the Ohio Valley, forcing some all the way to the Upper Midwest. The Ohio wilderness, its resident tribes forced out, became a huge hunting ground for the Iroquois. Some Iroquois eventually formed scattered villages in the Ohio Valley wilderness, where they became known as the Mingoes and, at least nominally, remained under central Iroquois leadership. Whites settling in eastern Pennsylvania and Virginia also pushed tribes such as the Delaware out of the Susquehanna Valley and the Shawnee out of the Shenandoah Valley. Moving westward over the mountains, these tribes took refuge in the giant wilderness haven of the Ohio. Here the migrating Delaware, Shawnee, and Mingoes lived in bark longhouses in widely scattered villages surrounded by cornfields, often together in the same settlement. But British traders, eager to acquire the abundant furs of the Ohio Valley, which would fetch steep prices in Europe's fashion markets, followed the migrating tribes from Pennsylvania and Virginia into the Ohio wilderness. Here they established trading posts at the Indian villages, where Indian hunters could trade furs for manufactured goods. The Ohio tribes had also begun to establish relations with the Pennsylvania government on their own. This could bring trouble for the Iroquois League. The league claimed the Ohio Valley as its own, and the Delaware, Shawnee, and Mingoes who lived there were subservient to it. The league wished to maintain its neutrality between the French and the British.

Seeing the potential for renegade action from the Ohio tribes, the Iroquois Council, based at Onondaga, near what is today Syracuse in upstate New York, appointed a regent to oversee the tribes and their settlements. His name was Tanaghrisson, and he was then about fifty years old. He had been born a Catawba in the Southeast but was captured as a child by Iroquois raiders, carried north to the Great Lakes, and raised as one of them. The British called him the Half King because they believed he was only partially empowered by the Iroquois Council and could negotiate on behalf of the Ohio Indians but had to clear all his decisions with the council.

Where the Half King's true loyalties resided in this mix of conflicting claims—with the Iroquois who appointed him, or the British, or the French, or mainly to his own interests—was not entirely clear, nor was his actual power over the Ohio tribes or whether he had simply stepped in and filled a power vacuum. One story says that he nurtured a lifelong hatred of the French because they supposedly had boiled and eaten his father. Whatever the case, he walked a delicate line between these European nations competing for the Ohio wilderness, whose representatives were seated or standing with him around the campfire on the bank of Sugar Creek. He may have simply wished that both British and French would keep their forts or settlers out of the Ohio Valley, and that the Iroquois League would leave him alone as the head chief of the tribes who lived there. Yet he had accepted from the French a wampum belt—a beaded band carrying a symbolic message—that signified friendship between the French and the Half King's followers. He had promised Washington, however, that he would break the friendship alliance with the French by returning the wampum belt when they reached the French commandant. He had also promised that he would make the other tribes in the Ohio—the Delaware and Shawnee—do the same.

So far, they had not. They had good reason to feel ill-disposed toward the British. In an infamous land deal sixteen years earlier, the sons of William Penn had contrived to slice off one and a half million acres of Delaware Indian lands and sell them to white settlers, leaving many Delaware wandering in search of a home. Some had come to the Ohio Valley. They had not forgotten the wrong committed by the British in Pennsylvania.*

Beyond the fire's orange circle of warmth stood the skeletal shapes of trees, or so one imagines, their bark and bare limbs wet with the day's rain. It had now paused. Perhaps a gust of wind sighed through the treetops, knocking droplets from high, swaying branches. They fell unseen through the night and pattered on the fallen leaves behind the men who faced the fire, like whispers from a spirit world. Soaked canvas tents, damp clothes, the smell of wet wool. Steam rose from wet breeches on the thighs of men who stood before the flames, drying themselves, turning around to expose chilled, wet backs to warmth and light. The firelight shone off the bald-plucked heads of the Indian sachems, feathers tied in their topknots, wrapped in their matchcoats and deerskin moccasins, cross-legged near the fire.

Young Washington stood taller than the rest, over six feet, with big

---

* In the notorious Walking Purchase of 1737, the sons of William Penn, after their Quaker father's death, had produced an old document which supposedly showed that the Eastern Delaware had given up a piece of land measured in part by how far a man could walk in a day and a half. The Penn sons apparently hired the fastest white runners around, had a trail cleared through the forest, and set them loose. One runner covered seventy miles, slicing off more than one million acres of Delaware lands in the Delaware Valley, which the Penn brothers proceeded to sell to white settlers.

bones, long limbs, and big hands and feet, although still on the skinny, gangly side of his eventual 175 pounds. He possessed broad, muscled shoulders but hips that appeared wide and a shallow chest relative to his other proportions, and wore his hair in a short queue. Next to the reddish-brown faces of the Indians around the fire, perhaps painted with bear grease and red, yellow, and black designs, and even against the French soldiers and British frontiersmen, Washington bore the notably pale, papery complexion of the Englishman, his cheeks slightly pitted with smallpox scars and his blue-gray eyes looking out from under a heavy brow. Although technically the Virginia officer heading the party, young Major Washington probably remained measured and even quiet, listening and watching to take his cues from wilderness expert Christopher Gist, wishing to be taken seriously by the older, more experienced members of the campfire gathering.

Perhaps the cold rain resumed, hissing on glowing coals and producing little puffs of steam from the white ash that ringed the campfire, an ancient circadian signal sending the men off toward their tents. Beyond the fire, the darkness was utter and encompassing, starless on this rainy night. Washington may have paused momentarily before crawling in. They had left towns and cities far behind—three hundred miles to Philadelphia and New York, even farther to the quiet, familiar streets of Williamsburg, and nearly five hundred to Montreal. Ahead to the west lay the great swell of the continent, of forest, prairie, river, mountain, extending a distance unimaginable to him, populated by unseen Indian tribes—painted, ornamented, roaming—and a strewn handful of French fur-trading outposts. Somewhere far, far beyond them—it was still unclear exactly where—lay the near shores of the Western Sea. Facing west into the dark night from where Washington stood on the banks of Sugar Creek, the nearest city lights would not appear until the coast of China.

Did he feel threatened out here, so far beyond the familiar? Did he sense euphoria, or rage, hidden out there in the continental darkness amid the scattered tribes? Perhaps from Christopher Gist he had begun to learn the shifting dynamics of native alliances and rivalries, now exacerbated by long-running power struggles imported from Europe to the

New World. Maybe he didn't care, or preferred to keep it simple: Britain and France vying for the Ohio wilderness and Britain the clear owner. One could reduce all these complications to that. He may have felt a thrill, and a sense of pride. Standing in the rainy darkness, still warm from the fire and his belly full of roasted game and perhaps relaxed by a tot or two of rum, with the French soldiers, painted Indian sachems, leather-fringed frontier guides crawling into tents or wrapping themselves in bearskins, and he in charge of them all—he may have wished that Lawrence could see him now. And the young ladies back at the Tidewater plantations. And his friend from earlier frontier travels, George William Fairfax, and Colonel Fairfax, his father, at the family's Belvoir estate. And especially George William's charming and beautiful twenty-three-year-old wife, Sally Cary Fairfax, two years older than Washington, who held a special spot in his heart. He would have stories to tell when he got back to Belvoir. He was far out in the wilderness defending the king's possessions and Virginia's integrity. Surely they would marvel at the exoticism of it all.

* * *

By the gray light of dawn they could see that Sugar Creek still surged high. Breakfasting in the rainy morning, they packed up their camp, folding or rolling the wet canvas tents, tying up bearskins and blankets. For their crossing of the creek, they chose a large tree that had fallen across it, bank to bank, dragging its branches in the surging current. They climbed out on the trunk with hatchets and chopped away the branches that protruded upward. With a path cleared, they balanced on the log and hauled their baggage across, piece by piece, stacking it on the far side. They then swam their horses, riding the puffing, chesty animals across the stream, chilling water splashing up the horses' shoulders and over their own boots and thighs, hooves spinning silently in the depths until the horses felt the solid bottom beneath them and heaved themselves with muscular lunges and sprays of water up the far bank.

"The major and I went over first," recorded Gist, "with our boots on."

It was growing colder. Flecks of white streaked down with the falling raindrops. It began to snow as they splashed through mires and jogged through meadows and forest groves. In the muddy low spots, the horses' hooves stamped out saucerlike puddles, pocked with falling rain and snow. The saturated air—just above freezing—seeped through their layers of wool and cotton, and the cold metal of the stirrups pressed against their leather boot soles so their feet ached. From beneath them rose the warmth of the horses, steaming in the dankness.

Late in the day, having traveled twenty-five miles, they arrived at Cussewago, an old Indian town founded by a Delaware chief. They unloaded their horses and pitched their tents. Children emerged from the scattered collection of single-family wigwams and multifamily bark longhouses. The Half King knew these people. In theory, he held power over them as a representative from the Iroquois Council, which considered these Delaware a tributary people. They did not necessarily agree, however.

Perhaps the Delaware sachems of the village came to call on the Half King, or perhaps he took Washington, Gist, and interpreter Davison to a longhouse to visit them. If it were a typical longhouse where families lived, Washington would have ducked to enter through a low door hung with a bearskin flap. His first impression probably was of dimness, and that it smelled of the aged woodsmoke that had stained its interior to a dark patina, as it would have been windowless except for smoke holes in the vaulted pole-and-bark roof. Sound inside was muffled by its swept earthen or mat-covered floor and walls of stitched sheets of bark, grass matting, and moss to chink the cracks. It may have stretched fifty, sixty, or one hundred feet or more in length and been about twenty feet wide. Several fires burned at equally spaced intervals down the center. From the rafters hung braided corn and dried squash and other foods. Sleeping compartments containing bearskin-covered pallets lined the walls, one for each family. Families facing each other across the central corridor shared a fire, the women propping pointed clay pots in the hot coals and boiling stews of corn, beans, squash—the "Three Sisters"—that they cultivated in nearby fields.

The men typically cleared land for the fields, constructed fish traps in the streams, and hunted for venison and bear meat in the forest to roast, sizzling, over the flames on sticks or lay directly on cleanly glowing coals. Inside the longhouse, the flickering firelight would have glinted faintly off their greased and painted faces—red, black, yellow, white—and their bald heads plucked with mussel-shell tweezers. They left a small patch of long hair protruding on top of their skulls, which they stiffened with bear grease and dyed red to stand upright in a fierce "scalp lock." Bear claws or shells perhaps dangled around their necks, while draped over their shoulders were mantles of bearskin against the season's cold, and to protect legs and feet against the wet or snow, deerskin leggings and embroidered moccasins.

The Indian women wore embroidered deerskin skirts and leggings, winter capes sewn of layered turkey feathers that repelled rain and snow, and headbands of wampum. They highlighted their eyelids with red. Women held tremendous power in both Iroquois and Delaware societies. Both societies were matrilineal, as were the Shawnee—meaning that family descent, property rights, and leadership positions were traced through the female line. They were also matrilocal—meaning that a newly married man and woman moved into the longhouse of the woman's family. And in the Iroquois especially, women wielded a great deal of political power. "Clan mothers" nominated the male chiefs. The women could approve or veto the chiefs' decisions. If necessary, the women could strip them of their chieftaincy. As the givers of life, they also held the power to decide matters about the taking of human life—including declaring war or making peace. Although young Washington might not have been aware of it, the Indian women of the Ohio Valley could play a major role in deciding whether the Ohio tribes sided with the French or with Washington, Governor Dinwiddie, and the British throne.

One horse had grown exhausted trudging over the swampy ground. They left it behind at Cussewago, and set out again on the narrow, crooked Indian trail along Rivière Le Boeuf. It was hard going—mud, roots, stream crossings, swamps. They looked across to the opposite side of the river where the going appeared easier and drier. When they reached a spot known to their Indian guides as a ford, they found the rain-swollen river still too swift and deep to cross and too broad for fallen trees to bridge. They made camp. The Indians looked for logs to lash into a raft. The party discussed the idea and gave up on it, figuring more flooding creeks would greet them on the far side. It would be faster to skip the trouble of building a raft.

Despite his urgency, Washington still assessed the land as he rode along as if he were surveying it for settlers and judging its agricultural potential. In his journal he wrote:

> *We passed over much good Land since we left Venango and through several extensive and very rich Meadows, one of which I believe was near four Miles in Length, and considerably wide in some Places.*

They camped after only eight miles that day, stopped by another deep creek that they balanced across on fallen logs while carrying their luggage. With the party's food supply dwindling, the Indians tracked and killed a bear, and they roasted fresh bear meat over the fire. It had now been a month since Washington had left the plantation houses of the Tidewater region. He had ridden to the northwest nearly three hundred miles, as the crow flies, from the Atlantic coast almost to the shores of Lake Erie, much farther than that if accounting for the winding trails. The temperature had dropped. Winter was coming. With no mountain barriers to stand between here and the Arctic, cold winds swept down from the broad reaches of Canada, across the big lake, picking up moisture and releasing it on the forested southern shore. Snow began to fall more steadily. Washington now brushed up against another realm where he had never traveled before—the north.

As darkness fell on December 11, nearly a week after leaving Venango—a week of rain and snow, swamps and creek crossings—they arrived on the stream bank opposite Fort Le Boeuf. It stood as a black hulk on a low rise, almost surrounded by a loop in a headwaters stream of Rivière Le Boeuf. It was a much more substantial structure than Washington had seen for weeks, formed by four long two-story buildings arranged in a square and enclosed by walls, or palisades, of logs twelve feet high buried upright in the earth, with the top ends sharpened to discourage any would-be climbers. Triangular structures called "bastions" projected from each of the fort's four corners, so it resembled a four-pointed star, and each bastion displayed dark portholes like eyes, where cannon lurked.

It would have been a grievous error to surprise those watchful eyes. Young Washington no doubt received advice from wilderness veterans Gist, the Half King, and La Force on how to make a proper diplomatic approach when emerging unannounced from the forest at night. Interpreter van Braam, who spoke both French and English as well as his native Dutch, was dispatched across the stream, probably accompanied by La Force and the three French soldiers, to inform the fort's officers of the presence of Major Washington and his party. Soon, canoes arrived from across the stream to greet the travelers.

WASHINGTON IN HIS TWENTY-ONE-YEAR-OLD URGENCY and impatience wanted to present himself as soon as possible to the fort's commander. One imagines the French subofficers telling Washington that he must wait until the proper moment for an introduction to occur, that French military formality prevailed here. They camped in their tents outside the fort's palisades, or some perhaps occupied a spare barracks, and the French subofficers may have supplied food to cook, or even provided supper. "[T]hey received us with a great deal of complaisance," wrote Gist.

It wasn't until the next morning that the fort's commander officially received Washington. A French subofficer met the young major and led him through the fort's front gates to the commandant's quarters within the stockade. The door opened. A fire in a fireplace dimly illuminated the low-ceilinged interior of the log structure. Washington was led in. Out of the dimness the French commandant stepped forward. "He is an elderly Gentleman, and has much the Air of a Soldier," recorded Washington.

One pictures an impeccable French uniform, graying hair, a lined face, a perfectly erect bearing, a certain precision to his movements, and a reticence to his person. He was missing one eye. He was an aristocrat and a knight—Jacques Legardeur de Saint-Pierre de Repentigny, Knight of Saint Louis. He came from an illustrious line of explorers, fur traders, and military officers in New France and was great-grandson of Jean Nicollet, the first European to explore what's now called the Upper Midwest. Fluent in several Indian languages and having served at many distant outposts, he was known for his wisdom and intelligence in diplomacy with the Indians and had only one week earlier arrived at Fort Le Boeuf to replace the previous commandant, who had recently died. Age fifty-two and heir to a century-old feudal estate near Montreal, he had just returned from a major exploratory mission for the governor of New France to find the shores of the Western Sea—the Pacific Ocean—when young Major Washington appeared at his door.

One can imagine Saint-Pierre's formal military greeting to young Washington. For his part, Washington—tall, broad-shouldered, still a bit youthfully awkward in body—no doubt drew himself up in a posture to match Saint-Pierre's. He announced that he was Major Washington of the Virginia militia. With some pride, he showed the one-eyed French knight his passport from the jowly Scottish administrator, Governor Dinwiddie. "Whereas I have appointed George Washington Esqr. . . . My express Messenger to the Comandant of the French Forces on the River Ohio, & as he is charg'd with Business of great Importance to His Majesty & this Dominion. . . ."

A heady document for a twenty-one-year-old to be carrying, the handwritten passport asked His Majesty's subjects and allies, and all others who abided by the Laws of Nations, to help the bearer on his journey. Also holding out his commission as messenger from Governor Dinwiddie—"I reposing especial Trust & Confidence in the Ability Conduct, & Fidelity, of You the said George Washington . . ."—he announced that he possessed an urgent letter to deliver to the commandant.

This all had to be worked out through the interpreter, van Braam. The answer came back from Saint-Pierre: Please hold on to the documents until another high-level officer, Captain Repentigny (probably a relative of Saint-Pierre), arrives. He was expected any hour. He commanded the Presque Isle fort, located about fifteen miles overland from Fort Le Boeuf on the southern shore of Lake Erie. This was the terminus of a supply route followed by large French cargo canoes from Montreal, up the Saint Lawrence River, across lakes Ontario and Erie, to the natural harbor at Presque Isle. Following an ancient Indian portage trail, they then carried canoes and goods inland the fifteen miles from the lakeshore to the headwaters of Rivière Le Boeuf, which flowed southward into the Ohio River system, giving access to that huge valley and the Mississippi beyond. The French military had recently built the forts at Presque Isle and Rivière Le Boeuf, creating a formal travel link between the Great Lakes and the Ohio River, separated by the fifteen-mile portage over a rise of ground.

At 2 P.M. Captain Repentigny arrived from Fort Presque Isle. Washington, again summoned to Saint-Pierre's headquarters, once more presented his credentials. This time, Commandant Saint-Pierre and Captain Repentigny accepted the documents and the sealed letter he carried from Governor Dinwiddie. Washington, mindful of the governor's instructions that he should not wait for more than a week for a reply, asked the French officers for a quick response.

They begged to be excused. Leaving Washington waiting in the main room of the headquarters, they entered a side room. In its privacy, Captain Repentigny, who knew some English, scratched away in ink on a blank page to translate the letter into French. Commandant Saint-Pierre then summoned Washington to enter the private room with his own interpreter, van Braam, and asked that van Braam correct any mistakes in the French translation of the governor's letter. Formality, protocol, and hospitality governed the diplomatic proceedings between these two rival empires wherever they bumped into each other, even out here in the wilderness, at least during times of peace.

This ended business for that first day. Commandant Saint-Pierre probably invited Washington, Gist, van Braam, and perhaps some of their companions to dine that evening. Again, one pictures the guttering candlelight inside the log structure, the roasted venison, the wine. The elderly veteran officer Saint-Pierre in his formal hospitality surely carried himself more guardedly than Captain Joncaire, the officer who had hosted Washington for dinner at Venango and bragged that the French, by God, would take the Ohio. Commandant Saint-Pierre, older and more experienced, and having arrived only a week earlier at his new post, would not discuss French intentions for the Ohio in this casual setting. With nothing resolved, they all retired from their dinner to their separate quarters.

The next day, December 13, Saint-Pierre and the other French officers held a council of war among themselves and also met with the Indian chiefs, Half King, White Thunder, and Jesakake. While the French held council, Washington walked around Fort Le Boeuf and did some spy work for Governor Dinwiddie, taking the dimensions of the fort and counting its cannon. In another heady moment for a twenty-one-year-old, and for the first time of many to come, he also dispatched his own spies, sending his companions to assess the many French canoes hauled up on the riverbank near the fort. They counted fifty birch-bark canoes and one hundred and seventy hollowed out of pine logs, plus many more "blocked out" of pine in preparation for final shaping. This fleet of canoes, Washington realized, could transport many hundreds of men down the Ohio to build French forts.

The Half King had warned Washington of this. He had earlier told the young major about his meeting with Marin, the late French commandant of Fort Le Boeuf, on a previous visit. The Half King, according to his own account, had vehemently objected to the French building "great houses"—forts—on Ohio Valley lands that belonged to neither French nor British but to the Indians. He was not afraid to evict the French, the Half King had told Commandant Marin. The French commandant, however, had angrily dismissed him and used the Indians' own imagery drawn from the natural world to show that he did not fear their threats:

> *I am not afraid of Flies, or Musquitos, for Indians are such as those; I tell you, down that River I will go, and will build upon it, according to my Command: If the River was block'd up,*

I have Forces sufficient to burst it open, and tread under my Feet all that stand in Opposition, together with their Alliances; for my Force is as the Sand upon the Sea Shore. . . .

As his men counted the hundreds of canoes at Fort Le Boeuf, Washington surely remembered that imagery. There was little doubt that when winter broke and spring arrived, these canoes would carry French troops—as numerous as grains of sand—down the Rivière Le Boeuf to the Forks of the Ohio and other key points. Here they would build forts and secure the Ohio Valley against the claims of the British and the Tidewater speculators of the Ohio Company, including the Washingtons.

Washington anxiously awaited the reply of Saint-Pierre and Repentigny. Every day he lost at Fort Le Boeuf would give the French a head start in the spring. He grew convinced that the French officers were stalling, deliberately delaying him and at the same time trying to convince "his" Indians to join the French side by bribing them with gifts and "fair Promises of Love and Friendship." While he waited, the temperature dropped daily. The snow drifted in on Arctic blasts and blanketed the earth around the fort. Through the swirling veils the forest's edge blurred to a hazy gray. Rims of ice appeared on the dark, looping creek. Washington grew concerned that his party's exhausted horses were weakening daily with the deepening cold and snow. He decided to send them, unloaded and in the care of the servitors, south to Venango to wait there. Once he had received a reply from Saint-Pierre, the rest of his party—Washington, Gist, the interpreters, and the Indian sachems and hunter—would paddle down the Rivière Le Boeuf to join them, using canoes that the French officers had offered to lend them.

In his anxiety to get moving, Washington probably did not appreciate the design of the birch-bark canoe. He may have seen it here for the first time, as it was a vessel of the north. The French had learned about it from the Indians, and the Indians had perfected it over uncounted centuries, using a spectrum of materials from the forest.

It started with its skin—the outer "rind" of a large birch tree. Working in July's heat when the bark peeled easily, the native canoe builders judged which side of the tree grew the thickest bark and, on the opposite side, cut a long vertical incision. This would give them a long sheet of bark whose thickest, toughest part ran down its center and could be used for the bottom of the canoe, which endured the most punishment from rocks in the river. Vertical incision made, they cut a horizontal incision around the trunk high up and another down low, then carefully peeled off the outer layer of bark in huge strips up to eighteen feet long and four feet wide while leaving intact the inner layer that kept the tree alive. White and papery dry on their outer surfaces, these big birch-bark strips had a thick orange undersurface with the flexibility and toughness of leather. They could be shaped easily for about a week.

The builder laid the large, flexible strip—or several strips stitched together if making a large canoe—out on the ground. On top of it he laid a simple template. This was made of two bowed strips of wood tied together at each end so that they had the shape of a canoe seen from above. He then bent the edges of the bark sheet so that they stood upright to make the canoe's sides, taking on the canoe shape of the template. He sandwiched each of the two upright edges between two long, flexible strips of white cedarwood, tightly binding each sandwich together to make the canoe's gunwales, using as cordage the long, skinny, flexible, and incredibly strong roots of the spruce tree, which, when stripped of their outer covering and split lengthwise, were called "watap." Removing the template, he now had a birch-bark canoe with vertical sides and a flat bottom.

To give the hull its round-bottomed shape, the native canoe builder fashioned ribs by splitting long, thin planks off a cedar log, steaming them for several hours, and bending them into a U shape. To make the ever-smaller U-shaped ribs that gave the canoe its end-to-end taper, the builder first bent the largest steamed rib—the rib for the center of the canoe—and nestled the next smaller steamed rib inside of it and the next smaller inside of that, like the layers in an onion slice. Once dried and hardened, the ribs would be removed from their nest, inserted into the flexible but square-bottomed birch-bark hull of the canoe, and fixed to the gunwales, giving the hull its rigidity, rounding out sides and bottom, and thus creating its handling ability and streamlined speed in the water. Sealing the birch-bark seams with a waterproof gum made from melted spruce resin, tallow, and pulverized charcoal, the builder also added the crosspieces called "thwarts" to stiffen the hull as well as provide a means of carrying the amazingly light vessel upside down on one's shoulders during a portage.

Except for the sealskin-covered kayaks of the Arctic's Inuit, no craft in the world equaled the birch-bark canoe's portability, maneuverability, and ability to carry weighty cargos.* It had allowed Native Americans to travel easily across much of the North American continent. Where Indians instantly saw the superiority of European muskets and metal knives, French explorers and fur traders quickly grasped the superiority of the birch-bark canoe. These became the workhorses of the French expansion in North America and emblematic of the way the French blended with the native population and borrowed its best aspects, adapting to the culture, landscape, and resources at hand, in contrast to the British attempts to impose a linear order over the landscape with survey lines and straight roads.

Young Major Washington did not take specific notice of the birch-bark canoe's attributes, or perhaps dismissed them, or was too distracted to care about them. He had to get going. Winter was closing in. Three days after Washington's arrival, Commandant Legardeur de Saint-Pierre had finally given him a reply letter to carry to Governor Dinwiddie. But now the French delayed him in other ways. As he put it, they employed "every Strategem that the most fruitful brain could invent" to detain the Half King, White Thunder, and Jesakake and win them to the French side. "I can't say that ever in my Life I suffer'd so much Anxiety . . . ," Washington recorded.

He clearly felt outmaneuvered by the veteran French officers, who

---

* The French-Canadian voyageurs in the fur trade, working in conjunction with the Indians, built huge cargo craft called "Montreal canoes" up to forty feet in length, able to haul four tons in addition to fourteen paddlers, yet light enough to be carried on a portage by four men.

possessed decades' worth of experience in the wilderness and Indian diplomacy. Without the advice of Christopher Gist, he would have found himself in deep trouble. He wanted to carry out his first mission for Governor Dinwiddie as conscientiously as possible. The governor had ordered him to employ all possible haste. If he did not leave Fort Le Boeuf very soon, snow and ice would block his path over the mountains to the Tidewater region, possibly delaying him for months. He needed to get out now. If he rushed his departure, however, and the Indians were not ready to go and decided to stay behind, he risked losing them to the French side. This would be a disaster, too. How was he to resolve this dilemma?

With the help of interpreters van Braam and Davison, and surely with the advice of Gist, Washington scuttled back and forth between two powerful forces in the Ohio Valley. On the one side was the gentleman officer and wilderness veteran with his eyepatch and soldierly bearing ensconced in his headquarters inside the fort; on the other, the regal Half King, painted and adorned, in barracks or tent just outside the fort's bastions along the wintry wilderness creek.

As Washington recorded in his journal:

> *I went to the Half-King, and press'd him in the strongest Terms to go. He told me the Commandant would not discharge him 'til the Morning. I then went to the Commandant, and desired him to do their Business, and complain'd of ill Treatment; for keeping them, as they were Part of my Company. . . . He protested he did not keep them, but was ignorant of the Cause of their Stay; though I soon found it out:—He had promised them a Present of Guns, &c. if they would wait 'til the Morning.*

Finally, on Sunday morning, December 16, after Washington persuaded the Indians not to indulge in the proffered French liquor, his party climbed into the borrowed French canoes to begin the long journey back to Governor Dinwiddie. Saint-Pierre had studied the French translation of the governor's letter. The governor had written:

> *Sir,*
>
> *The Lands upon the River Ohio, in the Western Parts of the Colony of Virginia, are so notoriously known to be the Property of the Crown of Great-Britain, that it is a Matter of equal Concern and Surprize to me, to hear that a Body of French Forces are erecting Fortresses, and making Settlements upon that River. . . . I must desire you to acquaint me, by whose Authority and Instructions you have lately marched from Canada, with an armed Force, and invaded the King of Great-Britain's Territories. . . . it becomes my Duty to require your peaceable Departure. . . .*

Washington carried Saint-Pierre's reply to this message buried deep in his satchel as he climbed into the canoe. Likewise penned in the felicitous phrases of diplomacy, the French commandant promised the Virginia governor to forward Dinwiddie's letter to his commander, the Marquis Duquesne, governor of New France, whose "Answer will be a Law to me. . . ." Then Saint-Pierre drove home the main point. "As to the Summons you send me to retire, I do not think myself obliged to obey it."

CHAPTER FOUR

THEY EMBARKED THAT SUNDAY WITH TWO CANOES CARRYing four men in each—one paddled by the three sachems and their hunter and one paddled by Washington, Gist, and the two interpreters. They may have used lightweight birch-bark canoes or perhaps the heavier but more durable wooden canoes hollowed from pine logs. The Indians' canoe pushed off first from the creek in front of Fort Le Boeuf, crossed a small lake just downstream from the fort, passed from the lake into a larger creek, and disappeared around a bend. After a mile or two, they entered the Rivière Le Boeuf. Washington's canoe soon followed, although the Indians were lost from sight.

It was cold paddling on a winter day. As they knelt or sat in the canoe, toes numbed and joints stiffened against the chilled hull, bloodless fingers clenched paddle shafts. Snow along the riverbanks accentuated the dark water twisting in U-shaped bends through forest and marsh under the gray sky. They paddled much farther than the crow flies. At times they looped back on where they had just been, perhaps twice passing the same skeletal dead tree protruding from the marsh, each time from a different direction. Steering the canoe through the tight bends, each dip of the paddle blade spun off little sucking whirlpools. The water appeared more viscous than a summer river, as if thickening and congealing with the cold. In the slack-water shallows, ice rimmed the riverbanks. The V wave from their canoe bow angled out across the narrow river and washed over the rims of ice along the banks, rocking them gently with a whispering sound.

Young Washington had never undertaken a canoe journey this long. A certain tedium set in, stroke after stroke, on this wintry day. Perhaps Gist knew some of the songs of the French-Canadian voyageurs—those laborers of the French fur trade who propelled the huge, fourteen-man birchbark cargo canoes along the Canadian rivers and lakes, thirty thousand or more paddle strokes covering fifty to ninety miles daily, with loads up to four tons.

Washington and Gist kept their eyes open for the Indians in their party. No sight of them, except perhaps the dark splash marks of their bow wave over the snow-dusted rims of ice. Apparently they had paddled ahead—unless they had somehow returned to the French fort.

In the shallow riffles, the canoe hull crunched into the gravelly bottom. They clambered over the gunwales and stepped into icy water that made their anklebones ache. They bent to pull the canoe over the shallows, often for half an hour at a stretch. Climbing back in, they dodged the canoe back and forth, skimming past rocks protruding under the surface that could stave in the hull.

Early darkness fell. They had paddled about sixteen miles without sighting the Indians. They pulled over to the riverbank, unloaded baggage, pitched their tent, and lit a fire to warm their stiff limbs. The fire's yellow light lapped out into the darkness. Washington worried about "his" Indians. Had the commandant somehow convinced them to switch sides? Should he have been more cautious in letting them paddle ahead?

After paddling some distance the next day, Monday, December 17, they spotted the canoe of the Indians lying on the riverbank and realized they had gone off hunting. They waited for their return. Three of the Indians eventually walked into camp hauling loads of bear meat. They had killed three, perhaps using the traditional Indian method of identifying a hollow tree where a bear lodged, climbing to the entrance hole, and dropping in smoldering bundles of punkish wood. Escaping the smoke, the bear eventually climbed from the hole and was shot by the hunter. Meat from the black bear provided a dietary mainstay for many of the Native American tribes of the forests of eastern North America, where there was an estimated population of some two million black bears.

They probably roasted the best pieces, cut from the legs and loins. After their bear meat feast, they prepared to pack up camp the next morning, but one of the Indians traveling in the Half King's canoe still had not returned from the bear hunt. Now there was another cause for urgency. With the end of the rains and the onset of snow and winter, the water level of the river was dropping quickly. If it dropped too far they would have to abandon the canoes and walk to Venango to retrieve the horses they had sent ahead. Unwilling to risk this major delay, and probably on Gist's advice, Washington decided to leave the Indian party behind at the hunting camp. He would travel on with Gist and the two interpreters.

It was the toughest travel he had yet experienced. The temperature dropped still further. The water level fell. With soaking boots and cold-stiffened fingers, they hauled the canoe over shallow, rocky bars. In the calmer stretches, thin sheets of ice reached all the way across the river, blocking their way. Leaning out over the gunwales to break a path, they struck at the clear, smooth sheet with thunking paddle blades, spiderweb cracks skittering off in all directions, leaving shattered ice triangles spinning in black water in the canoe's wake. In swifter stretches, the current snapped off big rafts from the ice shelves anchored to the shore, which bobbed freely downstream.

For three days after leaving the Indian hunting camp, they worked their way downstream, making scant mileage—seven miles one day, twenty the next. On the third day they came to a tight, U-shaped bend. As the river narrowed around the bend, the current squeezed together the freely bobbing ice rafts, piling them crunching atop each other, as if a giant set of unseen hands swept together the playing cards strewn across a tabletop.

They could not smash their way through the ice jam. They walked downstream, scouting for a passage. Where the river looped back on itself, a narrow neck of land only four or five hundred yards across separated the two sections of the U. Camping here for the night, they hauled their baggage and carried their canoe across this neck, launching it again downstream of the ice jam the next day.

As Washington's party was portaging across the narrow neck, three

French canoes and the canoe of the Half King and other sachems were appearing around an upstream bend and pulling over to shore above the ice jam. A fourth French canoe had overturned upriver, losing both the canoe and its cargo of gunpowder and lead, and its passengers now had crowded into the other three French canoes. Even for the experienced French, the Rivière Le Boeuf was not easy to navigate, especially in the low water of winter.

Now all five canoes—Washington's, the Half King's, and the three French canoes—started down these lower stretches of the Rivière Le Boeuf, twenty or thirty miles above Venango and the river's confluence with the Allegheny. It ran faster and rockier here. The stiff current shoved them toward the obstacles. They frequently had to jump out of their canoes and grab them to avoid catching on rocks and tipping. Water froze to their clothes. The French canoes were slightly ahead, with all parties struggling to wend their way down the shallow, icy river, when one of them caught sideways on a rock and capsized, spilling goods and flailing passengers into the frigid waters and delighting those in the British canoe. Here was one instance, at least, where Washington's party had outmaneuvered their French rivals. "[W]e had the pleasure of seeing the French overset," Gist recorded, "and the brandy and wine floating in the creek, and we run by them, and left them to shift for themselves."

SEVEN DAYS AFTER SETTING OUT from Fort Le Boeuf, they arrived at Venango and the river's mouth where it joined the Allegheny. It was now December 22. Washington had left the Tidewater region six weeks earlier, in early November. With Saint-Pierre's reply to Governor Dinwiddie in his satchel, he knew he must complete the return journey much more quickly.

More problems, however, beset Washington as he hurried to leave Venango. Reunited at the Indian village with his party's horses and servitors, he learned that the animals had grown "feeble" after weeks of cold, lack of forage, and the heavy loads on the long journey into the wilderness. Now, to get out quickly over the mountains, and with little chance to stop or rest en route, the party would have to load the horses even more heavily with food for the party and feed for the animals, especially if they did not expect to find grassy areas where the horses could easily graze. An average-sized horse of about one thousand pounds, if working intensely, especially in winter, needs as much as 33,000 calories per day. With hay providing about 900 calories per pound and oats providing about 1,250, the horse would need up to thirty pounds of feed a day, part forage and part grain. To carry a week's supply of its own food, the horse would be burdened with over two hundred pounds, or the weight of a substantial grown man.

Instead of riding weakened, overburdened horses, they realized that some of the party would have to walk, potentially slowing progress further. Another issue was White Thunder. He had arrived at Venango both ill and injured, perhaps from the bear hunt. Too infirm to continue on foot, and without a horse strong enough to carry him, he would have to be transported by canoe from Venango down the Allegheny River to Indian villages farther downstream. The Half King and Jesakake would accompany him. Washington worried that the Indians might tarry at Venango after he left and that the smooth-talking Captain Joncaire would persuade them to abandon their alliance with the British. Washington warned the Half King. "I told him I hoped he would guard against his Flattery, and let no fine Speeches influence him in their Favour," Washington recorded. "He desired I might not be concerned, for he knew the French too well, for any Thing to engage him in their Behalf. . . ."

Washington and his party and the riderless horses left Venango and headed into the forest on Monday, December 24—Christmas Eve, 1753. It snowed all that day, covering the trail. On Christmas Day, the temperature dropped again. The horses lagged. The streams froze, making it difficult to find drinking water for the horses. The trail grew icy. Three men in the party suffered frostbite and provided no help. The party trundled slowly along on foot.

Washington could feel the days slipping away and see in his mind's eye the hundreds of French canoes lined up on the creek bank at Fort Le Boeuf, waiting for spring when the ice would break and the French would set off down to the Ohio. If those canoes launched before Governor Dinwiddie had a chance to respond to Saint-Pierre's letter, Washington would have failed in this, his first mission, which he had hoped would launch an illustrious career as gentleman officer and Tidewater planter.

It troubled him that night. He was young and strong and not slowed by frostbite. He was convinced he could outpace the other members of the party and their agonizingly slow progress. If he broke off from the rest and pushed ahead bearing the crucial letter he might save days otherwise lost to the French. Governor Dinwiddie would at least know that he had given it his all. The governor might even regard this bold stroke as heroic, and so might the gentlemen and ladies back at Belvoir on the Potomac.

The next day, December 26, Washington proposed that he and Gist leave the rest of the party of horses and baggage and strike off alone, on foot, to deliver the message in the speediest way possible. Instead of following regular Indian trails, Washington proposed to his guide that they travel in the most direct line possible—through the trackless woods by compass bearing.

Gist wasn't so sure. He knew that Washington, as a barely out-of-his-teens member of the Tidewater horse-riding class, had never traveled on foot like this through the wilderness, especially in winter, despite his eagerness to attempt it. "The Major desired me to set out on foot, and leave our company . . . ," recorded Gist. "Indeed, I was unwilling he should undertake such a travel, who had never been used to walking before this time. But as he insisted on it, I set out with our packs, like Indians. . . ."

Equally surprising, Washington underwent a major change in fashion, either at his insistence or on Gist's advice. He stripped off his heavy British overcoat—and possibly his breeches and boots—and put himself in "Indian walking dress."

Washington was enthusiastic. Gist was worried.

As Washington and Gist stepped into the lonely forest that Christmas of 1753, three thousand miles away across the Atlantic, the city of London burst with exuberant life beneath a damp chill and a pall of coal smoke so heavy that at midmorning one might need a candle to read. Nearly seven hundred thousand people were crowded into greater London, which had recently surpassed Paris as the largest city in Europe. Merchant ships jammed the Thames River, bearing cargos—sugar, tea, tobacco, spices—from far-flung colonies. Dockworkers and sailors thronged the riverbanks and nearby slums. Merchants packed the coffeehouses, making deals, reading the day's news, debating issues, and creating innovative ways to finance and insure their valuable cargos at sea. In Tower Street, the patrons at Edward Lloyd's coffeehouse gave birth to the great shipping insurer, Lloyd's of London. Private banks sprang up. This newly thriving hub of transoceanic trade and finance, one essayist observed, spawned "agents, factors, brokers, insurers, bankers, negotiators, discounters, subscribers, contractors, remitters, ticket-mongers, stock-jobbers and . . . a great Variety of other Dealers in Money."

Forced off the land by the consolidation of small farms into large ones, tens of thousands of English country dwellers had moved to the great metropolis, swelling it beyond the winding streets of its old medieval center. Trades and crafts flourished—weaving, porcelain-making, shipbuilding. Other laborers sweated in sugar refineries, breweries, and gin distilleries, the imbibing of the latter becoming a drug epidemic, with an estimated two pints a week consumed for every man, woman, and child in London. There were bone-boilers, grease-makers, glue-makers, paint-makers, rope-makers, and soap-makers. The old street markets gave way to a new sales venue—the retail shop.

In the news that December of 1753 were the opening of Hampton Court Bridge across the Thames upstream of the city (the old London Bridge still stood, with hundreds of structures built on it) and the award of a gold medal by the Royal Society to Benjamin Franklin for his recent experiments with lightning and electricity. The aging King George II had now sat on the throne for twenty-three years. The realm was no longer called just England but Great Britain, after the parliaments of England and Scotland had merged several decades before. Since the English Civil War, a century earlier, a Bill of Rights had passed, inspired in part by the ideas of political philosopher John Locke, limiting the power of the throne and greatly expanding that of Parliament. Parliament now controlled domestic matters. George II retained influence in foreign affairs. There had been peace with France for the last five years, since 1748, although King George II still made vigorous appointments of officers to the military. Tensions between the two empires remained high both in Europe and in distant colonial outposts in India, Africa, and the Americas.

Londoners celebrated Christmas that year over the course of several days, the traditions having been revived after the Puritans in Parliament had banned them. Well-off families burned great hunks of wood called "Yule logs" in their fireplaces, and friends and relatives feasted on tables piled with roast beef, goose, turkey, venison, puddings, mince pies, ales, wines, and wassail bowls. Carolers sang. Revelers gave presents. Children joyfully received special treats and toys. The tradition of the Christmas tree had not yet arrived from Germany, but the city's residents hung up holly, laurel, and evergreens on houses and church pews. The generous-minded gave to the poor, and the day after Christmas the city's wealthy gave a "Christmas box" of gifts to their servants to take home to their own families, while the city's apprentices knocked on the doors of their masters' clients and begged for alms. Thus evolved the holiday Boxing Day. The celebration usually ended on the Twelfth Night—January 5—when households shared a special cake in which a bean and a pea had been baked. Whoever received the slices with the bean and pea would become the King and Queen of Misrule, in which the normal hierarchy of the household would be turned upside down.

Beyond Christmas celebrations, the bulging city, spilling over from its ancient, twisting streets, offered countless other forms of entertainment for the curious, the enterprising, the weak of flesh. Where the old gentry living on their rural estates had once disdained London and cities generally as sinful, they now took townhouses and shopped at luxurious retail stores, perused bookshops, visited their private clubs. Wealthy Londoners had moved into the fashionable West End and taken to strolling for entertainment in the landscaped public squares and gardens when the weather was fine. Theater thrived, and in that December of 1753, while young Washington slogged through the snowy Ohio wilderness, the great David Garrick, manager and actor at the Theatre Royal Drury Lane, was playing King Lear opposite a Cordelia played by the celebrated actress and contralto Susannah Cibber, for whom the composer Handel had specifically tailored solo parts in his original *Messiah*.

There were men's clubs, gambling houses, an estimated 207 inns, 447 taverns, 5,875 beerhouses, and 8,659 brandy shops, in addition to some 500 coffeehouses and countless brothels. One in every five women in eighteenth-century London was a prostitute, according to some estimates. A frequent patron of those prostitutes, the eighteenth-century diarist and biographer James Boswell once remarked to his literary mentor and friend, Dr. Samuel Johnson, that he preferred to visit London rather than live there, so as not to grow tired of it. "Why, Sir," Johnson famously replied, "you find no man, at all intellectual, who is willing to leave London. No, Sir, when a man is tired of London, he is tired of life; for there is in London all that life can afford."

IF LONDON WAS THE BULGING CITY of commerce and finance, Paris was the city of ideas and fashion, the intellectual capital of Europe, the beacon from which radiated the great revolution in thought called the Enlightenment, when the darkness of ignorance would be illuminated by the light of reason and idea of human progress. It was the era of the French philosophes—of Voltaire, Diderot, Rousseau—who became darlings of the French aristocracy. Great Parisian ladies hosted them to hold forth in glittering salons at their townhouses, or opulent *hôtels particuliers*—urban estates with courtyards and gardens—to partake in evenings of wit and intellectual conversation, fashionable dress, and lavish food and drink. That December of 1753, the movement was reaching a pinnacle of sorts. The philosophe Denis Diderot and the mathematician Jean d'Alembert had recently embarked on a revolutionary new publication—the *Encyclopédie*. Resting on the assumption that reason and science triumphed over blind faith and ignorance, and tolerant and liberal in its views and in places even hostile to the Church, this massive undertaking attempted to collect and publish all human knowledge. It was twice suppressed by authorities and clergy but continued to publish nevertheless, eventually reaching thirty-five volumes and 71,818 articles.

It helped that the mistress of the king, Madame de Pompadour, embraced Enlightenment thinking and promoted the *Encyclopédie*'s publication to Louis XV. In December of 1753, he had held the throne for thirty-eight years. When only five, and after smallpox had decimated his immediate family, he had succeeded his great-grandfather, the spectacular Sun King, Louis XIV, who had risen like a great, rainbow-hued bird surrounded by the fabulous luxury of Versailles. The Sun King had consolidated and centralized power in France and ruled, unlike the British king, now constrained by Parliament, as an absolute monarch who embodied all branches of government and whose word was law. A believer in divine right, the Sun King felt his mandate to rule had come directly from God and he was accountable neither to the aristocracy, the French people, or the Church. Personal glory was his highest ideal. "In my heart," he wrote in his memoirs, "I prefer fame above all else, even life itself. . . . Love of glory has the same subtleties as the most tender passions. . . ."

The Sun King's great-grandson and heir, Louis XV, carried on with the Sun King's belief in divine right and absolute monarchy, but with considerably less astuteness and success. The first whisperings against the French monarchy were heard during the course of his rule. The shift in thinking against absolute monarchy and divine right had been a long time in the making, and was slowly coming to a head. A century earlier, during the bloody chaos of the English Civil War (and about the time the ancestral John Washington left England for Virginia), English political philosopher Thomas Hobbes stated that humans in their primitive, "natural" state are inclined to hurt or kill each other. Life was "nasty, brutish, and short," as he famously put it. To prevent this bloody chaos, a "social contract" existed between a ruler and those ruled—each individual giving up some part of his natural rights and freedom in order to receive the protection of a strong sovereign against harm from fellow humans. Hobbes adamantly opposed the notion that monarchs have a "divine right" to rule. But he wanted a strong one.

Writing a few decades later, his fellow British philosopher John Locke took this concept of a social contract between ruler and ruled several steps further, asserting that humans in their primitive state of nature have an inherent moral code, which is commanded by God. Because all humans are

God's creations, and because all are "equal and independent," harming an individual is harming God's property. Thus, under the "Laws of Nature," individuals are prohibited from harming each other's "life, health, liberty, or possessions" but otherwise have freedom to pursue their own interests. Individuals, however, sometimes violate these natural laws by stealing another's property, or enslaving another, in effect declaring a state of war. While the wronged person may be morally justified under natural law in killing the transgressor, once a state of war begins it is hard to end. For this reason, humans willingly consent to give up to a body of government the power to punish those who harm them, and it is this civic body that will take action against the transgressor. The individual also consents to subject himself to the authority of that body and in turn receives its protection. By leaving the State of Nature and forming a governmental body, humans thus create laws, judges to rule on those laws, and an executive power to enforce those laws. By the same reasoning, if that governmental body or ruler no longer protects the life, health, liberty, or possessions of the individuals who have consented to join it, those individuals have the right to rise up against that government and break the contract.

In contrast to Hobbes's and Locke's measured reason, in which one could still hear the faint clink of the British aristocracy's silver and china, Rousseau, philosophically speaking, kicked over the table. He declared that all existing social contracts were fraudulent. Born to a respectable family of watchmakers in the Swiss Confederacy, Rousseau lost his mother at his birth and was raised partly by relatives. As a teenage apprentice to an engraver, he ran away from his home in Geneva, then a city-state republic ruled by an elected body. Rousseau lived a vagabond youth, including an idyllic stint in rural Savoy as the young lover of a minor noblewoman and naturopath who specialized in rescuing lost Protestant youth (such as himself) and converting them to a kind of mystical Catholicism. She encouraged young Rousseau to read deeply in philosophy and mathematics, to study music, to learn about plants. Eventually landing in Enlightenment Paris—socially awkward, self-educated, exquisitely sensitive, and highly excitable—he bluffed his way into a job teaching music and befriended philosophes like Diderot. In 1749, on a whim, he entered an essay contest sponsored by the Académie de Dijon, responding to the question "Has the Restoration of the Arts and Sciences had a Purifying Effect upon Morals?"

Decidedly not! Rousseau answered. He argued that man in his primitive natural state is good. It is civilization—including the arts and sciences—that corrupts him.

This won him the celebrated Dijon Prize and flung open the doors to the gilded salons of the great Parisian ladies, although, ever the contrarian, he had now taken up with a seamstress. In Paris, that December of 1753, as Gist and Washington trekked through the frozen woods, Rousseau had just turned down an annual pension from King Louis XV for a popular opera he wrote, condemned Paris and its society as corrupt, and prepared to return to the lakes and mountains of his native Geneva. It was to this republican city-state that he dedicated his next essay, his *Discourse on the Origins of Inequality.* The natural goodness of man, he argued, is corrupted by the advent of jealousy, private property, and inequality. All existing social contracts are fraudulent because they mainly benefit the rich. A few years later, in *On the Social Contract,* he would elaborate on these themes and propose a social contract that represented the will of the people rather than the interests of the rich and powerful, words that would help torch revolutions on both sides of the Atlantic in the decades to come.

"Man is born free; and everywhere he is in chains," Rousseau began his *Social Contract.* "Rousseau discovered nothing," it would later be said of him, "but inflamed everything."

In December 1753, there was little outward sign that France's Ancien Régime would one day be torn down or of the great spasm of violence that would follow. Rather, during the five years since the 1748 treaty ending the War of the Austrian Succession, a brief period of peace had prevailed in the centuries-old rivalry between Britain and France, however fragile it might be.

In the frozen woods, Washington was traveling between these two great empires, unwittingly carrying in his satchel the weight of seven hundred years, and hurrying with Governor Dinwiddie's instructions to use all possible haste. It no doubt sounded heroic to leave behind the weakening horses, frostbitten men, and winding Indian paths to strike out by compass bearing, but veteran woodsman Gist knew the young Tidewater gentleman had never before seriously traveled on foot, especially during the rigors of winter. Politely, he told Washington, in effect, that he didn't know what he was doing. But Washington insisted, and Gist gave in.

Displaying what would be a lifelong fascination with uniforms, Washington seemed very proud of his "Indian dress" for the occasion. "I . . . pulled off my Cloaths; tied myself up in a Match Coat; and with my Pack at my Back with my Papers and Provisions in it, and a Gun, set out with Mr. Gist, fitted in the same Manner, on Wednesday the 26th," he recorded.

A matchcoat was an Indian robe sewn of animals' skins, the soft fur turned inward to nestle against the wearer's body. After European traders arrived with their woven woolen cloths, many Indians fashioned it from a piece of heavy duffel about seven feet long, wrapped around the upper body and belted at the waist. Either version offered warmth in winter and freedom of movement, especially if worn—as Washington may have done—with the traditional leggings and moccasins. At night, the wearer could wrap himself in it to sleep, a great advantage to Washington and Gist. They could avoid carrying heavy blankets or skins or bedrolls in their packs, already heavily laden with food. They would burn through an enormous load of calories trudging through snow in cold weather with a heavy pack, exhausting even the hardiest traveler and slowing progress to one mile an hour or less. By some estimates, during this kind of extreme winter exertion an adult male needs 4,500 to 6,000 calories daily, or the equivalent of nine square meals a day.

Despite the new "Indian walking" outfits, Gist proved right in his skepticism about Washington's ability to handle the long journey. The day they left the horses, December 26, he and Washington walked eighteen miles, but it exhausted the young Virginian. "[T]he Major was much fatigued," Gist wrote, tactfully not mentioning that he had predicted exactly that.

They followed a known trail through the forest, not yet having cut across country. Gist knew the way. "It was very cold," he recorded. "[A]ll the small [creeks] were frozen, that we could hardly get water to drink."

No birds chirped. No water trilled over rocks. A frigid wind from the north might have creaked in the oak branches. Or the Arctic air may have settled in a silent, crystalline mass. Tree sap froze. Expanding, it popped, the hard, woody sound echoing through the silent woods. Boots crunched and squeaked, or the soles of Indian moccasins brushed softly over the snow.

As daylight faded, the prospect of camping looked overwhelming. Gist knew of an Indian shelter, or lean-to, along the trail. There they spent part of the night before a fire.

But the fire died, or they grew chilled wrapped in their matchcoats, unable to sleep. They rose very early, moving vigorously to warm themselves, and got back on the trail at 2 A.M., Gist recorded. They passed through a Delaware Indian village. Here an Indian called out to Gist, using his Indian name. He appeared glad to see Gist, as if familiar with him. Gist wondered if they had met at Captain Joncaire's at Venango. The Indian had questions for Gist: Why are you traveling on foot? When did you leave Venango? Where did you part from your horses?

Gist replied that he and Washington were heading for the Forks of the Ohio. Because the young major had "insisted" on taking the shortest route possible through the forest—as Gist pointed out again in his journal description of the meeting, as if to exonerate himself from putting the two of them in this situation—Gist did not know the way. "We asked the Indian if he could go with us, and show us the nearest way," Gist wrote, as if this was at Washington's insistence, and not altogether trusting the Indian.

The Indian appeared pleased to be asked. He took Washington's pack and slung it on his own back. Leaving the known path, the threesome set out into the snowy forest heading east—too far east and too far north, Gist thought—and walked fast for eight or ten miles. After the previous day's eighteen miles, plus the distance in the predawn hours to the Delaware village, plus another fast eight or ten miles with the Indian guide, the athletic Tidewater horseman whose servants normally hauled the loads acutely grasped the difference between covering distance on a mount and traveling quickly on foot in the exhausting cold carrying one's own baggage.

"The Major's feet grew very sore, and he very weary," wrote Gist. ". . . The Major desired to encamp." The Indian, however, did not. In response to Washington's weariness and urgings to stop, the guide began to act strangely. First he offered to carry Washington's gun for him, presumably to lighten his load. Washington refused. "[T]hen the Indian grew churlish," wrote Gist, "and pressed us to keep on, telling us there were Ottawa Indians in these woods, and they would scalp us if we lay out."

The Indian proposed they go to his lodge, where they would be safe. Gist and Washington pondered. It was very difficult in this wilderness to tell friend from foe. They started toward his lodge, which, he said, lay the distance away that one could hear a gunshot. After some time walking—again heading too far north, thought Gist, growing increasingly suspicious but saying nothing to Washington—they still hadn't arrived. The Indian assured the two that his lodge now lay at only two whoops distance. They went another two miles without seeing smoke or any other sign. Washington no longer trusted the guide, either. "The Major said he would stay at the next water," wrote Gist of the exhausted young Virginian.

By now it was dark. They entered an open meadow. Blanketed with snow, the meadow reflected enough starlight and moonlight to appear quite bright, surrounded by the dark shapes of trees. In the starlight, the dark figure of the Indian stood about fifteen steps away. Washington saw him turn and raise his gun. He fired. The muzzle flash lit the snow, and the heavy musket concussion ruptured the quiet, starry night. From Washington's perspective, he had appeared to fire at Washington or Gist.

"Are you shot?" called out Washington.

"No," replied Gist.

The Indian ran to a spreading white oak tree and began reloading his musket under it. They hurried to him. Had he been shooting at some small animal? When they saw that he loaded a lethal ball into his gun—as opposed to the pellets used to hunt small animals—they believed he had intended to shoot one of them. Without revealing their suspicions, as he stood nearby they deliberated what to do about the Indian. "I would have killed him," wrote Gist, "but the Major would not suffer me to kill him."

Gist's readiness to kill the Indian may have shocked Washington. He had never killed before. It was one thing to suspect that the Indian in the starlit meadow might have been aiming at them. It was quite another to kill him for it. Perhaps Washington suddenly doubted himself. Had he too hastily rushed to a conclusion? Did what he thought he saw really happen? Maybe the Indian was in fact shooting at an animal, or signaling his family, or accidently discharged his musket. What terrible consequences for a mistaken glimpse in the dark on Washington's part. Did he want blood on his hands for that? With Gist acting as the eager executioner? Or perhaps he was sure of what he saw and simply possessed a greater sense of mercy. In any case, he was less ready to kill than the seasoned frontiersman.

"As you will not have him killed," Gist said to Washington, "we must get him away, and then we must travel all night." The veteran Gist laid a strategy. At a small creek nearby, they pretended to make a camp. They had the Indian light them a fire. The two kept a close guard over the guns. After a time, Gist made the opening: "I suppose you were lost, and fired your gun."

The Indian replied that he knew the way to his lodge. Gist now tried to convince him to go home. "Well," Gist told him, "do you go home; and as we are much tired, we will follow your track in the morning; and here is a cake of bread for you, and you must give us meat in the morning."

Gist had diplomatically given him a way out of the tense situation. The Indian, Gist recorded, was glad to leave. Gist stealthily followed the Indian through the woods for a distance to ensure that he had gone. Then Gist made his way back to Washington. Should the Indian return, especially return with others, he did not want the two of them to be found in the same spot. They hiked a half mile or so and lit a small fire. Here they rested and set their compass. They would follow a compass bearing through the forest toward the Allegheny River and the Forks. At the Forks they would come to Shannopin's Town, an Indian village where Gist knew they would find Indians who were friends.

All night they walked through the snowy forest. They had to outpace any Ottawa Indians who might be tracking them or other Indians rallied by their former guide. They had now traveled nearly nonstop for twenty-four hours. Probably for the first time, young Washington felt his life genuinely in danger. "Very weary" even before they reached the starlit meadow where the Indian shot his gun, he now had no choice—keep walking, exhausting as it was, or risk being killed.

Spurring their pace was the knowledge that it would be easy for the Ottawa Indians to track them in the snow. He had wanted adventure and excitement, like that of his older brother Lawrence. Now Washington was getting it. Was this a heart-thumping thrill for the young Tidewater gentleman new to deep wilderness and Indian threat? Or simply frightening and exhausting, and he just wanted it to end? He had come to the wilderness in part to discover exactly that.

Washington was certain they were being followed. He recorded:

> *[We] walked all the remaining Part of the Night without making any Stop, that we might get the Start, so far, as to be out of the Reach of their Pursuit the next Day, as we were well assured they would follow our Tract as soon as it was light. . . .*

Washington had overestimated his own capabilities and understanding when insisting they head off into the pathless forest. He had also underestimated the disconcerting power of the unknown. With the possibility of danger behind every shadowy tree, the wilderness must now have appeared malevolent to him, moving wearily, pack on his back, Indian leggings and fur tunic wrapped close, snow swishing around his moccasined feet, trying to follow Gist's lead as he slipped through the forest like an Indian, silently bending past branches, avoiding protruding sticks, sliding over snow-covered logs.

What was that!?
Nothing, just the night.
What's that ahead? Ready your gun!
It's nothing, just your eyes.

They might pause, look around, hearts thumping from exertion and adrenaline. Starlit snow and the dark boles of trees—a cluster, like a silent party of Ottawa Indians.

It's nothing, just the forest, we must keep moving.

Or so one imagines. Whatever exactly took place, this night surely became a transforming experience for young Washington. Hour after hour they strode along, one leg after the other in the dark forest, endlessly. At times the taste of danger must have given way to boredom, and bone-deep weariness, and then to chest-thumping, ear-ringing danger again. It might have been all those things, thrilling and adventurous and frightening and utterly exhausting. Perhaps, like a woman's pain in childbirth, the fear and exhaustion would eventually recede into the deeper recesses of memory. But from this danger-charged night forward, one senses that Washington began to have a consciousness of himself as one of the dramatis personae—as if always keeping one eye on how his harrowing adventures would play in the sitting rooms of Belvoir Manor, in the chambers of the Governor's Palace, and eventually on the stage of history. This eye toward posterity, perhaps, underlay his youthful heedlessness, when his ambition for recognition far outpaced his capabilities or any sense of caution.

In this situation, it was very fortunate that he had Gist. When daybreak came, they found themselves at the headwaters of a stream Gist called Piney Creek. All day they followed the creek downstream. They had now traveled nonstop for thirty-six hours, having started at 2 A.M. two days before. Toward nightfall, as they wended their way down the forested banks of Piney Creek, they spotted Indian tracks in the snow. It appeared to Gist that the party had been hunting. Perhaps he saw the bright red spots of splattered blood from a wounded deer in the snow. In all events, they wanted to avoid the Indian hunters, having no idea who they were.

They split up. This surely was Gist's idea, the experienced frontiersman knowing how to confuse a tracker. They agreed to rendezvous at a designated landmark farther along and made the passage without incident. "We encamped," wrote Gist, "and thought ourselves safe enough to sleep."

Nothing bothered them in the night, except perhaps the cold. They surely slept close to the fire, rousing every so often and reaching out to throw more wood on the glowing coals, wrapped in their matchcoats, blankets, or animal furs. To insulate themselves from the frozen ground or snow on which they slept, which threatened to drain all their body heat away, they may have improvised sleeping pads of dry leaves or pine boughs. They were traveling light to move fast. The warmer the camping equipment one brought along—whether tent, sleeping pad, bedroll, or extra clothing—the heavier the pack and the harder to carry. This is the paradox of winter camping, an inescapable equation and logistical problem that Washington first encountered here, under Christopher Gist's guidance, and would encounter again.

The next morning, December 29, no doubt roused by the cold, they got an early start. Heading eastward through the forest, they soon reached the west bank of the Allegheny River, which flowed south to the Forks of the Ohio. Shannopin's Town, their immediate destination, sat on the far bank a few miles downstream. They had hoped the extreme cold would have frozen the surface of the river so they could cross it on foot. To their dismay, however, they discovered that instead of a smooth sheet from bank to bank, a solid rim had frozen along the shore but, in the broad center channel, the powerful Allegheny shoved along great floes of shattered ice, smashing, grinding, and hissing into each other. It had probably frozen over somewhere upriver but then fractured into pieces. "[T]he Ice I suppose had broke up above, for it was driving in vast Quantities," Washington recorded.

There was only one way to get across—make a raft. They possessed one "poor hatchet," as Washington called it, not wishing to carry a heavy ax while on foot. They went to work. They had to select suitable logs—preferably logs from pine trees or poplar, which are lighter and float more easily than dense hardwoods like oak. Standing dead timber makes the best, driest logs—trees that have died but not yet fallen to the forest floor and rotted. To fell a tree like this with one small, dull hatchet would require a tremendous amount of effort. One pictures Washington and Gist taking turns with the little hatchet, chopping away like beavers, woodchips flying. The young Virginia gentleman was not used to this kind of work. Having procured several large logs, they would lay them parallel to each other, place smaller logs perpendicularly across, and bind them together with rope to create a platform. It had to be hand-numbing work in the severe cold to lash the logs and tie the knots, but the alternative was to spend another night on the frozen ground rather than in front of a fire in an Indian lodge in Shannopin's Town.

After laboring on the raft the entire day, they finished it just after sunset. They slung their backpacks on board and slid it off the ice shelf and into the river. They jumped on board, grabbing the "setting poles" they had crafted—long wooden poles with which, standing on the raft, they could push against the bottom of the river, propelling the crude craft forward through the water.

They pushed out into the broad Allegheny in the twilight. The sudden power of the current seized the raft, jerking it downstream. Thick floes of ice swirling in the river pressed around them. The raft wedged between moving floes. It jostled and bumped, its logs driven underwater by the force of the current and floating ice, threatening to capsize. "[W]e expected every Moment our Raft to sink, and ourselves to perish," Washington recorded in his journal.

Determined to save himself and Gist and the important mission, Washington shoved his pole to the river bottom and leaned on it with all his powerful six-foot frame to steady the raft. The youthful strength and fine horsemanship that had distinguished him in the Virginia plantation country meant nothing here. The river's powerful current shoved the raft into the pole while Washington clung to it, suddenly snapping the pole forward like the arm of a catapult. The force flung Washington into deep, swirling water.

He surely emitted a reflexive gasp as he was immersed in a splash among the ice floes, arms reaching out. The human body can survive only a few minutes in water just above the freezing point before succumbing to hypothermia—dangerously low body temperatures. One reflexively hyperventilates and the heart rate jumps as the frigid water chills sensors buried deep in the skin. The body preserves its vital functions by closing capillaries in fingers and limbs, sending blood to the core to warm brain, heart, and lungs. Fingers quickly stiffen, then arms and legs, responding only clumsily to commands. Thinking becomes confused as core temperature drops below 95°F and brain enzymes slow. At a core temperature of 86°F, the hypothermia victim in frigid water becomes unconscious and drowns. Or, long before then, the current could sweep him downstream, pulling him beneath the floes, trapping him underwater, bumping along under the grinding ice, trying to surface for air, head hitting the underside of the floes, frantically seeking an opening, until finally his breath expires.

Tossed overboard, Washington churned his feet, seeking the bottom. The water ran too deep for him to touch the riverbed. He reached with his hands, seizing hold of one log of the raft. He pulled himself up, dripping with the frigid river, onto the wobbling platform. In the dimming light, he and Gist took up poles again and tried to shove the heavy, clumsy raft toward the far shore. The force of the current swept them straight down the river channel, surrounded by the hissing floes. Trapped in the current and floating ice, they could not propel themselves to the safety of either shore.

An island appeared in the twilight; the river swept the raft toward it.*

* This island today is known as Herr's Island or Washington's Landing and is located in the Allegheny River about two miles above Pittsburgh and the Forks of the Ohio. The floes may have originated at a horseshoe bend of the river miles upstream, today known as Brady's Bend, where ice jams frequently occur and, breaking loose, send masses of ice floes down the Allegheny.

Grabbing their backpacks, they abandoned the raft and clambered onto the island. The cold pushed in as night began to fall. They were not saved. Without a fire, they could easily perish overnight of hypothermia—especially Washington, his clothes soaked through, his body already badly chilled. But starting a fire at night on a snow-covered island, striking flint and steel with fingers almost unbendable with cold, was a daunting, painful, exacting task. Washington was probably too chilled, slow, and clumsy to perform it, and perhaps did not possess as much fire-starting skill as his partner. The task would fall to Gist, who would suffer for it, squatting barehanded in the snow to strike the spark and ignite the tinder and feed tiny sticks into the weak flame. "The Cold was so extremely severe, that Mr. *Gist,* had all his Fingers, and some of his Toes frozen," recorded Washington.

One emergency technique was to shove cloth lint into the barrel of a gun and shoot it out, the discharge of powder causing it to catch fire. Soon the fire blazed, and they felt the blood return to painfully throbbing fingers and toes.

The stars must have been bright that cold night on the lone island in the middle of the river, with the ice floes hissing by. It must have been tortuous to keep warm, to dry their clothing. Perhaps they stood by a large fire until deep into the night, drying their soaked Indian garments piece by piece, warming their bodies. Washington had again overestimated his strength in attempting to shove the raft against the current and moving ice, and had underestimated the power of nature.

Eventually they wrapped themselves in their matchcoats or bedrolls or skins. Slowly the flame would die with guttering hisses and pops. They may have slept, intermittently. The fire ebbed to orange coals. They tossed, lay still. The hissing river grew silent.

By morning the river had frozen solid. They slung on their packs—Washington's containing the letter he had been entrusted to carry—and started on foot across the ice toward the warmth and safety of Shannopin's Town.

CHAPTER FIVE

One can picture Governor Dinwiddie's outrage as he read the letter Washington had brought from the Ohio wilderness over the mountains to the Tidewater coast and small Virginia capital of Williamsburg. Heavyset, buttons stretching over his belly, white wig hugging his jowly face, thin lips, prominent nose, and dark circles under his eyes from a bureaucrat's frustrations, the sixty-year-old governor no doubt flushed with anger. Perhaps his jowls shook and the letter trembled in his plump, agitated hands. Young Major Washington might have stood before him politely, but also sharing his outrage. After "as fatiguing a Journey as it is possible to conceive," in Washington's description, he had ridden that day, January 16, 1754, down long, sand-covered Duke of Gloucester Street, with its scattered string of clapboard houses, taverns, and shops, to call at the Governor's Palace, a large, Georgian-style brick structure topped with a white cupola. Herein resided the British Empire's projection of power in colonial Virginia, in the person of the Scottish-born governor himself.

About half a mile away, at the head of the sandy street, stood another large structure that served as the embodiment of Virginia's wealthy white male property owners—the Virginia House of Burgesses. In the roughly two years that Robert Dinwiddie had served as acting governor, he had thoroughly alienated the one hundred and four burgesses, and they in turn had become a persistent and troublesome obstacle to his plans. The latest flap between the two powers at either end of the sandy street flared when Governor Dinwiddie, looking to bolster his salary, had legally, if unwisely, exercised his authority to impose a tax on land transactions, known as the pistole fee. This, of course, fell personally on many members of the House of Burgesses, who were plantation owners or land speculators. The money would be going directly from their pockets into Governor Dinwiddie's. "Liberty, property, and no pistole!" was the cry that broke out among the Virginia burgesses, in what amounted to an early protest against royal taxation.

A lifelong merchant and bureaucrat, an administrator of the far-flung British Empire, Governor Dinwiddie had a nose for profit. Born in 1692 to a family of prosperous Glasgow merchants, he had received an education at the University of Glasgow and learned the family business. He had then sailed to the colony of Bermuda and set up a substantial trans-Atlantic trade, married, and had children. Ambitious as well as profit-minded, he had joined the Council of Bermuda and captured the attention of authorities in London when he suggested ways Britain could improve trade with its American colonies. This advice earned him the post of inspector of customs for coastal colonies in British North America. While making inspection tours on the east coast in the early 1740s, the middle-aged merchant and administrator became particularly interested in the Virginia colony's busy seaports and thriving plantations. He also may have become acquainted with its extensive land claims—that Virginia claimed all lands to the west, over the mountains, to the Pacific. He could plainly see that, with its coastal and trans-Atlantic trade, fast-growing population, and virtually unlimited land in its interior, the Virginia colony had tremendous potential for profit.

He had returned to a merchant's life back in Britain in the late 1740s, but when he learned of the resignation of Virginia's Governor Gooch in 1749, he thrust himself forward to the London authorities. In 1751, after a great deal of maneuvering, Robert Dinwiddie won the appointment as lieutenant governor of Virginia and arrived in Williamsburg with his wife and two daughters, Rebecca and Elizabeth. (The aristocratic Earl of Albemarle won the actual governorship but left the act of governing to his lieutenant governor, Dinwiddie.) At a time when offices could be bought and sold, Dinwiddie had to pay a big chunk of his salary to the

Earl of Albemarle for the privilege of his office and attempted to make up for the financial drain on his resources by charging the pistole fee on patented land. Imposing a fee like this to supplement one's salary was not unusual among colonial governors, but Governor Dinwiddie charged a considerably higher fee than others. One pistole, a Spanish gold coin then in circulation and also known as a doubloon, was about the price of a cow and a calf.*

The dispute had come to a climax that past fall of 1753, with the Virginia House of Burgesses and Governor Dinwiddie appealing to authorities in London to decide the matter. At the same time that Governor Dinwiddie tangled with the burgesses, ominous rumors arrived in Williamsburg of the French moving into the Ohio Valley "in an hostile Manner, to invade His Majesty's Territories," as Dinwiddie reported it to the burgesses. The governor had sent his message to the French commandant by young Major Washington, hoping to receive a reply by December, while the House of Burgesses was still in session. But the House had adjourned—the pistole controversy still unresolved, both sides still aggrieved—a month before the exhausted Washington finally rode up Duke of Gloucester Street. The governor was already in an ill humor when he broke the wax seal on the French commandant's reply.

The elderly gentleman officer Legardeur de Saint-Pierre had begun his reply by stating that he wished that Governor Dinwiddie had sent the letter to Canada and the governor-general of New France, the Marquis Duquesne. The marquis could thoroughly acquaint Governor Dinwiddie with "the evidence and reality of the rights of the King, my Master, upon the Lands situated along the River Ohio, and to contest the pretensions of the King of Britain thereto." But the French commandant promised to forward Governor Dinwiddie's letter to the marquis and further prom-

* Britain prohibited the American colonies from minting their own money. Foreign coins were used instead. In Virginia, in the first half of the eighteenth century, common coins included the Spanish pistole, or doubloon, worth a little less than a British pound, and the Spanish peso, or piece of eight, worth 4 shillings 6 pence, or a little less than a quarter of a British pound.

ised that "His answer will be a law to me. . . ." The French commandant was there by order of the marquis, and Governor Dinwiddie could be assured that Saint-Pierre would follow the marquis's orders with "all the exactness and resolution that can be expected from the best of officers." Furthermore, the French commandant wrote, he was not aware of any action by the French in the Ohio that could be seen as an act of hostility, or contrary to the peace treaties prevailing between the two empires. If Governor Dinwiddie had specified his particular complaints, the commandant would have had "the honour of answering you in the fullest."

In the middle of the letter came the line that surely enraged Governor Dinwiddie the most: "As to the Summons you send me to retire, I do not think myself obliged to obey it."

It was a direct affront—a challenge to the authority of His Majesty the King, to the might of the British Empire, to the territorial integrity of Virginia, and to the honor of Governor Dinwiddie. And the governor was not going to stand for it.

Washington surely related the most galling of his discoveries: that the French considered the Ohio River theirs by right of the explorations of La Salle; that the French employed all manner of deceitful ruses to win Indian allies from the British; that they had already started to build a series of substantial forts along the river and its tributaries; that Washington had personally witnessed hundreds of canoes under construction on the banks of Rivière Le Boeuf; and that with winter's end and the melting of the river ice, the French planned to send hundreds of troops downstream to construct their plan's centerpiece—a substantial fortress at the key strategic spot, the Forks of the Ohio.

This last revelation perhaps alarmed the governor most. Virginia was very far from assembling something equivalent to a major French force descending the Ohio. The Ohio Company had a few weeks earlier sent a party to build a "stronghouse"—a reinforced warehouse—at the Forks, but it was a small party and hardly a military mission. Unless he acted fast, the governor realized, Virginia might lose to the French the huge expanse of the Ohio wilderness, including the Ohio Company's and his stake in it. Wasting no time, the governor immediately ordered young

Washington to write up his notes from the journey into coherent journal form. Washington, ever eager to prove his worth, over the course of a day hurriedly drafted a seven-thousand-word journal. He signed it with an obsequious flourish, writing that his chief aim had been to satisfy Governor Dinwiddie.

> *With the Hope of doing it, I, with infinite Pleasure, subscribe myself, Your Honour's most Obedient, And very humble Servant,*
> *G. Washington*

With its alarmist tone, Washington's journal did, in fact, satisfy his Honour—so much so that Dinwiddie immediately had it published. He ordered copies sent not only to Virginia's council but to newspapers in other coastal colonies, to warn them of the imminent threat of French takeover and the French plan to descend the Ohio by canoe in the spring. It would soon also be published in London. He demanded *action*—with funding. "I hope to bring [the burgesses] into a proper way of thinking when they meet . . . ," he wrote to an acquaintance.

A growing sense of urgency wracked Governor Dinwiddie. Already it was midwinter. The spring breakup of the ice was not far off. On his way out of the wilderness, Washington had passed on the trail the first of the pack trains sent by the Ohio Company to start construction of its stronghouse at the Forks. With Washington's report, the governor realized that he needed some way to protect the company's modest and fledgling British establishment. He called an emergency session of the burgesses for mid-February. This still wasn't soon enough, however. Feeling that the chance to secure the Forks for Virginia and Great Britain was slipping through his fingers, Governor Dinwiddie decided to take action single-handedly.

A mere five days after Washington returned from the Ohio wilderness, Dinwiddie authorized the young adjutant to raise one hundred troops for the Virginia militia and lead them into the Ohio. He ordered another hundred to be raised by William Trent, the Indian trader and agent for the Ohio Company who had been sent to build the company's stronghouse at the Forks. He made Trent an officer in the Virginia militia by sending him a captain's commission and, in effect, militarized the construction of the Ohio Company's outpost. (Trent was also the son of a wealthy merchant who in 1719 had founded Trenton, New Jersey, when he built a country home on the spot.) Dinwiddie planned to ask the Virginia burgesses to fund another four hundred militia troops when they met in February, making a combined force of six hundred from Virginia's militia. In addition, he hoped other colonies would raise troops and that the Indian allies of the British would help safeguard the British stronghold at the Forks.

It was a key moment, propelled by haste, or panic. Neither young Washington nor Governor Dinwiddie had any experience in military matters, nor, in a sense, did Virginia itself. Virginia had a law permitting a draft of every free white man over twenty-one, each of whom was required to equip himself or pay for a similarly equipped person. But it had no infrastructure for training officers or recruits and no troops with the expertise to cross a major mountain barrier into the wilds. This total lack of military experience seems to have deterred neither Dinwiddie, the merchant-administrator, nor young Major Washington. "Novices," as Washington biographer Douglas Southall Freeman put it, "were inviting war in a forbidding land."

The struggle to raise troops and money began at once. Frederick County lay under Washington's jurisdiction as adjutant, to provide troops. Lord Fairfax was in charge of finding the quota of fifty troops from the county's populace, either volunteers paid fifteen pounds of tobacco per day or, if necessary, draftees. He failed to do so. "You may, with almost equal success," Washington would later write to his brother, "attempt to raise the dead to life again, as the force of this County."

Other Virginia counties struggled to raise their quotas, too. Likewise, Virginia's neighboring colonies—some of which, like Pennsylvania, claimed large parts of the Ohio Valley—showed little of Governor Dinwiddie's urgency to protect their interests from the French. The governor of Maryland said that his assembly probably wouldn't vote any funds unless the French actually invaded Maryland. The governor of Pennsylvania responded that he couldn't convince the nonviolent Quakers in his assembly to spend money on military ventures. It looked as if Governor Dinwiddie was on his own.

The Virginia House of Burgesses convened in an emergency session on February 14. A castlelike brick structure built in the early 1700s, the grand old capitol building at the head of Duke of Gloucester Street had burned a few years earlier. It had recently been replaced by a plainer, two-story building with an elegant, Palladio-inspired portico that looked down the long street. The Governor's Council, or Upper House, met on the second floor, while the House of Burgesses, along with a courtroom, occupied the ground floor. With the Speaker of the House, John Robinson, sitting in a tall, thronelike wooden chair at the room's head and the burgesses taking dark wooden benches in galleries along the walls, Governor Dinwiddie presented his case before them. This must have been a heady, even intoxicating, moment for Washington, just shy of his twenty-second birthday: to see his account of his journey into the wilderness assume a place at the center of Virginia politics.

Governor Dinwiddie rose to the drama of the occasion. Speaking "with great application, many arguments, and everything I could possibly suggest," the governor told the burgesses, as he recounted later, that he had sent Major Washington to the French commandant deep in the Ohio wilderness. He related how Washington had returned with the report of hundreds of canoe loads of French preparing to descend the Ohio in the spring to build fortresses, and this on His Majesty the King's own lands. The governor—one can hear his voice rising in anger—stated that the French commandant had told Major Washington that the English must desist from trading in the Ohio. The French insisted that they had a right to seize any British traders who remained, such as the trader Fraser, who had built the log structure at Venango that French officers now occupied.

"That Man," the French officer had replied to Washington upon his query about Fraser, "was lucky that he had made his Escape, or he would have sent him Prisoner to Canada."

Of these "unjustifiable insults," Governor Dinwiddie saved the most powerful images for last. Washington had reported to the governor that on his way out of the wilderness he had encountered a party of friendly

Indians who described how they had recently come to a homestead where they found the corpses of seven white settlers scattered about and partially eaten by rooting hogs, which they claimed was the work of French-allied Ottawa Indians. All the bodies were scalped except for "one Woman with very Light hair." Drawing on the same massacre that Washington had heard about and recorded in his journal, the governor, standing before the assembled burgesses, now painted a graphic picture of a recent raid on frontier settlers.

> ... that poor unhappy Family! surrounded by a Crowd of Miscreants, dreadfully rushing on to perpetrate the most savage Barbarities, inexorable to the Parent's Intreaties, insensible to the Cries of the tender Infant. . . . Think you see the Infant torn from the unavailing Struggles of the distracted Mother, the Daughters ravished before the Eyes of their wretched Parents; and then, with Cruelty and Insult, butchered and scalped. Suppose the horrid Scene compleated, and the whole Family, Man, Wife, and Children (as they were) murdered and scalped by these relentless Savages, and then torn in Pieces, and in Part devoured by Wild Beasts, for whom they were left a Prey by their more brutal Enemies.

The French themselves, or so he had heard, instigate and accompany "their" Indians on these merciless raids, Governor Dinwiddie told the burgesses. He drove home the message: *These are the people who are moving into our neighborhood!* The burgesses could expect "[i]n the Bosom of your Country, these Evils, that you as yet have only the melancholy Tidings of from your Frontiers." He implored them to work together and "drive away these cruel and treacherous Invaders of your Properties, and Destroyers of your Families. . . ."

It was a frightening message. The burgesses, however, did not respond with the enthusiasm Dinwiddie had hoped. The Ohio Valley seemed remote. They didn't care about settling this deep interior. The Virginians were suspicious of him, with his Ohio Company connections. They didn't want to spend money. There was a question of which lands in the Ohio Valley belonged to Virginia and which to Pennsylvania. No clear border existed between the two, so who was responsible for evicting the French?

Some burgesses even doubted that Virginia had a claim to the Ohio and suspected that it was actually French territory, based on the work of a famed French cartographer. French fur trappers and explorers had traveled the interior of North America for many decades and brought back their knowledge to Paris-based mapmakers. Guillaume Delisle's 1718 map showed the interior of eastern North America in great detail—with geographical knowledge that far exceeded that of the British. In Delisle's rendering of the midcontinent, the French territory of Louisiana, splashed in large letters across North America's midsection, covered all the Mississippi Basin including the Ohio River and lands eastward to the crest of the Appalachians. In this view from Paris, the British colonies of Virginia, Maryland, and Pennsylvania were crowded together on the Eastern Seaboard, their boundaries extending no farther west into the Ohio than the backbone of the Appalachian Mountain chain, and Pennsylvania's western border not even that far.

"[One Burgess] pretended to ascertain the Right of the French to these [Ohio] Lands from Mons'r De Lisle's Maps," an indignant Governor Dinwiddie complained in a letter to a London acquaintance. "You may conceive how I fir'd at this, that an English Legislature sh'd presume to doubt the Right of His Majesty to the interior Parts of this Cont't, the back of His Dom's; how this French Spirit sh'd possess a Person of His Dis tinct'n and Sense, I know not."

The exasperated governor used every rhetorical trick he could summon to convince the burgesses to counter the French. He appealed to their sense of fear, to their loyalty, to their sense of shame. Washington watched this controversy unfold at close hand. It was a bitter first taste of partisan politics between a strong-willed executive and a skeptical House. He did not forget it. He later recalled with lingering resentment that some of the burgesses believed his entire journey to the Ohio was a "fiction and scheme" to further the interests of a "private company," an obvious reference to the Ohio Company.

"I assure You I have had a very up-hill Work of it . . . ," Dinwiddie said in his letter to his London friend, wiping the metaphorical sweat from his brow. "However, that is pretty well got over. . . ."

Spring lay only a month away. Tree buds were beginning to fatten. The warming sun soon would perforate the ice on northern rivers until the current broke it and swept it away. Every day's delay meant that the French canoes had a jump on the British. Governor Dinwiddie and young Washington waited anxiously for political action. The House of Burgesses finally voted what the governor called a "small sum"—£10,000—for the defense of the Ohio lands against the French and their Indian allies, as well as £50 reimbursement for Washington's expense and trouble on his journey to the French commandant. The funds would help Dinwiddie raise a force. The governor expected help not only from neighboring colonies like Maryland and Pennsylvania but also from British allies among Indian tribes.

Instead of drafting recruits to fill quotas, however, the governor would use the funds to help lure volunteers. While actual pay was a relatively modest fifteen pounds of tobacco per soldier per day (pounds of tobacco served as a medium of exchange), the offer of land at the completion of service provided a real incentive to would-be volunteers. Dinwiddie allotted two hundred thousand acres of land in the Ohio to serve as rewards for service.

Nor was this lure lost on young Washington. He now had two good personal reasons to defend the Ohio from French encroachment. From his family heritage and stint as a surveyor, he knew the value of land. He understood the relentless push toward the west from the growing population of the Eastern Seaboard. The real profit lay in acquiring large tracts of cheap land ahead of development and then selling it off in pieces to newly arrived settlers. His ancestors had strived for centuries in both England and Virginia to acquire large landholdings. He knew that the land at the Forks of the Ohio would have great value one day. He wanted to have a piece of it.

The acquisition of Ohio land was a theme that would run through the rest of his life. Like his brother Lawrence and many ancestors before him, he wished to propel himself into the upper aristocracy, to acquire a good name and reputation. In the absence of a guiding father and lacking the landholding status of a first son, the military offered Washington one clear avenue to rank and social standing. Large tracts of land and military promotion—these could provide the status and wealth he sought. Both were his to find in the vast wilderness of the Ohio. It was an opportunity he could not miss.

For a force to send to the Ohio, Governor Dinwiddie decided to recruit three hundred volunteers—six companies with fifty men in each. Washington immediately began to press his superiors for advancement. He rode five miles out of Williamsburg and called at the Green Spring estate of Philip Ludwell, who served as one of the twelve members of the Governor's Council, the upper house of advisers to the governor, usually wealthy planters appointed to the position. Here, at Ludwell's estate, young Washington had a conversation with Richard Corbin, another member of the council, about his future. Corbin suggested that Washington could receive a rank higher than his current major. He might even be named one of the commanding officers of the expedition to the Ohio.

Washington now carefully positioned himself. He gave promotion a great deal of thought. He decided he wanted a higher rank, but not *too* high. Cloaked in a veil of modesty and loyalty, his ambition shone through in a letter he wrote to Corbin soon after that conversation at Green Spring. He flatly told Corbin that he didn't want command of the whole Virginia force of three hundred.

> *. . . [I]t is a charge too great for my youth and inexperience to be intrusted with. Knowing this, I have too sincere a love for my country, to undertake that which may tend to the prejudice of it. But if I could entertain hopes that you thought me worthy of the post of Lieutenant-colonel, and would favour me so far as to mention it at the appointment of officers, I could not but entertain a true sense of the kindness.*

In other words, I don't want to be commanding officer, but lieutenant colonel certainly would be nice. He saw personal opportunity in the growing tension with the French. He knew now was his moment—his time to act. He carefully aimed for a position that elevated him but did not overwhelm him with responsibilities beyond his abilities. In a twenty-two-year-old, his focus on seizing an opportunity, his diligence in pursuing it, and his acute awareness of exactly what it was that he wanted for himself—as if picking out a set of clothes, or a uniform, or his Indian walking dress—was remarkable.

Nor did he hesitate to ask for what he wanted. From the Fairfaxes, aristocrats who lived in a world of connections and bloodlines and hierarchy, he may have absorbed the lesson to ask his superiors for advice and help in advancement. Always sensitive about being paid what he thought he was worth, Washington also focused hard on his potential salary. What might he expect for a salary, he asked Governor Dinwiddie, were he to be named an officer in the Virginia colonial troops for this wilderness expedition? About 15 shillings as a lieutenant colonel and 12 shillings 6 pence in his current rank as major, Dinwiddie estimated. "I then complained [the pay] was very much off," Washington recalled some weeks later, comparing the Virginia militia officer's pay with that of an officer in the regular British military. Dinwiddie assured Washington that the Virginia colonial officers "were to be furnished with proper necessary's" such as food. For this reason, the pay was lower than that of a regular British Royal Army officer, who had to buy certain "necessaries" out of his own pocket.

Washington got his first wish—higher rank—on March 15, 1754, when his commission as a lieutenant colonel came through. But pay was another matter. He was informed that his pay would be 12 shillings 6 pence—what Governor Dinwiddie had estimated for the lower rank of major. In joining the military, Washington had entered an institution with a closely defined, measurable hierarchy. From Lawrence, from the Fairfaxes, and from others around him, he had developed an acute sensitivity to his standing in the Tidewater world. Lawrence had died, however. He had no father to help give him an anchoring sense of belonging and worth, a sense of security about *this is who we are, and this is how you fit in.* Here, in his officer's pay in the Virginia militia, was one precise measurement of his standing. Too low—at least as he saw it. This was an insult more than young Washington could bear. Or, as an aspiring Tidewater gentleman following an unwritten code of behavior, perhaps he thought himself required to *take it* as an insult that demanded a response in order to protect his honor.

He went to his friend and mentor, Colonel Fairfax, to discuss it. "This, with some other Reason's, induced me to acquaint Colo. Fairfax with my intention of Resigning," Washington recalled soon thereafter in a letter to Governor Dinwiddie.

It would be the first of many times that young Washington threatened, almost reflexively, to resign. There was a touch of petulance to the threat. Colonel Fairfax soothed the young man's anger and sense of insult. Perhaps he convinced young Washington that this was not as great an affront as he might have thought. The colonel could see that to resign would be a mistake. He convinced Washington to stay and said he would personally speak with Governor Dinwiddie about raising Washington's pay. In the absence of a father connected to the Virginia aristocracy, young Washington had in Colonel Fairfax a powerful patron.

With the issue of pay still hanging, the fledgling officer's next set of problems arrived—one that would plague Washington in the years ahead. How to get arms, equipment, uniforms, and pay for his men, and how to get the men themselves. The responsibilities came down fast and hard on the twenty-two-year-old, and his learning curve was steep. At first, he seemed mostly concerned with outfitting his men in fancy, bright uniforms and the shiny trappings of the military. Even before his official commission had arrived, he wrote to Governor Dinwiddie with his urgent needs. First on his list was tents. This was followed by requests for the archaic, showy weapons of war carried by officers to lead and signal troops—"Cutlasses, Halbards, Officer's half Pikes"—as well as decent drums. In this first letter requesting goods, written from Belvoir in early March, and perhaps with an eye to the figure he would cut in front of an admiring Sally Fairfax when he returned in officer's dress,

Washington gave a great deal of attention to bright red uniforms. He also proudly displayed his knowledge of wilderness and Indians. He had made a study of the "Tempers" of Indians, as well as their "Customs and dispositions," he wrote to Governor Dinwiddie. It is their nature to be struck by showiness, he said, thus the need for brilliant uniforms for the men. We could win the Indians to our side, he suggested, if only we had the right uniforms.

> *If it was only a Coat of the Coursest red which may be had in these parts it would answer the Intention—red with them is compard to Blood and is look'd upon as the distinguishing marks of Warriours and great Men—The shabby and ragged appearance of the French common Soldiers . . . affords great matter for ridicule amongst the Indians. . . .*

Two days later, Washington wrote to Governor Dinwiddie again. He had now ridden from Belvoir Manor to Alexandria, a new town and fledgling shipping port a short distance upstream from the Fairfax estate and Mount Vernon, near the head of navigation on the Potomac. Founded a few years earlier by Washington's late half-brother Lawrence and other Ohio Company partners, the river port was intended to provide the land speculation company with the closest shipping point to the Ohio wilderness. At Alexandria in early March, Washington assembled troops in a camp at the edge of town. He acquired supplies with the help of Scottish merchant John Carlyle, official commissary to the expedition, who had married one of Colonel Fairfax's daughters and recently finished an ornate stone mansion in Alexandria that overlooked the Potomac. From his camp at Alexandria, Washington dispatched to the governor ever more urgent appeals for clothing, equipment, and pay.

Now he faced a "generl Clamour" from the twenty-five or so men he had recruited. They wanted to know how often they would be paid. The British troops received pay once a week; so should the Virginia troops, Washington believed. Here again arose a theme that would come up over and over in young Washington's world—a constant comparison between the Virginia colonial troops and the more privileged British Royal Army regulars, a ceaseless measuring of where he stood. With growing passion, Washington, a colonial leading colonial troops, would seek equality with British regulars and officers and would grow increasingly frustrated when it wasn't forthcoming. With dogged earnestness, and perhaps at first a youthful naïveté, he pushed to better the situation for himself and his men.

And yet another problem quickly became apparent in those first few days—the quality of the recruits themselves. "[W]e find the generality of those who are to be Enlisted, are of those loose, Idle Persons that are quite destitute of House, and Home, and I may truely say many of them of Cloaths . . . ," Washington wrote to Dinwiddie. "There is many of them without Shoes, other's want Stockings, some are without Shirts, and not a few that have Scarce a Coat or Waistcoat, to their backs. . . ."

Their lack of clothing caught his attention first; soon enough it would be their lack of discipline. Washington was mindful of not complaining too much to his superior. He apologized with a somewhat martyred air that he didn't want to be a burden and would bear the complaints of the recruits himself. "But I must here in time put a kirb to my requests, and remember that I ought not to be too importunate; otherwise I shall be as troublesome to your Honour, as the Soldiers are to me. . . ."

In these first weeks as lieutenant colonel, trying to assemble equipment and troops and disgruntled about pay, young Washington alternately reminded Governor Dinwiddie of his self-sacrifice and issued petulant threats to resign. Quickly, however, a sense of urgency overtook his disgruntlement. Governor Dinwiddie wrote to Washington that he was surprised to hear in the latest information emerging from the wilderness that the French were expected to launch their canoes down the Ohio much sooner than he expected, ". . . which I think makes it necessary," wrote the governor to Washington on March 15, "for You to march what Soldiers you have enlisted immediately to the Ohio, & escort some Waggons, with the necessary Provisions."

Now Washington truly had his hands full. Events had moved quickly. It was mid-March 1754. Only two months had passed since Washington had canoed, ridden, and walked out of the wintry January wilderness bearing the French commandant's letter. Only three weeks had elapsed since the House of Burgesses authorized £10,000 to protect Virginia's interests in the Ohio. In this letter from Dinwiddie of March 15, Washington received his official commission as lieutenant colonel in the Virginia militia. The post of overall commander of the Virginia Regiment went to Joshua Fry, a portly and aging former professor of mathematics and natural science at the College of William and Mary. Born in England, the multitalented Fry had studied at Oxford, moved to Virginia as a young man, married a landed Virginia widow, and, in addition to his teaching, run boundary surveys in the wilderness with Peter Jefferson (Thomas Jefferson's father) and produced a definitive map of Virginia. By Dinwiddie's orders, Colonel Fry would stay behind to gather more recruits and then follow Washington into the Ohio. Governor Dinwiddie wrote that he also expected help from a large force of British-allied Indians in the Carolinas to the south—one thousand Cherokee and Catawba warriors.

To Washington's elaborate request for red uniforms, the governor had no objections but thought time too short to have them made, unless they were sent along later. The governor said he was sending sloops to Alexandria carrying more recruits as well as twenty-four tents, as Washington had requested. As for the requested "Picks, Cutlasses, or Halberts," Governor Dinwiddie had none in the Virginia magazine at Williamsburg and, with what seems a roll of the eyes, wrote that Washington's officers would have to make do with small firearms. The operation clearly had to get under way immediately or all could be lost. "I wd gladly hope, as Capt. Trent has begun to build a Fort at [the Forks of the Ohio], that the French will not immediately disturb us there; & wn our Forces are properly collected we shall be able to keep Possession, & drive the French from the Ohio. . . . Pray God preserve You, & grant Success to our just Designs. . . ."

Working in tandem, Governor Dinwiddie and young Colonel Washington made for a volatile mix. Both were utterly lacking in military experience. Washington hungered to establish himself in Tidewater society and climb in officer's rank, while Dinwiddie deeply understood the commercial potential of the vast Virginia colony and, at age sixty-one, with retirement looming, stood to gain a fortune in land speculation if Virginia settled the Ohio.

While deeply loyal to the Crown and defending lands they considered rightfully the king's, both Dinwiddie and Washington had personal stakes in evicting the French from the Ohio—Washington to boost his name, and Dinwiddie to boost his retirement account. As they launched this rush to the Forks, neither Governor Dinwiddie nor young Colonel Washington grasped its continental—or global—significance. Having warred in Europe for seven hundred years, with empires that nearly circled the earth, Britain and France had now woven a fragile peace. With much of the North American continent still unexplored and uncharted by Europeans, however, the boundaries delineating their vast claims in America appeared much blurrier on the map than their smaller, more distinct—if sometimes shifting—borders in Europe. Both had staked out but neither had settled the Ohio wilderness, where Indians were few but the lands rich enough—in furs to sell and soil to till—to pull both empires in. They slid into the power vacuum hovering over the Ohio Valley like warm and cold weather fronts colliding. This vast and rugged region would become—as a later historian described another remote and contested region of the globe—a "storm center of the world."

CHAPTER SIX

THE RACE WAS ON. THE RIVER ICE HAD BROKEN, AND ON THE banks green shoots poked up through dead leaves in the warming sun. For the French, it was about two hundred and fifty river miles from Fort Le Boeuf downstream to the strategic Forks of the Ohio. Paddling on the high, swift meltwaters of the Rivière Le Boeuf, they would sweep away from Lake Erie southward through countless bends until the small river poured into the powerful Allegheny. Then they would ride that river in a looping arc southeast to its confluence with the Monongahela at the Forks, where together they formed the Ohio. They would have to dodge rocks, haul canoes over logjams, and contend with any lingering ice. But, augmenting their paddle strokes, the flow of the current would help propel them, and their big craft would easily carry their food and supplies. In essence, they were making a downhill run in a canoe.

Washington faced an uphill run, on foot, and farther. Starting at the port of Alexandria staging grounds, at sea level on the Potomac, his recruits would have to push about three hundred miles along crude, muddy forest paths, climb mountain ranges to three thousand feet, and ford several rivers to reach the Forks. For the early going, they could purchase food and fodder for their horses at frontier settlements. But beyond the last white outpost, at Wills Creek—an Ohio Company warehouse and fledgling settlement at the base of a tall, formidable ridge known as Wills Mountain—they would jump off into wilderness. They could not hunt enough game to feed over one hundred troops marching on foot. The party had to carry its own food, in addition to tents, medicines, ammunition, and other supplies. Washington would need wagons instead of canoes to haul his supplies—many wagons—as well as horses to pull the wagons plus grain to feed the horses. In this race to the Forks, Washington faced thorny logistical problems utterly new to him and daunting for anyone, much less a twenty-two-year-old.

Problems delayed him from the start. It wasn't until April 2, two weeks after receiving Governor Dinwiddie's order to get under way immediately, that newly commissioned Colonel Washington began the march from the Tidewater region into the continent's interior. Mounted on horseback, the young officer led one hundred and twenty men in two companies—one under Peter Hog, a fifty-one-year-old Scotsman appointed by his countryman Governor Dinwiddie, and the other by Jacob van Braam, the former Dutch military officer who had served as interpreter on Washington's mission to the French commandant. As they were recruited, more troops would follow under Joshua Fry, the retired mathematics professor appointed by the governor to head the Virginia Regiment. Washington had secured only two wagons to haul the supplies but could delay no longer. He hoped to hire more when he reached the settlement of Winchester beyond the Blue Ridge.

The party only made six miles that first day. The next few days, the pace picked up—slightly. The column marched northwest from the Tidewater farms into gentle, wooded rises lifting toward the first of the Appalachians' corrugated ridges. Staying at settlements and homesteads, they made on average of eleven miles a day. The rough track, notched through a low, forested rise, emerged in flatter terrain on the far side, where they could see the first prominent ridge—the Blue Ridge—painting a hazy bluish line across the horizon. Seeking a passage through the barrier, the crude road bent northward, angled upward, and cut through a notch in the ridgeline called Vestal's Gap. They now stood nine hundred feet above sea level. Following the track, they dipped down into the broad Shenandoah Valley, cupped between the Appalachian ridges. Homesteads dotted the landscape, and through the valley wound the shining course of the Shenandoah River. Reaching its banks, the column paused. A small ferry

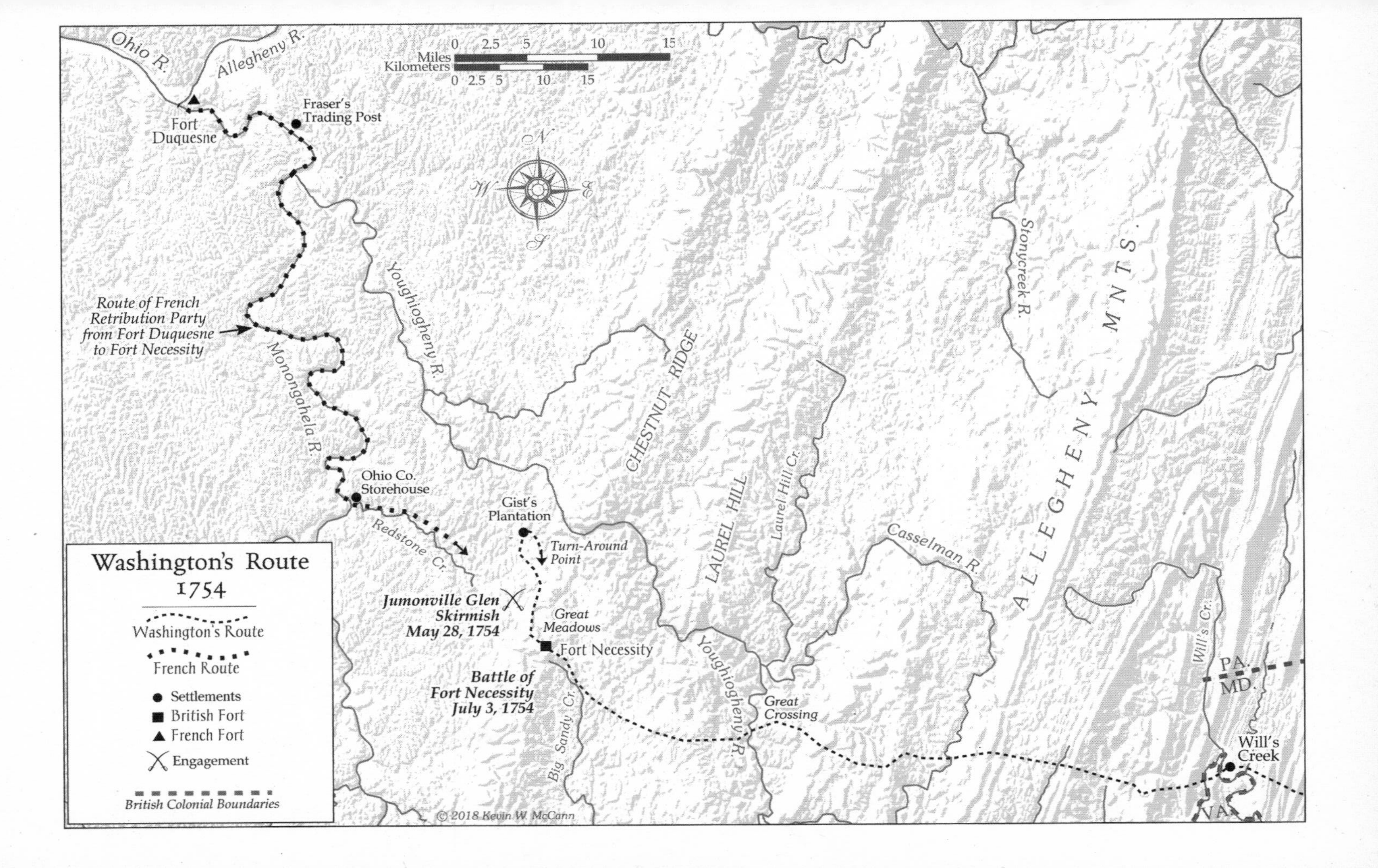

Washington's Route
1754
Washington's Route
French Route
Settlements
British Fort
French Fort
Engagement
British Colonial Boundaries
Miles
Kilometers
0 2.5 5 10 15
0 2.5 5 10 15
N
S
E
W
Ohio R.
Allegheny R.
Fort Duquesne
Fraser's Trading Post
Route of French Retribution Party from Fort Duquesne to Fort Necessity
Monongahela R.
Youghiogheny R.
Ohio Co. Storehouse
Redstone Cr.
Gist's Plantation
Turn-Around Point
Jumonville Glen Skirmish May 28, 1754
Great Meadows
Fort Necessity
Battle of Fort Necessity July 3, 1754
Big Sandy Cr.
CHESTNUT RIDGE
LAUREL HILL
Laurel Hill Cr.
Youghiogheny R.
Great Crossing
Casselman R.
Stonycreek R.
ALLEGHENY MNTS.
Will's Cr.
PA.
MD.
VA.
Will's Creek
© 2018 Kevin W. McCann

floated troops, wagons, and horses across, load by load. It was April 8. Nearly a week had elapsed since leaving Alexandria.

After his former mission traveling with a much smaller and quicker party, mostly mounted, Washington discovered that a column of one hundred and twenty men on foot and two horse-drawn wagons moved slowly and deliberately. One can imagine the frustration. He knew the river ice was melting to the north. On tree branches all around him sprouted tender green buds gathering light and warmth to open into leaf. No word had arrived about whether French canoes had launched on the Rivière Le Boeuf. Stroking in unison in their big canoes, French voyageurs could move with remarkable speed—easily capable of fifty miles a day on calm water, or three hundred miles in the six days since Washington had left Alexandria. He had covered only a fifth of that distance. If they had started when he did, the race was over. The French would have secured the Forks of the Ohio, and he would have missed his perfect opportunity to shine as a subject of the king. He would be known as the Virginia officer who lost the race. It was not the start of his reputation he had imagined.

Crossing the Shenandoah River on the clumsy ferry, he hurried his troops toward the village of Winchester in the valley's center. Here he was to meet troops recruited by Captain Adam Stephen, a thirty-six-year-old Scottish doctor who had served aboard a Royal Navy hospital ship before emigrating to Virginia. Stephen had enlisted an additional forty or so men from the Shenandoah area. Washington desperately needed wagons to carry supplies and hoped to hire them in Winchester, but wagons, as it turned out, were very difficult to find. One option was to "impress" the wagons—force the owner by law to hire the wagon out for military use. Governor Dinwiddie had given "full power" to his agent at Alexandria, Scottish merchant John Carlyle, to impress all manner of conveyances—"Boats, Sloops, Waggons, Carts, Horses or any Thing else that is necessary for the safe Conveyance of Provisions or Stores." But did the power to impress wagons and horses extend to young Colonel Washington, too?

Washington's march to the Forks stalled for a week. Still not knowing whether the French had launched, he approached local farmers in the Shenandoah Valley. They did not part easily with their wagons. The horses to pull them were in such poor shape that the recruits would have to help push the wagons uphill. In his determination to get to the Forks before the French, Washington did not allow impediments such as the law to stand in his way. "Out of Twenty four Waggons that were impress'd at Winchester," Washington wrote to Governor Dinwiddie, "we got but Ten after waiting a week. . . . I doubt not but in some points I may have strained the Law, but I hope as my sole motive was to Expedite the March, I shall be supported in it should my authority be Questioned. . . ."

In fact, his authority *was* questioned. During that week in Winchester, Frederick County issued a warrant for his arrest for trespassing—presumably trespassing on the property of some farmer who didn't want his wagon and team seized for a journey into the wilderness to counter some abstract threat. Brandishing firearms, exerting military authority that was less than a month old, young Colonel Washington refused to surrender to the sheriff. "He kept me off by force of arms," the sheriff wrote on the back of the arrest warrant.

It was turning out that Washington possessed a stubborn doggedness and, at times, when thwarted, an explosive anger. Still, his defiance of local authorities did little good. He acquired only twelve wagons. He had covered only about one hundred miles in seventeen days. The Forks still lay nearly two hundred miles away, across major mountain ridges. This progress was far too slow. A frustrated young Washington changed strategy. He decided that when he reached the final jumping-off point for the wilderness, the Ohio Company storehouse at the base of the big ridge at Wills Creek, he would employ packhorses for the steep mountain ridges ahead. He sent a message ahead to Thomas "Big Spoon" Cresap, who had a trading post near Wills Creek, to procure horses and have bags made from sixty yards of coarse linen fabric to strap on the animals and carry the party's supplies. "The difficulty of getting Waggons has almost been insurmountable," he wrote to Cresap. ". . . I am determined to carry our provisions & c. out on horseback. . . ."

After a week of struggling with the wagons at Winchester, and joining Adam Stephen's company with his other two, Washington finally left for

Wills Creek. He made his way out of the Shenandoah Valley and over the tall ridge beyond it, the North Mountain or, as some knew it, the "Devil's Back Bone." Many other mountain ridges, some much taller, still lay ahead to the west. He had traveled a distance past Winchester when a mounted messenger rode up the path.

It was Christopher Gist, Washington's former guide on his mission to the French commandant. He carried an urgent message from the Half King. The sachem was at the Forks, where William Trent, initially working for the Ohio Company but made a captain in the Virginia Regiment by Governor Dinwiddie, was building a fort. "[Gist] informed me that the Indians are very angry at our delay, and that they threaten to abandon the country; that the French are expected every day at the lower part of the river; that the fort is begun, but hardly advanced. . . ."

Two days later, on April 19, still worse news arrived. Wrote Washington, "Received a messenger from Mr. Trente, Captain-lieutenant at [the Forks], who urges that reinforcements be sent with all possible diligence, because a corps of 800 French are hourly expected."

The French, it appeared, had launched.

MARQUIS DUQUESNE, the new governor-general of New France, was descended from an exalted French naval family and Norman gentry. Appointed only two years earlier on the death of the previous governor-general, and criticized by his detractors for a lordly and self-important demeanor, Marquis Duquesne quickly grasped the strategic situation of the French in North America. He saw the blurriness of boundaries between the rival empires and the threat of British traders moving into the Ohio, which could sever communication between Canada and France's territories farther west along the Mississippi and south at the river's mouth at New Orleans. His predecessor, La Jonquière, had started an effort to secure the Ohio, sending the explorer Céloron down the river in 1749 to bury lead plates reaffirming its possession for France, this after La Salle had explored it for France in the late 1600s. Now the new governor-general, Marquis Duquesne, resolved to take action and build a series of forts from the Great Lakes down the Ohio—or Le Belle Rivière, as it was known to the French. He was taking an aggressive approach to France's territorial interests in North America.

Together with the great portage at the Niagara Falls between lakes Ontario and Erie, the Forks of the Ohio formed "the key of the Great West," as historian Francis Parkman put it. "Should France hold firmly to these two controlling passes, she might almost boast herself mistress of the continent."

Within just a few months of his arrival in New France in the summer of 1752, the Marquis Duquesne had sketched out plans for an enormous expedition in the spring of the following year, 1753, to embark as soon as the winter's ice had left the lakes and rivers. That fall, he wrote to the commandant of the French fort at Niagara, notifying him of the huge expedition of two thousand people that he would have to portage around the great falls: "In the course of the month of May I shall send from Montreal 300 soldiers, 1700 *habitants,* and about 200 Indians whom I am assigning to go and seize and establish themselves on the Belle Rivière, which we are on the verge of losing if I do not make this hasty but indispensable effort."

In April of 1753, under Commandant Marin, the main body of troops journeyed from Montreal, up the Saint Lawrence and along the length of Lake Ontario, arriving at the Niagara gorge in mid-May. A crusty sixty-year-old veteran of New France's wilderness outposts, Marin was described by a superior as "half-ferocious." The enormous task of portaging so many men, supplies, and equipment around the falls was aided by horses and hired Indians. In canoes and boats, they paddled, rowed, and sailed along the exposed southern shore of Lake Erie to the natural harbor protected by a long, hooking sand spit at Presque Isle. Here they built a substantial fort of squared chestnut logs, designed by the European-trained head of artillery and engineering, Le Mercier. They constructed a portage trail from the harbor at Lake Erie over fifteen miles, much of it swampy, to the headwaters of the Rivière Le Boeuf. Here they erected a simpler fort and started to build hundreds of canoes and pirogues to descend that autumn to the Forks of the Ohio. Food shortages arose. Kegs did not contain as much salt pork as marked. Some troops subsisted on nothing but grain and water for weeks on end. Under heavy traffic, the portage trail collapsed into a mud rut. Tremendous labors on a diet solely of salt pork and biscuit caused their health to deteriorate, including that of the hard-driving Marin. One veteran officer wrote:

> *The labor of our troops was excessive. The soldiers, sunk half-leg deep in the mud, and weakened by the recent fatigues of the first portage [at Niagara], succumbed under their burdens. It was impossible to use the few horses that remained. It was an afflicting spectacle to behold these debilitated men, struggling at the same time against the bad season and the difficulties of the road, broken down by the weight of their weapons and of the loads which they had to carry.*

Scurvy set in. At Presque Isle, they were burying four or five soldiers a day, remembered the commissary who had served there.* For centuries, the disease had been the scourge of long-distance seafarers and European explorers of northern regions who relied on preserved foods and were ignorant of native remedies such as berries, herbs, fresh animal organs, or undercooked flesh, all containing vitamin C. It is now known that vitamin C tightly weaves amino acids together to create a tough, flexible protein, collagen, which forms the connective tissue found in ligaments, arteries, and elsewhere. Lacking vitamin C for ten or twelve weeks, one's connective tissues literally begin to unravel. Teeth loosen in one's gums, old scars reopen, and finally organs, such as one's arteries or heart, hemorrhage. Centuries ago, explorers observed that depression often

---

* The expeditions of early French explorers to North America had suffered greatly from scurvy. In the 1530s, Jacques Cartier's ship became locked all winter in the icebound Saint Lawrence while scurvy decimated the crew, until twenty-five had died and only three or four had the strength to go above deck. They were saved by the local Indians, who showed them how to prepare a tea from the needles and bark of what is thought to have been the white cedar tree, which cured them in just a few days with its infusion of vitamin C. Apparently the remedy had been forgotten.

appeared in conjunction with the debilitating disease, and that those in "bad spirits" were more susceptible to it.

As the French soldiers struggled with poor food and heavy loads along the muddy portage trail, their spirits collapsed. Little rain fell in late summer and into the fall. The water level of the Rivière Le Boeuf dropped until it finally became too shallow to float their canoes and pirogues downstream to the Ohio.

Marin, now very ill, sent troops back to Montreal. He canceled the expedition down the Ohio for the year, halting fort-building farther downriver. It crushed him. In his mind, he had failed as the commander of this tremendous thrust. As the summer sun faded to the wan light of autumn, and cold winds blew from the north, Marin gave up in spirit. He wanted to die in the field. At the end of October 1753, depressed and hemorrhaging, Marin breathed his last.

In early December, Legardeur de Saint-Pierre, just returning from his explorations in search of the Western Sea, arrived as new commandant to replace Marin for the French thrust into the Ohio. Only a week into his command, Christopher Gist and young Major Washington showed up at Fort Le Boeuf with the letter from Governor Dinwiddie. Despite the setbacks that had destroyed Marin, the Marquis Duquesne had hardly given up on his plan for the Ohio. When Saint-Pierre forwarded him the letter from Governor Dinwiddie, he called Dinwiddie's claims to the Ohio "sheer imagination." He ordered a new expedition to set out for the Ohio in early 1754.

By April 1754, forty-one men were either laboring on Governor Dinwiddie's British fort at the Forks or were soldiers protecting the workers there, under Captain William Trent and his brother-in-law Ensign Edward Ward. On April 13, Captain Trent had journeyed back to Wills Creek, apparently for provisions, leaving twenty-eight-year-old Ensign Ward in charge of the fort under construction, when rumors first arrived that the French were only a few days away. After dispatching a messenger to inform Captain Trent at Wills Creek of this alarming development (Trent would in turn send the message to Washington at Winchester),

Ensign Ward consulted with the Half King as to what might be done about the advancing French. The Half King advised that a stockade be constructed immediately around the fledgling fort. With the commanding officer, Trent, away at Wills Creek, another officer, the trader Fraser, who had been made a lieutenant, should have been in charge, but claimed he had business to attend to back at his trading post. As Washington's column was still many days away, Ensign Ward was left to lead the British defense of the Forks of the Ohio by himself.

He remained resolute—at least in principle. "[I will] hold out to the last Extremity," Ward told the homebound Fraser, "before it should be said that the English had retreated like Cowards before the French Forces Appeared."

Construction began instantly on a stockade around the small fort. This sat near the point where the Allegheny and Monongahela rivers joined together in a V, each channeled by steep, wooded bluffs three hundred feet high, touched with the first shimmering, minty green of spring. In a symbolic act, the Half King laid down the first timber of the stockade, conveying that he had thrown in his lot and that of his people with the British against the French—although it remained unclear exactly who and how many his people were. After several days of furious work, and with the help of the Half King's Indians, Ensign Ward and his workers had erected the stockade and hung the main gate. But on Wednesday, April 17, the forty-one workers and soldiers looked up from their labors of hewing timbers and erecting log palisades to see an incredible sight: sweeping down the Allegheny, paddles flashing, oars stroking, came the Marquis Duquesne's massive flotilla of three hundred canoes and sixty bateaux.

It must have been stunning to see this flotilla on this wilderness river where the white man had scarcely paddled before. If they paddled five abreast with each canoe separated on all sides by three canoe lengths, this flotilla would cover the bluff-lined river with a dark mass of vessels for a mile in length. In addition, the sixty bateaux were laden with eighteen cannon and additional supplies. Each vessel held several men, for a total force that Ensign Ward estimated at one thousand. The forty-one workers and soldiers hewing timbers for their stockade must have dropped their axes in awe.

The canoes touched the riverbank near the fort. Hundreds of soldiers climbed out, pulling the canoes onto the river beaches in a nearly endless row, as if some great pod of river-swimming whales had run aground on the banks of the Allegheny. Assembling in groups, drawing themselves to attention, they marched in formation toward the little fort, trailed by work parties that hauled cannon. When the soldiers had approached to a point just beyond musket range from the stockade, the French officers called a halt. A messenger was sent to the fort. The French wanted to parley.

The practice of laying siege to castles, fortresses, or fortified towns had an ancient history in Europe's endless wars and had evolved into a very specific set of rules—a delicate, deadly minuet—by which the besieging army could offer terms of surrender, which the besieged castle or town was obligated to accept or be overrun and granted no mercy. (For an instance of this specificity, the besieging army's artillery master had the right to carry off the surrendered town's best bell, presumably to forge its metal into more armaments.) Although now deep in the Ohio wilderness, the French followed this strict siege protocol.

An officer, Le Mercier, chief of artillery and engineering, who had studied mathematics in France, stepped forward from the assembled ranks of the French, along with an Indian interpreter, known as the Owl. For the British, young Ensign Ward, by default commander of the fort, stepped forward for the parley. Le Mercier carried a message from the leader of the flotilla and commander of the Ohio forces, Contrecoeur, who waited somewhere to the rear of the group. "Nothing can surprize me more than to see you thus attempt to settle on the Territories of the King my Master," Contrecoeur's message began.

The two Crowns were now in peace and harmony with each other, he had written, and he very much wished to respect that. He cited the provocations of the British—the British traders (like Fraser) operating on French soil, the attempts by the British (i.e., Washington) to persuade the Indians to "take up the hatchet" against the French, and now the construction of this British fort on His Majesty's territory. The British must withdraw immediately, Contrecoeur insisted. He concluded ominously, "I hope, Sir, you will not prolong the Time, and thereby force me to Extremes."

Having delivered Contrecoeur's message, Le Mercier looked at his watch. It was two o'clock. By three o'clock, he said, Ensign Ward had to make his wishes known. Did he wish to surrender? Or fight? In the background, meanwhile, Le Mercier's cannoneers were busy setting up their eighteen pieces of artillery and aiming them at the little stockade.

Ward returned to the fort for counsel. For half an hour, until two thirty, he discussed the situation with the Half King. Using the same ploy that the French had used on Washington during his winter mission, the Half King advised Ensign Ward to tell the French officers that he wasn't qualified to give an answer and would have to wait for the arrival of a superior.

Ensign Ward, the Half King, and Indian interpreter John Davison walked to the French camp on the riverbank. There, instead of parleying with Le Mercier, they were received by Contrecoeur himself. Like Saint-Pierre, whom he had replaced, the forty-eight-year-old Contrecoeur came from a distinguished line of military officers in New France and held a feudal lordship, or seigneury, near Montreal. Deeply experienced in wilderness, military, and Indian affairs, he had most recently served as commander of Fort Niagara, the other "key to the West," where he had left his wife and nine children.

Their delaying tactic failed to impress the veteran French officer. "[Contrecoeur] absolutely insisted on [our] determining what to do that instant," Ensign Ward later testified, "or he should immediately take Possession of the Fort by Force."

Ensign Ward now found himself in a very tough position. On the one hand there was his hastily erected stockade and his force of forty-one, only thirty-three of them soldiers. On the other, one thousand French troops and a battery of cannon amassed on the riverbank nearby, threatening to attack at any minute. All his superiors were absent—Lieutenant Fraser literally tending his store, Captain Trent gone home to Wills Creek, and Colonel Washington struggling with wagons and logistics far away. While Fraser and Trent had conveniently made themselves scarce,

Washington was desperately trying to make his way to the Forks. Perhaps his delay was for the best. Had he been there, Washington might have valiantly defied Contrecoeur's demand to leave and tried to hold the fort. As it was, he lived to fight another day.

Ensign Ward was a sensible man. The math was simple. Despite his pledge to fight to the last extremity, he sized up the French troops—at least a thousand—took account of his own force—thirty-three soldiers and a handful of workmen—and promptly surrendered the fort. Contrecoeur gave him until the next day at noon to leave. He could take all the belongings he could carry.

That night Ensign Ward stayed about three hundred yards outside the fort at the Iroquois camp of the Half King. Commandant Contrecoeur, meanwhile, was encamped a short distance away. Extending wilderness hospitality, combined with a desire for information, the commandant invited Ensign Ward to dinner. One assumes they supped well in the commandant's tent and the wine flowed liberally. But fine dining and French wines notwithstanding, the commandant found Ensign Ward not inclined to be helpful, if Ward's later deposition is any indication.

> [Contrecoeur] ask'd many Questions concerning the English Governments, which [I] told him [I] could give no Answer to, being unacquainted with such affairs. . . . [T]he French Commander desired some of the Carpenters Tools, offering any money for them, to which [I] answer'd [I] loved [my] King and Country too well to part with any of them. . . .

One imagines the commandant plying Ward with another glass of wine and trying to pry some tidbit of information out of him, but the ensign held firm. Then Ensign Ward retired to bed.

Ward apparently accepted the loss, but not the Half King. He had thrown in his lot with the British, betting that the British would hold the Ohio Valley where the Half King and his people—the Ohio Valley Iroquois known as Mingoes—would thrive with their British alliance. Now, however, the British, in the form of Ensign Ward, were surrendering this key locale in the Ohio Valley and with it, in his mind, much of the Half King's and his people's future.

"[T]he Half King stormed greatly at the French at the Time [we] were obliged to march out of the Fort," Ensign Ward recounted later, "and told them it was he Order'd that Fort and laid the first Log of it himself, but the French paid no Regard to what he said."

With the fort surrendered to Commandant Contrecoeur, Ensign Ward and company then headed east along wilderness paths to find young Colonel Washington and the reasons he hadn't arrived.

HAVING RATTLED WITH HIS CUMBERSOME WAGONS as far as Wills Creek and now waiting there to hire enough packhorses to cross the mountains to the Forks, Washington first got word of the fort's surrender on April 20, three days after it occurred. Ensign Ward soon arrived at Wills Creek with his men and told Washington of the French, with their three hundred and sixty canoes and bateaux coming down the river, aiming a battery of cannon at the stockade, and offering him an ultimatum.

Washington had been running a race against the French since his mission to Fort Le Boeuf the previous December, when he learned of French plans to launch their canoes down to the Ohio in spring. Now, five months later, he had lost the race. The French had won the key strategic spot in the Ohio Valley—the Forks. It had happened while he struggled with wagons, and baggage, and logistics, and recalcitrant frontier farmers many miles and mountain ranges away. He had hurried as fast as he could. He and Governor Dinwiddie, however, had not understood crucial variables. For each additional man in the wilderness, he would need a great deal more support. This would require hiring wagons, which would require local cooperation. Without understanding just what he was doing, he had rushed ahead.

He surely wondered whether the fort would have fallen so easily if he had arrived in time to defend it. Now the entire Ohio Valley lay in danger. This was not how he had imagined his rise through the ranks—to lose the race to the French and disappoint the expectations of the gentlemen and ladies of Belvoir, whose warm firesides and captivated audiences he had enjoyed only six weeks before. He was determined to win the fort back. Washington charged ahead.

Now, speaking on behalf of the Mingoes in the Ohio wilderness, the Half King sent a message from his camp somewhere near the Forks to Washington at Wills Creek. It was also directed to the governors of Virginia and Pennsylvania: *Come immediately, or you will lose our allegiance to you British.*

> *My brothers the English . . . We expected for a long time that [the French] would come and attack us; now we see how they intend to treat us. We are now ready to attack them, and are waiting only for your aid. Take courage and come as soon as possible, and you will find us as ready to fight them as you are yourselves. . . . If you do not come to our aid soon, it is all over with us, and I think that we shall never be able to meet together again. I say this with the greatest sorrow in my heart.*

The pressure was on—from the king, from the colony of Virginia, from the governor, from his Indian allies, and, perhaps most of all, from Washington himself. Instead of backing off in the face of overwhelming odds, Washington now made plans to push forward toward the thousand French with his meager body of troops. Holding a council of war with his officers, he devised a plan to advance to an Ohio Company storehouse located on the Monongahela River about forty miles upstream from the Forks. Here he would erect fortifications. He would clear a road over the mountains from Wills Creek to transport heavy cannon to this riverfront storehouse. From there, the Virginia troops could construct rafts or boats to float the powerful weapons downstream to the Forks to besiege the French-held fort. While making these preparations, Washington would await fresh orders from Governor Dinwiddie and additional Virginia Regiment troops under Joshua Fry, possibly supplemented by troops from other colonies.

He wrote a flurry of letters—to Governor Dinwiddie requesting heavy cannon "to attack the French, and put us on an equal footing with them";

to two other governors he did not know personally, Governor Hamilton of Pennsylvania and Governor Sharpe of Maryland, to implore them to send troops with all possible haste; and to the Half King, thanking him for his "steadfast adherence to us" and reaffirming the British—or at least Washington's—commitment to him and his Mingoes. The small contingent of one hundred and fifty that Washington had under his command represented only a foretaste of what was to come, he wrote in his letter to the Half King, promising the immediate arrival of "a great Number of our Warriours" as well as "our Great Guns, our Ammunition, and our Provisions."

He signed the message "Go: Washington" as well as "Connotaucarious." In the late 1600s, Indians had supposedly bestowed this name on Washington's great-grandfather, John Washington, who had taken part in punitive expeditions against Doeg Indians who were suspected of killing white settlers. Washington, on his winter journey to the French commandant, had apparently told the Half King this story, and he, in turn, had bestowed this Seneca name on George. It meant "Town Taker," "Town Destroyer," or "Devourer of Villages." He wore the name proudly.

Washington showed little sign of trepidation at the prospect of this advance with a tiny body of troops toward a large French force. Rather, he brimmed with "glowing zeal," as he said in an impassioned letter to Governor Sharpe of Maryland. "[This cause] should rouse from the lethargy we have fallen into, the heroick spirit of every free-born Englishman to assert the rights and privileges of our king . . . and [rescue] from the invasions of a usurping enemy, our majesty's property, his dignity, and lands."

As he railed about the usurping enemy, Washington's glowing zeal cast a glare that blinded his broader vision. Britain and France, after endless wars, were finally at peace. A state of relative calm had settled over Europe. But not for Washington. His trumpeting rhetoric sounded as if he were ready to single-handedly declare war on the French. The loss of the fort while he rushed to its aid landed as a personal blow to him and to his ambitious vision of his future. The French may have shifted in status in his mind—from rival power at peace with his nation to personal enemy who blocked his path. His original orders from Governor Dinwiddie stated that once he reached the Forks he should help the men of the Ohio Company complete the fort they had begun. Governor Dinwiddie explicitly spelled out that Washington was not to be the aggressor, unless faced with aggression, obstruction, or resistance to removal: "You are to act on the Difensive, but in Case any Attempts are made to obstruct the Works or interrupt our Settlements by any Persons whatsoever, You are to restrain all such Offenders, & in Case of resistance to make Prisoners or kill & destroy them."

Both Dinwiddie and Washington were now getting into a situation over their heads. When he wrote those orders, could Governor Dinwiddie have imagined he had set the stage for two superpowers to collide in the Ohio wilderness?

Over the next days, young Washington's frustrations mounted rapidly, exacerbated by the ongoing issue of low pay. Still in Wills Creek with the main ridges of the Appalachians rising before him, he ordered a party of sixty of his men to widen the wilderness footpaths into crude roads in order to transport wagons and cannon to the Ohio Company storehouse on the Monongahela River. They "slaved" their way along, Washington reported, chopping and digging and yanking up stumps to hack a passage through "woods, rocks, mountains." Drenching rains fell. The trail turned to mud. Rivers and creeks swelled with rainwater. Their slow progress stopped for two days while they built a crude bridge across one swollen stream.

"The great difficulty and labour that it requires to amend and alter the Roads," reported Washington, "prevents our Marchg above 2, 3, or 4 Miles a Day, and I fear . . . we shall be detaind some considerable time before it can be made good for the Carriage of the Artillery with Colo. Fry."

This felt agonizingly slow for someone used to traveling thirty or forty miles a day on horseback. Washington had never before participated in a task of this magnitude, much less led one. It was far easier to plan on paper than to execute. Adding to this frustration were near-daily reports of the French advances in the Ohio, brought by British traders fleeing out of the wilderness with their packs of furs. His journal entries detailed a cascade of bad news:

May 4. We met Captain Trente's [business agent] who informed us that 400 more French had certainly arrived at the fort and that the same number were expected in a short time. He also informed us that they were busy building two strong houses. . . .

May 5. We were joined by another trader coming from [the Allegheny] who confirmed the same news. . . .

May 7. We met a trader who informed us that the French had come to the mouth of [Redstone Creek], and that they had taken possession of it with about four hundred men.

May 8. This report was contradicted by some other traders who came directly from there.

Other frustrations boiled up at the same time. The men who had served under Captain Trent were a rough lot whom Washington couldn't control properly and who finally took off on their own. Colonel Fry, commander of the Virginia Regiment, was expected to show up at Wills Creek at any time with more recruits but did not arrive. And then there was the matter of pay—it wasn't enough, not for the other officers, not, especially, for Washington himself, not to endure all this.

He had plenty of time to ruminate on his lack of pay while he and his men chopped and picked their way uphill through rocks and mud and dense, dripping forest. At times he may have wondered why he subjected himself to this. His thoughts traveled over the mountains to green Tidewater plantations and the admiring, long-skirted ladies who graced them. In a quiet moment, he wrote a letter to Sally Fairfax's sister-in-law, Sarah Fairfax Carlyle, enumerating his travails and perhaps asking her to reply. (The letter does not survive.)

After three weeks of mounting frustrations and torturously slow progress, Washington could not contain his bitterness over pay. Governor Dinwiddie, he felt, had no idea of the ordeal he and his men were undergoing. British officers received far more pay. Why not Washington?

Why should the Virginia Regiment receive less? Especially when enduring such hardship?

Bearing the weight of his ancestors and their—and his—aspirations to gentlemanly aristocracy, it was as if he were trying to work out within himself at what point, or at what level of transgression, a gentleman's pride and sense of honor took offense. But how to separate an offended sense of honor from simple anger and frustration arising from a task going badly, such as this road into the wilderness? Washington had not yet calibrated a trigger point, as he would in later years, for either his anger or his honor. He was prone to feel unappreciated. The last time this had happened, a few weeks earlier, he threatened to resign. He had gone to call on Colonel Fairfax, who calmed him and persuaded him not to quit. That was not possible now—Colonel Fairfax was many miles away. Instead Washington wrote to Governor Dinwiddie and tore into the issue of his status and pay, or lack of it. One can picture him scrawling this letter—in a tent in the rain, in a muddy wilderness camp—and unleashing his frustration to his superior.

> *Giving up my commission is quite contrary to my intention. . . . But let me serve voluntarily; then I will, with the greatest pleasure in life, devote my services to the expedition without any other reward, than the satisfaction of serving my country; but to be slaving dangerously for the shadow of pay, through woods, rocks, mountains,—I would rather prefer the great toil of a daily laborer, and dig for a maintenance . . . than serve upon such ignoble terms; for I really do not see why the lives of his Majesty's subjects in Virginia should be of less value, than of those in other parts of his American dominions; especially when it is well known, that we must undergo double their hardship.*

Enclosed with this angry letter was a list of complaints from the other officers and from himself—about low pay, the need for more sergeants and corporals, the need for funds to buy finer foods for an officers' table, and lack of an allowance to purchase beer and wine, among other issues.

Having vented his frustrations, Washington tried to push on. Traders had repeatedly told him of the impossibility of clearing a road good enough that one could haul a wagon—and certainly not heavy cannon—over these Appalachian Mountains. The traders appeared to be right. He now switched strategy. A river, the Youghiogheny, cut through the spiny mountain ridges to the Monongahela, slicing through the formidable obstacles of what are now known as Laurel Hill and Chestnut Ridge. Washington calculated that he could follow this watercourse in canoes or boats to cross the mountain barrier.

While his men toiled on the road, he outfitted a small expedition to explore the Youghiogheny River. His party of six set off on May 20—Washington, a lieutenant, an Indian guide, and three soldiers, in canoes. Within half a mile, a trader motioned them over to the riverbank and discouraged them from going forward, but Washington ignored the advice and kept paddling, as he had ignored the expert advice of Christopher Gist the past December about the shortcut through the frozen forest. Mountain ridges slanted straight up from the riverbank, their slopes thick and green with leafing forest. After camping on a gravel bar that night, they continued downstream the next day, scouting on foot, wading through the river.

The river flowed more swiftly between the ancient ridges of the Appalachians, steepening, tumbling over water-smoothed boulders, spinning in eddies. About ten miles downstream from the gravel bar, the river fell into a set of rapids—"rough, rocky, and scarcely passable," Washington noted. They kept going. The river bounced down wide stair steps of shallow rock ledges, each increasing in drop. They heard the thunder of whitewater. Ahead, mist rose. Looking down the riverbank, they saw only distant, hazy treetops far downstream as the river's horizon suddenly disappeared.

Working down the bank, they discovered, to Washington's distress, that the river plunged over a waterfall. "[It] then fell, within the space of fifty yards, nearly forty feet perpendicular," he wrote. It was now clear that, despite his best efforts at river exploration, Washington could not use the Youghiogheny to transport men and heavy cannon through the mountains to the Forks. He abandoned that effort and returned to the encampment near Wills Creek, where his men continued to hew a wagon track over the ridges.*

THE FRENCH SEEMED TO OCCUPY more of the Ohio wilderness each day. Reports came in about the French at the Forks building earthwork defenses twelve feet thick, raising yet another fort downstream at the Falls of the Ohio, sending out reconnoitering parties.

As the French pushed eastward in the wilderness, Washington's party pushed westward. It appears inevitable, in retrospect, that Washington's party and a French party should at some point collide. Washington's plan was that it would be on his terms, when all his men and all Joshua Fry's reinforcements and cannon had gathered at the Ohio Company warehouse on the Monongahela and constructed boats or rafts to float downstream and lay siege to the French fort.

More reports came in of French troops approaching through the forest. Washington now caught up with his main party, which had hewn the road to a valley between two tall Appalachian ridges. On May 24, a few days after Washington had quit explorations of the Youghiogheny, an Indian came into camp with a message from the Half King, somewhere near the Forks. The message, written in interpreter John Davison's crude English spelling, warned that the French had sent troops from the Forks to meet Washington. "[T]herfor my Brotheres I deisir you to be awar of them for [they design] to strike the forist Englsh they see. . . ."

Was the Half King to be believed? How would the sachem know the French intent? Another vague report arrived that the French had reached a crossing of the Youghiogheny, about eighteen miles away from Washington's camp in the forest. With the French rumored so close, Washington resolved to set himself up in a camp with some defensive potential. He rushed with his troops a few miles to a place called the Great Mead-

* Today the Youghiogheny Gorge is one of the most popular stretches of whitewater in the eastern United States for rafters, kayakers, and canoers. Ohiopyle Falls is clearly the drop that turned Washington back.

ows, near the base of Chestnut Ridge. Between the wooded hills lay a long, marshy strip of grassy meadow. A stream looped through the meadow, creating natural ditches that could act as defensive earthworks. To Washington it looked like a fine place to make a fortified camp. "We have with Natures assistance made a good Intrenchment and by clearing the Bushes out of these Meadows prepar'd a charming field for an Encounter . . . ," he boasted to Governor Dinwiddie, now at Winchester.

Washington sent out scouting parties to try to locate the rumored French parties. They returned with nothing. He was getting jumpy. They all were. It was as if two blindfolded persons were trying to find each other in a dense wood, their arms reaching out, their ears attuned, uncertain what would happen at the first startling touch of the other.

That night, May 26–27, sentries around his camp at the Great Meadows thought they noticed something suspicious out in the dark. They fired, the boom of their muskets rolling through the nearby hills, as if mocking this sudden aggression. Nothing happened. No return fire. No movement. No noise. The entire camp was ordered to arms. Under Washington's direction, the troops stayed on alert from 2 A.M. to sunrise. With the dawn, they discovered that six men had deserted. Washington and his officers conjectured that the suspicious activity in the night had been made by the escaping deserters.

The tension rose again that morning, May 27. Christopher Gist, who had built a cabin nearby, rode into the camp at Great Meadows. He reported that La Force, the French officer who had accompanied Gist and Washington on part of their journey to the French commandant, had approached his house along with a contingent of fifty or so French while he was away. He had seen their tracks. They would have broken everything inside it and killed his cow, Gist reported, but they were turned away by two Indian guards Gist had left. Washington immediately sent seventy-five men in pursuit of the rumored French party. What his troops would do if they caught up to La Force is not clear.

That evening of May 27 the tension notched still higher. At 8 P.M., as the darkening sky grew heavy with rain clouds, an express messenger from the Half King rushed into Washington's Great Meadows camp. The

Half King was on the trail en route to join Washington at Great Meadows and had camped about six miles away. Along the trail, the Half King, with his expert Indian tracking skills, had spotted the tracks of two men that descended into a "gloomy hollow." "[H]e imagined the whole party was hidden there," Washington recorded.

Washington sprang. Fed by uncertain rumors of French intentions, determined to win back what he had lost in the race to the Forks, both to the French and to his reputation, untutored in the rules of war or the arts of diplomacy, he appears to have convinced himself, with the Half King's help, that the French were about to attack him. He would go himself to head off this possibility.

It was now late in the evening. He summoned forty of his men. At midnight they set out on foot up the narrow forest path toward the spot where the Half King waited, leaving behind a contingent to guard the Great Meadows camp and ammunition should that be the French target. As they marched, a heavy rain began to fall. Raindrops pattered onto the thick, leafy forest canopy overhead, and the night, Washington reported, was "black as pitch." The trail wound through the undergrowth, barely as wide as a single man. They could not see each other. Wet branches swished against their clothes, drenching them. When one man stopped, the man behind bumped into him in the darkness. Sometimes they wandered off the trail altogether. It took fifteen or twenty minutes to locate it again in the heavy darkness, feeling for the earthen path with their feet.

Toward dawn, as the wet gray skies lightened, they arrived at the forest camp of the Half King. Washington and the Half King consulted. In what would prove a fateful decision, they decided that Washington and his troops, together with the Half King and his handful of warriors, would pursue the French tracks.

The Half King sent two Indians ahead to trace the tracks. They found that the French had set up camp in the "gloomy hollow"—"a half mile from the [trail] in a very obscure place surrounded with Rocks."

Led by the Indian guides, Washington and his men, along with the Half King, another Iroquois sachem named Monacatoocha, and their small party of warriors, marched in single file, Indian fashion, up a mountain toward the forest glen. Just after dawn, they quietly approached the defile from above. In another fateful decision, Washington chose to deploy his troops in combat mode. He ordered his men to spread out so that they surrounded the U-shaped glen on top of its rocky walls, in a "disposition to attack on all sides," as he later put it. His group held the curve of the U. Another group under Captain Thomas Waggoner held the left side. The Half King and his warriors crept down through the woods to come up through the open end, while Monacatoocha led warriors to the glen's right.

Below, a party of about thirty-five French were just waking from sleep in the bark lean-tos that had sheltered them from the night's rain. Some had started to cook over fires and eat their breakfast. Some had gone off in the woods to relieve themselves. Some still slept.

There is a great deal of disagreement about just what happened next. Did Washington give the order for his forty troops to open fire on the party of breakfasting French? Or did one of the Frenchmen spot Washington's troops atop the rock walls first, warn his comrades of an impending attack, and reach for his gun? Or did one of Washington's men, in the excitement of the moment, pull the trigger without orders, or the Half King or one of his men? Who fired first?

Whoever it was, by sneaking up on the French in a hostile manner, Washington had positioned himself at the epicenter of what would prove to be a global turn of events.

PART TWO

DEFEAT OF GENERAL BRADDOCK, IN THE FRENCH AND INDIAN WAR, IN VIRGINIA, IN 1755.

CHAPTER SEVEN

THEY WERE THE FRENCH, AND, IN WASHINGTON'S MIND, they were the enemy. At some point, he gave the order.

Fire!

White billows of musket smoke shot out from among the leafy trees rooted above the glen's rocky walls. The crack of gunfire reverberated among the rocks. Washington himself fired the first shot, an Iroquois warrior with Washington later testified.

The French, just roused from sleep, only fifty yards away, scrambled from their lean-to shelters. Others sitting around the smoldering campfires dropped their breakfasts, grabbed their muskets, and took cover as best they could behind trees and rocks. Shouts and cries in French tried to coordinate a response to the firing from the top of the walls. Musket balls thwapped down into the rain-soaked earth around them, zinged off boulders. The French aimed up toward the haze of smoke and trees, sighting on trembling leaves and crouching figures who reloaded on top of the walls. The French fired. The floor of the leafy glen erupted with flashes of yellow and puffs of white, and an instant later the ear-thudding roar reached young Major Washington and his men on the rock rim. Bullets tore through leaves. A man dropped.

Again Washington gave the order to fire. The volley exploded from the rim top. French soldiers in buttoned white coats fell to the earth. One, two, three, and still more. More French muskets aimed up toward the rim. When they pulled the triggers, the guns emitted only a small, metallic click. The night's rains had dampened the gunpowder in the muskets' flash pans, so when flint struck steel the spark fell harmlessly on the wet charge in the pan. The French realized they had little chance if unable to fire volleys at the British troops. They scattered, running down the slanted bottom toward the glen's exit, the opening of the U. As they dashed between trees toward the mouth, the Half King and his warriors suddenly emerged from the woods, armed with muskets and tomahawks. They blocked the way.

The French spun around and scrambled back up the glen to their encampment, dropping their muskets, their hands raised. The firing from the rim went silent. What remained were the groans of the French wounded. The fight had lasted fifteen minutes.

The French leader had fallen to the ground, having taken a musket ball. A thirty-five-year-old ensign in the colonial regulars, he belonged to a large French-Canadian military family that descended from French nobility. He made it known who he was—Ensign Joseph Coulon de Villiers de Jumonville. He also communicated that he carried a message to deliver to Colonel Washington—a summons. Washington worked his way down through mossy boulders and drooping ferns. Unlike on his previous travels into the wilderness, he now lacked the advice of that expert in Indian and frontier matters, Christopher Gist. He still surged with adrenaline from his first firefight, the sudden, violent eruption in the peaceful woods, the explosion of noise and smoke, shouts and groans. He had never before seen crumpled bodies like this among the dead leaves, bloodstains seeping across white and blue coats. He could not dwell on that now. He stepped forward in the glen to meet an emissary from the French camp, perhaps La Force, the officer who had accompanied him to Fort Le Boeuf and spoke many Indian languages and some English and now was with Jumonville's party.

The message was read aloud in the glen, either by La Force or another interpreter. It was from Commander Contrecoeur at Fort Duquesne. It warned that Contrecoeur had heard from the Indians that Colonel Washington at the head of an armed force had entered the lands of His Majesty, the King of France, and emphatically requested, in the name of the king, that he depart over the mountains "*in Peace*." Claims by Virginians to have purchased Indian lands at the Forks of the Ohio were so weak as to require, if necessary, eviction by force, he wrote, and, as it was Commander Contrecoeur's intention to keep the peace between the two Crowns, Colonel Washington would have to answer for any act of hostility. "Whatever your Schemes may be," concluded Contrecoeur, "I hope, Sir, you will shew Mr. Jumonville all the respect that Officer deserves, and that you will lend him back to me again with all Speed, to acquaint me with your Intentions."

Washington could not understand it. He took the message from the French party. He began to walk back to his own men to study it more carefully.

As Washington withdrew, his ally and guide, the Half King, stepped forward. He hovered over the wounded Jumonville. He addressed him using the traditional Indian term that acknowledged the paternalistic relationship between the French "father" and the Indian "children."

"Thou art not yet dead, my father," he said.

The Half King raised his tomahawk overhead and slammed it down with patricidal force on Jumonville's skull. The French officer slumped. The Half King lifted it again and slammed its sharp edge down again with a dull crack, and then again. The blows split open Ensign Jumonville's skull. Letting go the bloody tomahawk, the Half King reached down, scooped out the gray mass of Jumonville's brain, held it up, and kneaded it through his fingers, as if washing his hands in Jumonville's essence.

Washington was helpless to stop this, nor could he stop what happened next. The Half King's warriors fell upon the dozen or so wounded French scattered about on the ground. They pulled back the heads of the helpless soldiers, planted a knee in their back, sliced around the skull, grabbed a lock of hair, and yanked the scalps off them.

In both his first combat and his first taste of Indian-style warfare, the twenty-two-year-old probably watched dumbfounded in horror. He had no control of the situation that he had initiated and pursued. His stealthy approach to the breakfasting French had spun first into a massacre and then, by any standard rules of war, a murder scene, as the bloody sacrifice of the wounded unfolded in the leafy glen. He could have approached the French far more peacefully. He could have sent an emissary. He could have stayed along the main trail and waited for the French to arrive. He could have approached in an open manner. But he had chosen to approach stealthily, arranging his men along the glen's rim in a position "to attack on all sides," virtually guaranteeing that fighting would break out. Ambitious, impetuous, paranoid, and inexperienced in combat, Washington had just ignited a war.

The eighteenth-century British commentator Horace Walpole would famously remark, "The volley fired by a young Virginian in the backwoods of America set the world on fire."

THE SHOOTING WAS ONE ISSUE. The murder of Jumonville—for that is what it appears to have been—in the aftermath of the gunfire was another one entirely. The Half King split open the wounded French officer's head during a negotiation, contrary to all European rules of warfare, usurping from Washington this wilderness encounter between two rival European empires and turning it to his own ends. The Half King never revealed his true motives, at least not to anyone who ever wrote them down. Some historians have speculated that as the French established their presence in the Ohio, the Half King felt his power over the Mingoes and other Ohio tribes seeping away, and he intentionally provoked an outbreak of fighting between the two empires with the hope that his ally, the British, would win and restore his power. Other historians believe that the Half King targeted Jumonville and his party simply because he wished to protect his own family from the French, having heard they were pursuing him. Whatever the Half King's motivations—personal defense, tribal power struggles, revenge against the French for reputedly boiling and eating his father, or a combination of factors—Washington appears to have been blindsided by the Half King's sudden eruption of violence and the chaos it triggered.

After tearing off the scalps, the Half King and his Indians hacked off the head of a dead French soldier and skewered it atop a wooden stake. The gory Indian trophy-taking at an end, Washington and his remaining thirty-nine men took as prisoners the twenty-one surviving French soldiers, including La Force, leaving a dozen scalpless French bodies sprawled in the dead leaves of the glen's floor, and one headless corpse.

As the party marched with their captives from the glen, along the ridgetop and down the forested east slope toward the Great Meadows camp, did Washington speak to his former traveling companion, the prisoner La Force? Did he reflect on what had just occurred? Had it unfolded as he might have hoped, or did it seem wildly out of control and shockingly violent to him?

By body count, he and his Virginians and the Half King and his Indians had indisputably won the engagement against the French, the British suffering only one killed and one or two wounded. One imagines the Virginia troops celebrating British resolve over French decadence. Washington, however, may have needed the march back to the Great Meadows to make sense of what had just happened. How had the moment exploded in such violence and chaos? Had he acted properly? Jumonville had camped in a secluded glen off the trail. Washington may have ruminated on this during the trek back to the Great Meadows camp. By the time he arrived at his tent and sat with quill in hand and portable writing case in lap, he had convinced himself of a heroic version of the incident, that he had made a bold—even courageous—preemptive strike against a party from an enemy empire that was about to commit a sneaking, hostile act.

WAITING FOR HIM at the Great Meadows camp was a letter from Governor Dinwiddie expressing a great deal of displeasure. Now visiting Winchester in hopes of a meeting there with the Half King and other Iroquois sachems, the governor had received Washington's letter of May 18 with its long list of complaints from the officers and Washington's personal grousings about pay.

Dinwiddie countered them point by point: The "gentlemen" officers knew the pay before they signed up, so how can they complain now? It wasn't true that the British officers of an expedition to Canada received better pay and free wine and beer. Your personal allowance for expenses as a Virginia colonial officer is no less generous than that of British Royal Army officers. What you consider hardships, furthermore, His Majesty's officers see as "Opportunities of Glory [rather] than Objects of Discouragement."

But Governor Dinwiddie's most painful remarks expressed his personal disappointment in his young protégé for agreeing with the officers' complaints: "Now Colo. W: I shall more particularly answer wt relates to YrSelf, & I must begin with expressing both Concern & Surprize, to find a Gent. whom I so particularly consider'd, & from whom I had so great Expectats. & Hopes, appear so differently from himself. . . ."

It was a harsh rebuke, and it stung. It bothered Washington so much that he immediately sat down to respond to the governor, and, when he touched quill to paper, defense of his own reputation took precedence over the startling news of the incident with Jumonville and the French. "I am much concerned that your Honour should seem to charge me with ingratitude for your generous, and my undeserved favours," wrote Washington, "for I assure you Honble Sir, nothing is a greater stranger to my Breast, or a Sin that my Soul more abhor's than that black and detestable one Ingratitude."

Alternating in tone between obsequiousness and a prickly self-defense, Washington expounded his points to Governor Dinwiddie, page after scrawled page. He wished to receive pay on the same generous scale of a British Army officer or, alternatively, serve as a volunteer without any pay at all. But he could not honorably accept the meager pay offered by Virginia. He gave considerable space to the lack of proper uniforms and to an officer's expense account for food and drink, known as a "table allowance." Rather than sitting at the table of Colonel Fry, the Virginia Regiment's soon-to-arrive commander, he wished to invite his friends and officers to dine at his own table. It was hard to live in the wilderness, especially for gentlemen officers accustomed to a more luxurious life. "We are debarr'd the pleasure of good Living, which Sir (I dare say with me you will concur) to one who has always been used to it; must go somewhat hard to be confin'd to a little salt provisions and Water: and do duty, hard, laborious duty that is almost inconsistent with that of a Soldier. . . ."

Washington self-righteously excepted himself from this softness. He wrote to the governor that he, unlike his fellow officers and soldiers, for whom it came as a surprise, knew beforehand the difficulties of this rough country. His experience the previous winter trekking through the frozen Ohio wilderness had convinced him of his own toughness: "for my own part, I can answer, I have a Constitution hardy enough to encounter and undergo the most severe tryals, and I flatter myself resolution to Face what any Man durst, as shall be prov'd when it comes to the Test. . . ."

It was only after Washington had sketched out in the minutest detail where he stood, or thought he stood, on these issues—pay, honor, ingratitude, wilderness hardship—that he gave his commanding officer the briefest possible account of the firefight with the French the previous day. It was a whitewash. He described how he and his soldiers had tumbled over each other in the rainy darkness, reached the Half King's Indian camp along the trail, and decided to "strike" together. "I thereupon in conjuncton with the Half King & Monacatoocha, formd a disposion to attack them on all sides, which we accordingly did and after an Engagement of abt 15 Minutes we killd 10, wounded one and took 21 Prisoner's. . . ."

That was it for an actual account of what had happened. For his letter's next several pages, Washington attempted to justify his actions. He adamantly challenged whether Jumonville's party was on a diplomatic mission, as the French prisoners had claimed. "[T]he absurdity of this pretext is too glaring," Washington wrote.

He asserted that the French party was, in fact, a pack of spies "sculking" about in low, hidden places, gathering intelligence so Contrecoeur at the fort could launch an attack on Washington. But this was all Washington's after-the-fact justification for his ambush of Jumonville. That rainy night, with the Half King's urging, Washington had grown convinced that the French planned to attack him, so he had prepared to attack first, sneaking up on Jumonville's party at daybreak.

The French prisoners swore that, when Washington attacked them, they were on a diplomatic mission and they should be treated as diplomats rather than prisoners. "The Sense of the Half King on this Subject is, that they have bad Hearts," Washington wrote to the governor, "and that this is a mere pretence, they never designed to have come to us but in a hostile manner. . . ."

Besides, he went on, one prisoner, the officer La Force—"a person of [great] subtilty and cunning"—spoke the Indian languages fluently and wielded a great deal of influence among them. He was worth fifty Frenchmen, Washington wrote. He could not see turning La Force loose just because he claimed to be on a diplomatic mission.

Washington mentioned nothing of the nature of the massacre, the mutilation of the bodies, the taking of scalps, except to say that the Half King of the Mingoes dispatched the French scalps along with a hatchet to other Ohio tribes such as the Delaware to summon them to abandon ties to the French and ally themselves with the British. The Iroquois sachem Monacatoocha carried this message to tribes as far north as Lake Erie. The French might attack him at any time, Washington wrote, but he and his men would be ready whatever the hour. He closed his letter to Governor Dinwiddie with heroic and martial ardor. ". . . I doubt not but if you hear I am beaten, but you will at the same hear that we have done our duty in fightng as long as there was a possibility of hope."

He sounded ready to die for the cause. He could not help adding at the end, however, a last complaint about his out-of-pocket expenses and deficient wardrobe. He had given some of his own shirts to the captured officers La Force and Druillon, by which he had "disfurnished" himself, as he was traveling light and had only brought two or three shirts into the wilderness from the last outpost at Wills Creek.

He wrote a kind of addendum to this letter the same day, sending the shorter note by express messenger on horseback, warning Governor Dinwiddie to "give no Ear" to the "smooth stories" of the French prisoners now en route to the governor at Winchester. The prisoners were claiming that they had called out from the bottom of the glen to the British on the rock rim not to fire. "[B]ut that I know to be False," he wrote, "for I was the first Man that approach'd them & the first whom they saw, and immediately upon it [they] ran to their Arms and Fir'd briskly till they were defeated."

Furthermore, he emphasized to the governor, don't give them the honor of an embassy "when in strict Justice they ought to be hang'd for Spyes of the worst sort. . . ." Apparently it did not occur to him that he, too, had spied while on his first diplomatic mission to the wilderness.

This all came as bold talk for the inexperienced young colonel after his first victory—or what appeared a victory. He seemed unsure of how to present himself after the event, as if trying to strike the right pose in his communications to the outside world. In less than twenty-four hours, he had swung from an aggressive pursuit of the French, to defensiveness about his reputation in his letter to the governor, to rationalizations about his ambush of the French breakfasting in the glen. But he oscillated again in another twenty-four hours or so—perhaps after congratulations from fellow soldiers and officers—and performed a kind of macho strut.

Writing to his younger brother Jack, he breathlessly described the "sharp firing on both sides," the French dead, and the capture of prisoners. He wrote that he and his men on the right flank took the brunt of French fire but escaped relatively unscathed, with only one dead and two or three wounded. "I can with truth assure you," he added in a final boast and self-congratulating affirmation of his bravery, "I heard Bulletts whistle and believe me there was something charming in the sound."

Writing a day or two later again to Governor Dinwiddie, he went so far as to suggest that the French, generally, might be cowards and easily defeated. "If the whole detachment of French behave with no more resolution than this chosen party did, I flatter myself we shall have no great trouble in driving them to the [damned] Montreal."

The remarks would come back to haunt young Washington.

UNDERSTANDING THAT A FRENCH ATTACK might be imminent, Washington had his men reinforce his camp by erecting a log palisade and digging trenches. It sat on the narrow, four-mile-long forest opening called the Great Meadows, which could provide both rich forage for the horses and cattle and a clear field of vision to spot an approaching enemy. He wrote to Colonel Joshua Fry, the portly former math professor leading the Virginia Regiment over the mountains from Alexandria, urging him to bring reinforcements quickly. Washington pledged to Colonel Fry that, even if he should have to fight against a much greater force, he would not give up "one Inch of what we have gaind."

Reveling in the ecstatic flush of his first victory and with defensive preparations hectically under way, Washington now projected a certain invincibility: "We have just finished a small palisaded Fort in which with my small Number's I shall not fear the attack of 500 Men," he boasted by express letter to his boss, Governor Dinwiddie.

This hastily built fort of Washington's in the Great Meadows was in reality a claptrap affair. It consisted of a circular stockade a little over fifty feet in diameter—about the size of a small lawn—constructed of ten-inch-thick oak logs that had been split in half, sharpened to a point, and planted vertically in the earth, like a picket fence twelve feet high. At its center stood a shack about fourteen by fourteen feet roofed with a crude covering of bark and animal pelts to keep food and ammunition dry. A ditch and earthworks surrounded the stockade. In this tiny, ramshackle fort, young Colonel Washington had convinced himself that he and his small body of men could stave off half a thousand French.

Nor had he allowed for French battle tactics. He assumed that the French, and whatever Indian allies they might have recruited, would approach in regular formation, marching across the field of battle like the clashing armies of Europe. He had constructed his fort at the narrowest part of the meadows, where low, wooded hills fringed it nearby. These offered a ready source of oak trees to build the stockade, but Washington was oblivious to the help those trees might also provide to the French and their Indian allies.

His own recruiting efforts for Indian allies had met with little success. The Half King had arrived at the little fort on the Great Meadows with twenty-five Mingo families consisting of about eighty people, but most of them were women, children, or elders. This was a very meager turn-out of support for the British from the Ohio Valley Indians, despite the Half King's effort to rally them by sending French scalps and a symbolic hatchet to take up in war. Out of two hundred thousand square miles of Ohio Valley wilderness, he had managed to rally only about two dozen warriors. Governor Dinwiddie, at Winchester in the Shenandoah, where he hoped to meet with Iroquois sachems and assemble other Indian allies, expected a force of Cherokee warriors from the Carolinas to arrive at any moment on the Great Warriors Path. But so far, nothing.

IT MUST HAVE BEEN A MUDDY, MOSQUITO-RIDDEN PLACE during the moist month of June, this crude fort on the low, boggy ground of the Great Meadows surrounded by wooded hills. The men had spaded up earth everywhere—to plant the log uprights of the stockade, to dig the surrounding ditches and earthworks. One imagines the meadows' tall, thick grasses stamped flat by the shoes of the laboring soldiers, then trodden into a soggy mash of grass and mud. A scattering of cows that have been driven over the mountains graze about—the supply of meat. The men eat this tough, lean beef along with bread baked from precious flour. They sleep in tents pitched outside the small stockade. The plumes of smoke from dying fires climb into the moist night air. Mosquitos whine in tossing soldiers' ears.

Thwack! Hands slap cheeks. Soft curses drift through the sleeping camp.

Young Washington endured this alongside the men. By June 6 they had been encamped a week at the Great Meadows since the death of Jumonville, preparing for a possible French attack. They had completed the stockade. But the hungry men and the twenty-five Mingo families had eaten their way through the supply of flour. Food had become a critical issue. This day the flour ran out. At the same time, startling news arrived. Christopher Gist, Washington's guide from his winter mission in the wilderness, rode into camp and reported that Colonel Joshua Fry had taken a bad fall from a horse at Wills Creek. Appointed by Governor Dinwiddie to overall command of the Virginia Regiment, Fry had been en route from Winchester to the Great Meadows with reinforcements for Washington's camp. Fry was known for his knowledge of mathematics and natural science and his exceptional mapmaking abilities but not for his military prowess.* Well into his fifties, he was overweight and out of shape for vigorous riding.

A few days later, he died of his injuries. Governor Dinwiddie, commanding temporarily from Winchester, now wrote young Washington with stunning news. He was to replace Colonel Fry and command the Virginia Regiment, which could soon number in the hundreds. More troops were expected to arrive from North Carolina under the leadership of a seasoned older officer, Colonel James Innes, a fellow Scotsman from the Highlands whom Dinwiddie appointed to take overall command of the expedition to the Ohio. Colonel Washington and his Virginia Regiment would serve under Colonel Innes, whose Royal Army commission came from the king (of a higher order than a commission from a colonial governor) and who had seen bloody action and devastating tropical disease in the Battle of Cartagena during the War of Jenkins' Ear. Dinwiddie expected that Colonel Washington would find Colonel Innes's leadership "agreeable."

Now head of the Virginia Regiment, young Colonel Washington, only twenty-two, had commanded no men at all until eight months before.

THE MONTH OF JUNE hung in a kind of stasis. Anxious questions lay heavily on young Washington's mind. What were the French up to? When would reinforcements cross the mountains and arrive at the Great Meadows? Would his men run out of food? Should he stay put at the Great Meadows camp? Still flush with victory, Washington now had to deal with the uncertain aftermath. It was a game of waiting.

No one had declared war, yet he was impatient for action and outraged at the French. Governor Dinwiddie was complicit in all this. He reinforced Washington with congratulations for his "Success" in killing and capturing the French party and showing the Indians that the French could be defeated. But he also urged his young colonel to use caution. "[I] hope the good Spirits of Yr Soldiers will not tempt You to make any hazardous

* Joshua Fry and Peter Jefferson made groundbreaking surveys of remote parts of Virginia in the 1740s and in 1752 produced a widely acclaimed map, *A Map of the Inhabited Part of Virginia,* which became the standard of the era.

Attempts agst too numerous Enemy," he wrote. "When [reinforcements] join You, You will be enabled to act with better Vigour."

It seems Dinwiddie felt some unease about his newly made commander. While eager to take on the French, Washington had also been personally and exhaustingly defensive when chastised by the governor for his complaints about pay and elaborately self-justifying in explaining how he ended up pursuing the French to their leafy glen. In the emotional swings of Washington's long letter, and in his recent threats to quit or serve without pay to protect his honor, Dinwiddie may have sensed an uncertainty, a lack of self-possession, an insecurity about his place in the world, as if he had not learned how to comport himself when challenged or questioned in any way. This was hardly surprising in a twenty-two-year-old. An astute judge of character after his years as merchant and administrator, Governor Dinwiddie may have feared that his young protégé might choose a rash course of action in an attempt to show his bravery or to win respect and a hero's adulation from the Virginia aristocracy.

Why, then, did Dinwiddie choose Washington to command the Virginia Regiment? After two journeys into it, Washington, young as he was, had some familiarity with the Ohio wilderness, its terrain and its Indians, and had participated since the beginning in the governor's attempt to evict the French. While waiting for Colonel James Innes to arrive and take overall command of the Ohio expedition, the governor had few choices for a more senior officer with wilderness experience.

To encourage his young officer, Dinwiddie sent Washington a special medal he had minted to commemorate the "success" over Jumonville, along with a gift of rum from the governor's private stock. Washington effusively thanked him for the promotion to head of the Virginia Regiment, the commemorative medal, and the gift of rum, promising to think of the governor whenever he drank it. He "rejoiced" over the appointment of Colonel Innes as the commander to whom he would answer, but, characteristically, he could not help thinking about how the appointment of a superior officer would detract from his own chance to prove his worth. "[N]ow I shall not have it in my power to convince yr Honr, my Friends, and Country of my diligence, and application to the Art Military, as a Head will soon arrive to whom all Honour and Glory must be given."

ON JUNE 9, A SMALL BODY of reinforcements arrived at the Great Meadows under Captains Robert Stobo and Andrew Lewis and Lieutenant George Mercer. This boosted Washington's numbers by about one hundred and eighty—not a great many if it came to a face-to-face showdown with the French, but helpful. The flamboyant son of a Glasgow merchant, twenty-seven-year-old Stobo would become a legend in his own right. After the death of his parents in Scotland, he traveled to the American colonies to trade but enjoyed Virginia and its society life and stayed, becoming friends with fellow Scotsman and former Glasgow merchant Dinwiddie. Charmed by a romantic boyhood notion of military life, Stobo, commissioned as captain by Dinwiddie, joined the Virginia Regiment as it prepared to push into the Ohio wilderness. Crossing the mountains from the Tidewater region, he arrived at Washington's Great Meadows camp luxuriously outfitted with ten personal servants and his own horse-drawn covered wagon, which carried a barrel containing 125 gallons of Madeira to soften life in the woods. Stobo, Lewis, and Mercer also brought nine swivel guns—small cannon mounted on swivels to allow a wide range of direction and usually firing loads of small pellets called "grapeshot," able to mow down several enemy soldiers at once. These were welcomed by Washington to defend his fort on the meadows.

Also arriving at the swelling Great Meadows camp were two experts in Indian negotiations. Sent by Dinwiddie to assist the soon-to-arrive Colonel Innes, they left deep impressions on everyone who met them. One was the hard-to-miss trader Andrew Montour, three-quarters Indian and one-quarter French, whose father was an Oneida chief named Big Tree. His mother, Madame Montour, half French and half Huron, had proved a great friend to the English. Her mild-mannered but fearless thirty-something-year-old son Andrew, likewise friendly to the English, nimbly trod between Indian, white, and frontier worlds. He was a fluent speaker of English and French as well as various Indian languages, and his exotic dress reflected the peculiar amalgam of worlds that he inhabited.

"His cast of countenance is decidedly European," recorded a traveling German nobleman and Moravian minister, Count Zinzendorf, "and had not his face been encircled by a broad band of paint applied with bear's fat, I would certainly have taken him for one. He wore a brown broadcloth coat, a scarlet damaskin lappel waistcoat, breeches over which his shirt hung, a black cordovan neckerchief decked with silver bugles, shoes and stockings and a hat. His ears were hung with pendants of brass and other wires plaited together like the handles of a basket. He welcomed us cordially and when I spoke to him in French he replied in English."

More expert help in Indian diplomacy arrived with an entrepreneurial and wide-ranging fur trader, George Croghan. Born in Ireland, Croghan, about thirty-five years old, had emigrated to the colonies and in the 1740s become a fur trader in the Ohio Valley, building trading posts at Indian villages, learning the native languages, and attempting to acquire Ohio Valley land directly from the native peoples. A half-brother of Ensign Ward, whom Contrecoeur and the French had forced to surrender at the Forks, Croghan was commissioned to provide flour for Washington's troops at the Great Meadows. The veteran of a great many risky ventures into the unknown, Croghan appeared unconcerned that the French had offered a bounty for his scalp due to his trading incursions into the Ohio and trading posts he had built on what they considered their territory.

On June 12, two days after the arrival of Montour and the reinforcements, another party stumbled into the Great Meadows camp—French deserters. Indian scouts had reported a French party in the woods rumored at ninety strong. Washington eagerly set out at the head of one hundred and thirty men and thirty Indians. Fervently hoping to make another "present" of French prisoners to Governor Dinwiddie, as he put it, his expectations were crushed by wildly inaccurate intelligence and miscommunication. He had marched only half a mile before discovering that the French party in its entirety consisted of nine deserters. "I was as sensibly disappointed when I met these persons today as ever I was in my life," Washington lamented to his boss.

Despite the dimmed prospects for glory, Washington questioned the escapees. He wrote to Governor Dinwiddie that they corroborated his suspicions that the Jumonville party was sent to spy on the British and had instructions to show the diplomatic summons they carried from Contrecoeur only if discovered or overpowered. Whether the deserters knew this as a fact, or said it simply to please the big and very young Virginia colonel who interrogated them, is unclear.

Meanwhile, an Indian messenger arrived at the Great Meadows camp from Logstown and the village of Delaware chief Shingas to give the Half King important news. Rather than joining the British, the Shawnee and Delaware had "taken up the hatchet" against them. Confirmed by the nine French deserters, this was not good news, especially as Governor Dinwiddie had counted heavily on Indian support to stymie the French in the Ohio wilderness. On the advice of the Half King, Montour, and Croghan, Washington invited the Delaware and Shawnee to meet with him directly, hoping to persuade them to the British side.

On June 14, a few days after the discovery of the French deserters, more reinforcements wended into the marshy Great Meadows. With them came—in young Washington's mind, at least—a big problem. This was in the form of Captain James Mackay. He rode at the head of a company of one hundred regular British Army troops arriving from distant South Carolina and Georgia. Back in January, immediately after Washington returned from his winter mission to the French commandant, Governor Dinwiddie had written to authorities in London asking for military assistance to protect the Ohio from the French. In London, Lord Holderness, Britain's then secretary of state for the Southern Department, was willing to commit two independent companies from the New York colony and one from the Carolina colonies who were regular British Royal Army troops. The South Carolina company under Captain Mackay had sailed up the Atlantic coast, up the Potomac to the river port of Alexandria, and marched over the Appalachians to Washington's camp at the Great Meadows.

Now the question arose: Who was in charge of whom?

The issue apparently reared its stubborn head the very first night. One can imagine the thirty-six-year-old Captain Mackay riding into the Great Meadows camp that day at the head of his one hundred troops and attended by all the trappings of his King's commission. Son of a Scottish laird, James Mackay had emigrated to Georgia, founded his own sprawling estate, and served in the War of Jenkins' Ear as an officer in the British Royal Army. His captain's authority came directly from His Majesty the King, while Washington, though technically outranking Mackay as a colonel, derived his authority from the colony of Virginia and its provincial governor. Washington's lifelong preoccupation with uniforms, appearances, and rank may have started here. He may have looked on enviously as Captain Mackay's disciplined company of British regulars chosen from Georgia and South Carolina marched across the Great Meadows, no doubt with flags flying, drums beating, saddles creaking, and Mackay dressed in his best uniform. The glory of His Majesty's royal forces had arrived in the marshy meadow. Colonel Washington's Virginia men, by contrast, looked tired, dirty, and scratched from days of attempting to chop a road through dense forest, their drab uniforms torn and muddy. A soggy patch of stomped grass surrounded Washington's tent.

The two officers, Washington and Mackay, greeted each other cordially but eyed each other warily. Captain Mackay went off with his troops to pitch their own camp of twenty-five tents nearby. As a standard security procedure, that evening Colonel Washington issued a sign and countersign—password or phrase—to both camps so sentries and soldiers could recognize friend from foe in the night.

Captain Mackay balked. He refused to accept that Washington had authority over him, even in so routine a matter as this. Officers in the regular British army generally did not follow commands from officers in the colonial forces.

Governor Dinwiddie had been unclear on this point in his instructions, saying only that Captain Mackay was "an Officer of some Experience & Importance," and Colonel Washington, along with Colonels Innes and Fry, should not allow small procedural points to hinder the great mission to the Ohio. This warning, however, did not give Washington pause. He immediately sat down to write the governor an impassioned addendum to a letter he had written earlier that day, scrawling almost obsessively in endless sentences. He signed it, sealed it, and dispatched it by mounted express rider to Winchester.

He was confident that His Honour would see the absurdity and problems should Captain Mackay direct the Virginia Regiment, Washington wrote. For example, Captain Mackay's subofficers might employ the British regular army style of battle and order a direct frontal attack against a French force who used Indian fighting methods—in which the Virginia Regiment had some experience—and would expose them all to almost immediate death. Nor would Captain Mackay's regular army soldiers labor on clearing the road over the mountains unless they received extra pay, Washington noted, while the Virginia Regiment, which had done far more for the cause than any other troops, was finding it very hard to do double or triple duty for little pay and have their rank undermined, too. To achieve rank was what the officers of the Virginia Regiment signed up for, Washington wrote, and if Captain Mackay took command and they lost rank, an "Evil tendancy" would result.

Washington ended his long, self-promoting missive with a warning directed to any prying eyes, which reflected both his lack of grammar and an awareness that he stepped on uncertain ground in advocating so strongly for himself at the expense of a British regular army officer with a King's commission. "The Contents of this Letter is a profound Secret," it read.

Governor Dinwiddie had no doubt feared just this—that his young provincial officer, unconfident of his standing in the world but fiercely ambitious to better or at the least protect it, could not simply let the matter of rank rest. Colonel Washington would not back down. Captain Mackay would not budge. The two met and discussed the matter cordially but ended where they began—deadlocked.

With more people coming into the Great Meadows camp, food remained a critical issue. Captain Mackay had driven over the mountains with him sixty head of beef cattle but had brought only about five days' worth of flour, which the soldiers craved to accompany the beef. Including Mackay's men and the twenty-five Mingo Indian families with the Half King, the Great Meadows camp now had about three hundred people to feed each day.* (Governor Dinwiddie also insisted that for morale the

---

* Daily rations for a British soldier (1767): "1 pound bread or flour; 1 pound 9 ounces

troops receive a generous daily rum ration of a quart per four men, or the equivalent of four stiff drinks each.) The Virginia Regiment's commercial agent back in Alexandria, merchant John Carlyle, struggled to supply flour and other goods such as shovels to the Great Meadows camp. Even when he could locate enough flour to purchase, he could not hire enough wagons to carry the goods on the wilderness "road" over the mountains. The wagon drivers found the way far too rough. "*Can not you spare a few Men, to blow up any rocks . . . ,*" Carlyle wrote to Washington.

The wilderness rendered the already difficult logistics of military supply far more complicated. Washington now began to grapple with a problem that he would confront for the rest of his military life—how to keep his men supplied. And a still larger issue loomed—a strategic one. Should he stay put at the Great Meadows? Or move forward closer to the French? Or retreat to a safer point farther east, perhaps even over the mountains?

The latter option, to retreat farther east, does not seem to have occurred to the young Washington. Nor did he give much consideration to staying put at the Great Meadows. Rather, he was all about pushing ahead toward the Monongahela River and the French fort at the Forks of the Ohio.

Whether this was rashness or bravery depends on one's perspective. Surely Washington saw it as the latter. But there was so much he did not know: The number of French. Whether more supplies and ammunition were coming over the mountains on the terrible road, and more men arriving under Colonel Innes. Whether the British could attract enough Indians to help them. And the biggest question of all: After the bloody incident with Jumonville, would the French and Indians attack in response?

Despite the uncertainties, Washington moved ahead with the plan he had hatched before the Jumonville encounter, although with a twist. His original plan called for pushing the last twenty-five miles or so from the

---

meat or pork; 6/7 ounce butter; 3/7 pint peas; 11/7 pint rice or oatmeal. When the above is not available, the daily ration is as follows: 1/2 Pound flour or bread; 1 Pound Beef or 10 Ounces Pork. If no meat is available, the following is a complete daily ration: 3 Pounds flour or bread; or 1-1/2 pounds rice." The standard daily rum ration was a quart per six men, and beer was also widely distributed.

Great Meadows to the Monongahela River to build rafts and float cannon down to the Forks. Now he decided to march, for the time being, only about halfway there, to Christopher Gist's newly built homestead. This lay about twelve miles away on the other side of the last great barrier of the Appalachians, Chestnut Ridge. Here, at Gist's tiny settlement, he would hold a council, or meeting, inviting Indian leaders who had recently taken up the hatchet against the British—notably the Delaware and Shawnee.

With the help of the Indian experts Montour and Croghan and the presence of the Half King, leader of the friendly Ohio Iroquois (or Mingoes), he would convince these Delaware and Shawnee chiefs to abandon the French and join the British. Council completed, he would then proceed the last thirteen miles or so to the Monongahela River, establish his forward base as planned at the Ohio Company warehouse at the mouth of Redstone Creek, and await the arrival of cannon, supplies, more troops, and, he hoped, many more Indian allies. When a sufficient force had gathered at this forward base, he would raft the cannon the thirty-seven miles down the twisting Monongahela to where it joined the Allegheny and capture the prized French fort that stood at the Forks.

It was a bold plan, but young Washington saw himself as a bold leader. He had so easily vanquished the French in that first encounter with Jumonville and had boasted to Governor Dinwiddie that if all the French were that cowardly, he would have no problem in driving them back to "Damned Montreal." The Half King, nevertheless, advised Washington against advancing, according to one participant, the wealthy Virginia planter Colonel James Wood, who also noted, as if to mark the overconfidence of Washington and his men, that some were "heard to say how glorious it would be to take the [French fort] without the assistance of Cap. McCay."

On June 16, a few days after Captain Mackay's troublesome arrival, Washington left Mackay and his Carolina troops behind to guard the Great Meadows camp. With his Virginia troops and several rickety wagons hauling the nine swivel guns brought into the wilderness by Mackay, his column strained over rugged Chestnut Ridge and dropped down to

Christopher Gist's settlement in the valley below. His three crude log structures were surrounded by a split-rail fence and cupped in a grassy meadow in a mountain valley with a crystalline stream tumbling through it. Known to white settlers as an "old field," this meadow was one of the ancient clearings burned in the forest by Indians to encourage grass to grow for deer and to open up land for corn planting. With good soil, water, and few trees, the old fields offered prime spots for a settler's homestead.

A soft light suffused through small windows and an open door and dimly revealed the adze marks on the logs of the cabin's interior. In a circle on the flooring's split planks sat or squatted Gist, Croghan, Montour, Washington, and representatives, chiefs, and sachems from the various tribes of the Ohio. These included the Six Nations or Iroquois, whose headquarters were at Onondaga, outside the Ohio Valley, but who claimed control over the valley; the Shawnee and the Delaware, the latter represented by the powerful Delaware chief Shingas. Traveling from the Great Meadows camp at Washington's request, the Half King also attended, representing his band of Ohio Valley Mingoes.

In the measured cadences of formal speech, the Iroquois spokesman figuratively addressed the governor of Virginia, Robert Dinwiddie, who was not present. He told Dinwiddie's representative at Gist's settlement, young Colonel Washington, that they now feared the British would destroy them should they not actively join the British cause against the French. "My brothers, we your brothers of the Six Nations," began the Iroquois spokesman, ". . . have heard it said that you threaten to destroy entirely all your brothers the Indians who would not come to join you on the road. It is for that reason that we who stay in our villages expect every day to be cut in pieces by you. We should like very much to know from you yourself the truth of this news. . . ."

As in many preliterate cultures, oratory among North American Indians was a prized and highly refined skill and one that deeply impressed the early white explorers and frontiersmen who heard it. Speeches given by sachems at councils followed a formal pattern, usually delivered to a group sitting in a circle around a ceremonial fire. It opened with polite acknowledgment of the relationship between the gathered parties, moved slowly to the substance of the message, often expressed metaphorically, and concluded with an emphatic point.

Among the Iroquois and other tribes of the Eastern Woodlands, wampum played a powerful role in both long-distance communications and up-close oratory. Painstakingly embroidered with beads fashioned of seashells or glass, these beaded belts contained abstract designs and figures that conveyed various messages—the making of peace, the declaration of war, or the solidifying of bonds between groups. Speakers presented a belt of wampum during their speech to emphasize their main point and their sincerity—codifying in a physical, immutable object the speaker's message and avoiding misinterpretation if delivered long-distance. In a way, the belt of wampum represented the Native American counterpart to the European legal system's urge to "put into writing" in the form of a contract or letter of agreement a solution to a disputed matter. The intended recipient could either accept that belt of wampum—sign the contract or letter—or reject it outright, sometimes violently, by kicking it away or flinging it in the speaker's face.

Concluding his speech with a symbol of his sincerity, the Iroquois sachem presented Washington with a beaded belt of wampum. Then came Washington's turn to speak. The trader and negotiator Croghan had probably prepared him, sketching out his speech and style of delivery, perhaps with the help of Montour and the Half King. Young Washington started by disparaging the French. For decades, the French had represented themselves as "father" to the Indian "children," reflecting the patriarchal relationship that French priests, traders, and military officers maintained to the tribes. They encouraged the Indians to refer to the governor of New France as "Onontio," meaning father. As Washington later described the speech in his journal:

> *[The French] have a beautiful speech and promise the most beautiful things. . . . but all this is from the lips only, while in his heart there is only corruption and the poison of the serpent. You have been their children, and they have done everything for you, but scarcely have they believed themselves strong enough before*

they resumed their natural haughtiness and drove you from your lands and declared that you have nothing [in the Ohio].

The British, he promised, were different. The British opposed the French in order to give the Ohio lands back to the Indians. "[The British] at your repeated requests sent an army to maintain your rights, to restore you to possession of your lands and to guard your wives and children, to dispossess the French. . . . It is for this that the arms of the English are actually employed; it is for the safety of your wives and children that we are fighting."

This lay a very long way from the truth. Some Washington biographers have called it an outright lie.* Croghan may have advised the young colonel to state so emphatically that the British fought for the safety of the Indians. Or Washington may have believed it was partly true. Or he may have willingly told an untruth or half truth in an attempt to win allies to the British side. Or he may have felt that lying to Indians was not quite the same as lying to white men.

The speeches lasted for hours, then days. In the evenings, the Virginian troops retired to their canvas tents pitched on the meadow around Gist's settlement, and the Indians to bark lean-tos erected at the forest edge beneath Chestnut Ridge. The complex web of relationships among the region's tribes resembled a family tree. The Delaware called the Iroquois "uncle," and called the Shawnee "grandsons," while the French were "father," reflecting the diplomatic hierarchy between the tribes themselves and the tribes and Europeans. The British stood on a more equal-footed basis, the tribes and British addressing each other as "brother." Throughout this diplomacy conducted in a circle on the rough-hewn floor of Christopher Gist's cabin, the tribes each wondered with whom to side

---

* The famed story about six-year-old George Washington admitting to chopping down one of his father's prized cherry trees because he "could not tell a lie" was apparently concocted—and certainly wildly embellished—by Parson Weems, who wrote a worshipful biography of Washington a year after his death and sprinkled it with moral fables for the edification of the era's youth.

if war erupted between the French and British. Likewise, both the French and British realized that Indians would play a pivotal role in this contest of empires and that whichever side had the help of the tribes possessed an enormous advantage in determining the possession of the Ohio Valley.

Aided by his advisers, young Washington employed every stratagem he could to win over the tribes to the British. The giving of gifts informed a great deal of the diplomacy between whites and Indians, but Washington, who had urged Governor Dinwiddie to send plenty, had few to give. He apologized for the paucity of gifts, assured them that good gifts would arrive shortly, and blithely assured them that the British would feed and clothe the warriors' families, at the personal invitation of the governor of Virginia himself, without mentioning that Virginia was struggling to feed its own troops.

No bribes worked. Chief Shingas promised to help recruit Delaware warriors but could not do so openly, he said. He departed from Gist's settlement. Likewise, the Half King and his Mingo warriors, instead of pushing forward with Washington toward the Monongahela, left Gist's and backtracked over Chestnut Ridge to their families at the Great Meadows camp. The Shawnee simply disappeared. None of this boded well. Nor could the British get any sure intelligence about French strength at Fort Duquesne or be certain when Colonel Innes would arrive with reinforcements.

Having failed to recruit Indian allies during his stay at Winchester, Governor Dinwiddie returned over the Blue Ridge to the Tidewater region and his Governor's Palace at Williamsburg and now entertained some second thoughts about his gung-ho interim leader. On June 27, he wrote to Washington to proceed cautiously. He wished the young colonel had postponed marching toward the Monongahela River and the Ohio Company warehouse at Redstone Creek until Colonel Innes and his reinforcements had arrived. "You know the French act with great Secrecy & Cunning," the governor warned. But the letter did not arrive until too late.

CHAPTER EIGHT

WASHINGTON'S JOURNAL ABRUPTLY ENDS WITH HIS ENtry of June 27, 1754, four weeks after the Jumonville incident. The last entry describes sending sixty men out from Gist's settlement to enlarge the thirteen-mile trail from Gist's to the Ohio Company warehouse, which sat where Redstone Creek flowed into the Monongahela. Once wagons, artillery, and reinforcements had finally arrived over the mountains from the coast, this widened trail would allow them to travel on to the Monongahela River and thence downriver to the Forks to besiege the French fort.

The next day, June 28, everything changed. After a series of confusing rumors about French strength at the Forks, a message arrived from Monacatoocha, the Iroquois sachem sent to the Ohio to oversee the Shawnee and considered friendly to the British. En route by canoe from a Logstown meeting with the Shawnee to the council at Gist's, Monacatoocha had paddled by the French fort at the Forks. With his own eyes, he had observed French reinforcements arrive down the Allegheny, pulling their canoes and bateaux to the riverbank underneath the fort's thick earth-and-timber bulwarks. The Forks now teemed with French soldiers and Indian warriors. Monacatoocha had heard the French officers at the Forks say they were sending a force of eight hundred soldiers and four hundred Indian warriors to attack the English.

This was stunning, sobering news. Washington immediately abandoned the plan to proceed forward to Redstone Creek and the Ohio Company storehouse, at least for now. Instead, he would make a stand at Gist's settlement. Gathering his scattered troops, he summoned Captain Lewis and his crew of sixty trail clearers to return to Gist's. He sent a message to Captain Mackay at the Great Meadows to bring his Carolina independent troops over Chestnut Ridge and immediately join him at Gist's settlement. Combined with troops already at Gist's, this amounted to a total of perhaps four hundred men who would be at Gist's settlement by noon the next day. In the meantime, he ordered his own men to tear up Gist's fences and set the split fence rails on end to build a stockade.

The Indians who remained encamped in Gist's pasture did not like the look of the vulnerable situation. They threatened to abandon Washington unless he returned to the Great Meadows camp.

The inexperienced Colonel Washington found himself in a difficult situation. It appeared that the French planned to attack him imminently. He had assembled all his forces, which still amounted to only one-third of the French number. His troops were running out of food. His few potential Indian allies were backing out. This would give pause even to someone as eager and ambitious as young Washington. He did not want the planter aristocracy and ladies at Belvoir to know him for a retreat he led or, worse, a defeat. Unsure what to do, squeezed between a large French force, starvation, and the mountains, Washington, perhaps with George Croghan's advice, wisely called for a consultation—a council of war.

Made up of his officers, the war council assessed the situation and listed certain facts—that reliable reports indicated a French force of twelve hundred planned to attack their much smaller force, that the French could paddle by canoe to within a mere five miles of the current British position at Gist's, and that the British troops, already in a "Starving Condition," had only twenty-five milk cows and a single quart of salt to feed their four hundred men.

Given all this, the council of war unanimously decided that the troops should return to the Great Meadows. "The Reasons for so doing," read the minutes, recorded by Colonel Washington himself, "were very Weighty."

As Governor Dinwiddie had feared, Washington had pushed ahead too far and too fast without enough reinforcements or food. It must have come as a shock to find himself moving enthusiastically forward one day and, almost the next, in headlong retreat. It's easy to imagine that a good deal of second-guessing was going on from fellow officers and soldiers and perhaps especially from Captain Mackay, who, at Washington's request, had just rushed from the Great Meadows toward Gist's. As if refusing to show self-doubt or vulnerability and to affirm the correctness of his decisions and the sureness of his hand, Washington now threw himself into the retreat.

His men, however, had grown weak from meager rations, road cutting, stockade building, and the hard labor of crossing Chestnut Ridge to Gist's. The retreat to the Great Meadows back over Chestnut Ridge would require twelve miles of very rocky, hilly terrain to be traversed by weakened men, few horses, and two broken-down wagons. Loading up at Gist's meadow, the troops could not fit all their ammunition into the two wagons. Taking the lead, Washington loaded ammunition onto his own horse, paying four pistoles to soldiers to carry his personal effects. Instead of riding like a tall-booted officer, he would walk the muddy trail like a common soldier.

It was a remarkable moment, and a fateful one. In the strict hierarchies of the British military and of Virginia society, one did not often, or willingly, step down from one's privileges or position. When they joined the colony's military, the gentlemen members of the aristocracy or Virginia's planter class were expected to become officers, and officers were expected to behave in ways befitting their gentleman's status. They rode on horses while the common soldiers walked. Servants frequently attended them. (Washington's personal servant, apparently a slave, accompanied him on the expedition.) They dined on better food, had abundant liquor, drank wine, slept in larger tents, had portable beds. In this moment of need, Washington quite literally stepped down from all that.

Other officers followed his example and gave over their personal horses to carry ammunition, too, marching on foot like their leader and the troops. The men had brought the nine swivel guns in horse-drawn wagons from Great Meadows to Gist's settlement but now did not have enough horses and wagons to haul the heavy guns back to Great Meadows. Instead of horses, the council of war opted to use manpower to muscle the guns up and down the mountain trail—"the roughest and most hilly Road of any in the Allegheny Mountains," as Captain Adam Stephen described it. The men of Captain Mackay's independent company of British Royal Army regulars from the Carolinas, however, perhaps taking a cue from their own commander, refused to help Washington's Virginia troops haul the guns or ammunition back to the Great Meadows or clear the trail. "[This] had an unhappy Effect on our Men," wrote Stephen, "who no sooner learned it was not the proper Duty of Soldiers to perform these Services, than they became as backward as the Independents."

Leading by example, Washington plunged into the task, perhaps hoping to put the recalcitrant Captain Mackay and his British Army regulars to shame for placing selfish pride and pay above commitment to the common cause. One can imagine Washington striding along in breeches and boots beside his column of men, most of whom wore crude leather shoes. They grunt and haul while he—young, athletic, tall, powerful—perhaps lends a hand at crucial moments, leaning his powerful frame into the spoke of a wagon wheel or tugging on a rope as they heave the guns up and over a large boulder. By giving up his horse, he reached down through social strata directly to his men, as if to say, *we're all in this together.* He learned to make that connection here, at the age of twenty-two, while manhauling swivel guns and ammunition over Chestnut Ridge. The camaraderie he engendered with his troops would prove invaluable in the years and decades ahead.

Still, it was a crushing twelve-mile slog. For lack of flour, they had eaten no bread in eight days, subsisting on the tough, lean beef from the fresh-killed cows and at times only parched Indian cornmeal. Struggling uphill and down, they no doubt fantasized about the bread, bacon, salt, and rum that awaited them at the Great Meadows, and the relief of having fresh troops to reinforce them.

After the exhausting trek, they stumbled into the Great Meadows camp on July 1. They had expected to find that a large convoy had arrived bearing thousands of pounds of supplies, but were met by a few paltry flour sacks that could not remotely feed four hundred men. Nor had reinforcements or big cannon arrived. The men were exhausted, disappointed, and, reported Stephen, too weak to manhaul the swivels any farther.

Washington now faced another crucial choice—whether to stay at the Great Meadows camp or retreat farther east. He considered his weakened troops. He took stock of his disappearing food. He expected a supply convoy and two companies of troops from New York to arrive shortly. In his victory flush after the Jumonville incident, he had pledged not to give up one inch of ground. Abandoning the camp at the Great Meadows and retreating farther east over the mountains would mean giving up a great deal of it.

He resolved to stay. He would prepare to meet a French force here at Great Meadows, reinforcing the little fortification and sending express messengers to summon the extra troops and supply convoys with all possible haste. He named the fort, to express his situation, Fort Necessity. His resolve to make a stand may have been bolstered by correspondence from home. He received a reply to the letter he had written six weeks earlier to Sally Fairfax's sister-in-law, Sarah Carlyle, which apparently had referred to times spent with her family at Belvoir. She urged him to forgo his pleasant memories and concentrate on the much greater task at hand:

> *[T]hose pleasing reflections on the hours past ought to be banished out of your thoughts, you have now a Noblier prospect that of preserveing your Country from the Insults of an Enimy. and as god has blessed your first Attempt, hope he may Continew his blessings and on your return, Who knows but fortune may have reserved you for Sum unknown She, that may recompence you for all the Tryals past. . . .*

Washington ordered his men to reinforce the little circular stockade of upright logs by building around it a bulwark, or breastwork, of horizontal logs and spaded earth and dig a trench just behind the breastwork in which to stand protected from enemy fire. On July 2, they raised the earth-and-log breastwork, but dug the trench behind it only about two feet deep. To more experienced eyes, neither this nor the stockade offered much protection from an attack. Sensing impending disaster, the Half King and his warriors, supposedly allies of the British, took their wives and children and drifted away. The Delaware and Shawnee had long disappeared. "After [the Jumonville incident] Col. Washington never consulted with us nor yet to take our Advice," an Iroquois warrior recalled later, "as we knew the French were strong, and now it was dangerous Times we wanted a place of Security of Fort for our Wives and Children. . . ."

Only eight months had passed since Washington, as a young adjutant, had volunteered to serve as messenger to the French in the Ohio wilderness. He had risen rapidly from mere messenger to commander of a substantial body of troops—a commander who had pushed forward too far too fast and had to retreat, lacking supplies and reinforcements. If the young Washington was all about rapid, even heedless, advancement, it's not surprising that the Washington of later years would be known for his deliberation, planning, and caution.

"Perhaps the strongest feature in his character was prudence," Thomas Jefferson would write of the Washington of many years later, "never acting until every circumstance, every consideration, was maturely weighed. . . ."

That George Washington, however, still lay far in the future.

AS WASHINGTON'S TROOPS STUMBLED into the Great Meadows camp, some fifty miles to the northwest a man named Denis Kaninguen showed up at Fort Duquesne bringing choice news to the fort's commander, Contrecoeur. A deserter, he told Contrecoeur that he had been chased by a horseman from the British camp, fired a shot at the rider and broken his thigh, seized the horse, and galloped to the Forks. As Washington had feared, deserter Kaninguen gave over detailed intelligence about British numbers, movements, and condition—information that Commandant Contrecoeur found very useful.

Since mid-April, when Contrecoeur's canoe flotilla had paddled down the Allegheny and run off Ensign Ward and his small Virginia contingent, the French had been busily constructing a stronghold at the Forks. In honor of the governor-general of New France, they named it Fort Duquesne. The French engineer and artillery expert Le Mercier had designed the massive fort with mathematical precision in a star shape, guarded by bastions, cannon, and earth-and-timber ramparts that were twelve feet thick. Fort Duquesne, in short, was to Fort Necessity what a stone castle was to a thatched hut.

When news of Jumonville's violent end reached Fort Duquesne, Commandant Contrecoeur, outraged at what appeared to be an ambush of his diplomatic party followed by an assassination of his officer, readied his force of about six hundred French troops and one hundred Indian warriors to punish Washington's action. On June 26, as this force, led by Le Mercier, prepared to pursue the audacious young Virginia colonel, then camped at Gist's settlement, a canoe party of another one thousand French troops and Indian warriors arrived at Fort Duquesne. Sent from Montreal, they had rushed up the Great Lakes, portaged Niagara, and paddled down the Rivière Le Boeuf in a matter of days. They happened to be under the command of Jumonville's older brother, forty-three-year-old Captain Louis Coulon de Villiers. Like his father and brothers in his aristocratic family, he had served for years as a distinguished officer at remote interior posts and was thoroughly knowledgeable about Indian ways. He deeply grieved the loss of his younger brother. With Captain Villiers's arrival at Fort Duquesne, Commandant Contrecoeur postponed the pursuit of Washington for a few days and, in place of Le Mercier, who graciously stepped aside, appointed Villiers as its new leader to avenge his brother's death.

To spur the Indians to the French cause, Contrecoeur addressed the assembled Indian warriors whom Villiers had brought to Fort Duquesne in an outpouring of grief and lamentation for Jumonville's death and Washington's misdeed. "The English have murdered my children," he called out to them, "my heart is sick; to-morrow I shall send my French soldiers to take revenge. And now, men of the Saut St. Louis, men of the Lake of Two Mountains, Hurons, Abenakis, Iroquois of La Présentation,

Nipissings, Algonquins, and Ottawas,—I invite you all by this belt of wampum to join your French father and help him crush the assassins. Take this hatchet, and with it two barrels of wine for a feast."

"Both hatchets and wine," writes historian Francis Parkman, "were cheerfully accepted."

The Delaware, whom Washington had tried to woo to the British side, were also among the warriors at Fort Duquesne. They symbolically picked up the hatchet, too.

The French burnished muskets and the Indians sewed moccasins. On June 28, they embarked together in scores of canoes and pirogues up the Monongahela toward Redstone Creek and the Ohio Company storehouse located at its mouth. The French officers—Commandant Contrecoeur, Villiers, Le Mercier, and Longueuil—had coauthored the mission statement: "to avenge ourselves and chastise them for having violated the most sacred laws of civilized nations." After the British had been thoroughly punished and withdrawn from French territory, the French would inform them that in accordance with King Louis XV's wishes, as Parkman puts it, "the French looked on them as friends."

For two days they paddled up the Monongahela. On June 30 they pulled their canoes and pirogues ashore at the Ohio Company storehouse. They found no one around at the stout structure of squared-off logs, measuring about twenty by thirty feet. The loopholes cut for guns stared back empty. The forest around it remained silent, and the sweep of the river offered no clues. After a council with the Indian chiefs for advice, Villiers left the fleet of canoes at the creek's mouth and, with his great mass of troops and Indians, headed through the forest on foot toward Gist's settlement. Scouts probed ahead.

"The path was so rough," writes Parkman, "that at the first halt the chaplain declared he could go no farther, and turned back for the storehouse, though not till he had absolved the whole company in a body. Thus lightened of their sins, they journeyed on. . . ."

They reached the three cabins of Gist's settlement on July 2, finding the fence rails torn up and planted in a stockade and the site recently abandoned by Washington's men. They camped there that night in a drenching rain and at dawn marched on the steep, rocky trail over Laurel Ridge. In the downpour, Villiers and a small party detoured a half mile from the main path to the rocky glen where his brother was killed. Scalped bodies still lay strewn in the wet leaves of the glen's floor where Washington had left them a month before, unburied and no doubt gnawed by scavengers, and a French head was impaled on a stick. Turning away, Villiers knew that the young Virginian responsible for this scene of rotting carnage was now encamped only a few miles off.

WASHINGTON'S MEN AT FORT NECESSITY were still digging shallow trenches and reinforced breastworks on the morning of July 3. Mist-draped hills surrounded the Great Meadows, and the small creek twisted quietly past the little fort. Colonel Washington and Captain Mackay had about three hundred able men and another one hundred sick, malnourished, and exhausted by the last march over the difficult trail from Gist's.

Early that morning, according to Adam Stephen's account, word came that an outlying sentry had been shot in the heel by an advancing French party. As the men frantically dug the ditches deeper and the embankments higher, creating two defensive V shapes in the marshy ground to protect the circular fort, more information arrived in camp.

"[A]bout nine," Stephen reported, "we received Intelligence, by some of our advanced Parties, that the enemy were within four Miles of us, that they were a very numerous body, and all naked. We continued to fortify and prepare ourselves for their Reception."

At midmorning, a sentinel spotted the French. The French soldiers marched in three columns across the wooded hills that surrounded the Great Meadows. While not wholly naked, the Indian warriors who accompanied them perhaps had stripped down to leggings or breechclouts, and some of the French may have followed suit.

The sentinel fired his gun. The French columns paused, still about a thousand feet from the fort, and leveled their muskets. White smoke billowed into the gray morning, the quiet rain splitting with gunfire, but the shots fell harmlessly short.

Washington ordered his officers to draw up the men in formation in front of the knee-deep trench. They assembled, muskets at the ready, and began their march forward. Washington told them to hold their fire. Marching his men forward, he waited for the French to approach nearer, expecting to meet them in orderly rows on the open meadow he had cleared of brush, his "charming field for an Encounter," in the style of warfare conducted on the battlefields of Europe.

But instead of behaving predictably, the French troops suddenly broke from their columns along the edge of the forest, joined by scores of Indian warriors. With bloodcurdling Indian death screams and French battle cries, they surged forward. Major George Muse, leading one body of Washington's troops, called a sudden halt. "*The French would take the fort!*" he exclaimed and, turning about, ran with his men back to the protection of the shallow trench and fort, later to be accused of cowardice. At about the same time, Adam Stephen, realizing that the fort lay largely unguarded as the French and Indians swarmed downhill, ordered two platoons of his troops in the field to turn to the right and return to defend it. Suddenly the whole body of troops, on its own accord, wheeled right and turned back to the trenches.

"Had not this lucky Mistake happen'd not a man of us could have liv'd above an hour," Stephen later remembered. Had they followed Colonel Washington's lead into the field, Stephen's remark implied, they would all have perished.

As the Virginia and independent Carolina troops took to the trenches, Colonel Washington again ordered his troops to hold fire until the French neared. The screaming Indian warriors and French soldiers flowed over the grassy contours of the landscape, downhill but not toward the little circular fort and its protective trenches in the meadow's center. The swarming mass ran instead toward a low hill that projected into the open meadow, covered by a tongue of forest that reached to within sixty yards of the fort. Abandoning any pretense of European warfare, the combined French and Indian force assumed the Indian style of fighting and hid behind oak and ash and tulip trees and fallen logs, within easy shooting range of the trenches and fort but protected from British fire by this patch of hardwood forest. To have an easy source of timber, Washington had built his fort near the trees. This now looked like a big mistake.

The French discharged a second volley, smoke erupting from the projecting hill and its tongue of forest. Washington ordered the British to fire. The battle was joined, with the British huddled in shallow trenches or behind the low embankments or inside the little stockade, while across a short patch of grassy meadow and up a slope, the Indians and French, as Washington later put it, "from every little rising—tree—Stump—Stone—and bush kept up a constant galling fire upon us. . . ."

Hunkered in the trench, the British returned the fire, loading and reloading, musket shots booming toward the wooded hill, smoke rolling over the green meadow. Their grazing horses and cows dropped to their knees, then toppled over heavily, targeted by French and Indian sharpshooters to starve the British troops. They shot the British camp dogs, too. With thudding roars, two British swivel guns replied, blasting loads of grapeshot at the wooded tongue, the lethal clusters of small metal balls whistling and slamming through the trees. French officers now targeted the British cannoneers. French musket balls ripped into the fort's wooden palisade, wounding several men with flying splinters.

Their fire "extinguished" by his French musketeers, as Villiers later phrased it, the swivel guns fell silent. In misty rain the shooting crossed the patch of meadow all afternoon, gunfire rolling down from the woods and up from the trenches. Rainwater seeped into the trenches from the marshy earth. British soldiers dropped. Toward evening, the British quickened their firing. As if shaken by the chaos below, the skies suddenly let loose a deluge of rain that lowered a gray veil over the field of fire, "the most tremendous rain that can be conceived," Washington later remembered.

The British dead slumped in their defensive ditches, half submerged in knee-high water, mud, and blood. Everything in the fort was afloat, reported the Scottish doctor, Captain Adam Stephen. Wearing shoes but stockingless and wading through mud up to his calves, Stephen dashed among the unfinished fortifications ordering swivel cannon fixed and delivering gunpowder to the men, until his hands and face, as he put it, "were as black as a Negroe's." Still the Indians and French fired down upon the mire while sheltered from the upward whizzing of English musket balls. The downpour splattered over ammunition boxes, over the flash pans of muskets, trickled down the gun barrels. Guns misfired or did not fire at all. The British tried to clear their muskets. Startlingly, they possessed only two "screws," the devices that they could shove down the barrel to extract the wet powder and ball. With so many guns jammed, they would have to fight with bayonets, but they had few of these.

The British struggled under incoming fire, wet powder, rain, lack of food, the mud and blood and fallen bodies, the mounting chaos. As darkness fell, desperate men broke into the rum casks and lurched about, drunk. Washington's own servant took a French musket ball and fell, lethally wounded. Except for the brief skirmish at the glen, this was Washington's first real taste of battle, and he had never witnessed such violent confusion. His bold course of action, his relentless push forward, had led not to glorious subjugation of the French but rather to this disaster, with the French and Indians blasting down on this dimming muddy hole. How could it have happened? How to make it stop?

A shout came out of the rain and gathering darkness. "*Voulez-vous parler?*" The French were offering to talk.

Was this a ruse? The French and Indians had vastly superior numbers. They had hidden themselves well. They held the higher ground and every advantage. Why would they ask to talk?

The French offered to send an emissary from their position in the woods to the British fort. It was a trick, perhaps. Washington refused. There was a pause. The French called out again. The British could send an emissary to the French side.

Washington hesitated. The fallen lay all around him. What did the staring eyes and blank, mud-splattered faces of his dead and dying men ask of him, men whose names he knew? To fight onward for their honor against the French? To make them heroes remembered for all time? Or did they tell him to waste no more lives, to save their comrades who still survived? Did they plead with him to parley with the French, whatever the risks to the battle's outcome or his personal honor? What did his own dead want? And his commander, Governor Dinwiddie, and His Majesty thousands of miles across the ocean, what did they want?

Nothing was clear, nothing was glorious, certainly not on the battlefield or, probably, in his own mind. He had boasted to Governor Dinwiddie a month earlier that he would drive the French back to "Damned Montreal." He now realized that he faced a choice for both himself and his men, between negotiating with the enemy or martyrdom. He decided to parley.

He chose Jacob van Braam. The Dutchman van Braam had served only the previous fall as Washington's French-speaking interpreter during his first foray into the Ohio wilderness. To accompany Captain van Braam, Washington selected William La Péronie, who also spoke some French.

Van Braam stepped out into the rainy, darkening evening. Apparently wounded, La Péronie could not walk the short distance to the enemy position. French officer Le Mercier, the mathematician and artillerist, strode down from the woods into the meadow to meet van Braam, followed by Captain Villiers. After the formalities of introduction, Villiers gave van Braam intimidating news. The French were expecting four hundred more Indian warriors to arrive in the morning. He strongly advised the British to surrender. As France and Britain were not at war, Captain Villiers told Captain van Braam, he and the French forces were very willing to save the British soldiers and officers from the "Cruelties"—the torturing, dismembering, and scalping—they risked from the Indian warriors. If the British stubbornly refused to surrender, however, they had no hope of escaping the Indians' hands. Captain Villiers said he came only to avenge his brother's assassination and remove the British from the Ohio. He wished to respect the peace between the two Crowns, he told van Braam, so he would offer very favorable terms of surrender. The British troops could march out of the fort with honor, bearing their arms and all the provisions and ammunition they could carry. In return, the British must promise not to return to the waters of the Ohio for one year plus one day.

Van Braam walked back to the fort on the meadow. He relayed to Washington the terms that Captain Villiers had offered. The cocky young

Washington of a few weeks before, who had pledged not to give up one inch of ground, now found himself in a swampy little fort, food low, powder wet, men dying, and being asked to give up the entire Ohio Valley or face a cruel death at the hands of Indians.

He agreed to negotiate.

Word was sent back to the French camp. Negotiations began on just what the terms would be. The French officers drew up a document. Van Braam returned with it. He, Colonel Washington, Captain Mackay, and other officers retired to the shacklike log structure at the center of the fort or an officers' tent to read it by candlelight. The flame guttered. They unfolded the two-page piece of heavy paper covered with words penned in French and patches of ink blotted by rain drips.

> *July the 3d, 1754, at 8 o'clock at Night.*
>
> *As our Intention have never been to trouble the Peace and good Harmony subsisting between the two Princes in Amity, but only to revenge the Assassination commited on one of our officers, bearer of a summons . . . and to hinder any Establishment on the Lands of the Dominions of the King my Master: Upon these Considerations, we are willing to shew Favour to all the English who are in the said Fort, on the following Conditions:*

So it began. But these first words would prove crucial to later events, turning on the translation of the French words *l'assasin* and *l'assasinat du Sr de Jumonville.* The document went on to list seven articles of "capitulation," allowing the English to leave the fort with drums beating, carry what arms and ammunition they could except for the cannon, and leave behind two hostages to ensure the return of the French taken prisoner in the Jumonville incident. The British must give their word of honor not to build any establishments in the Ohio Valley. The French, in turn, promised that they would restrain the Indians from going after the British as best they could.

At around midnight, two officers scrawled their signatures by candlelight on the rain-splattered document: *James Mackay, Go. Washington.*

They did not realize that the French were not faring much better in their rainy tongue of woods. Captain Villiers had offered to parley because French ammunition was running low, their Indian allies planned to leave in the morning, their soldiers were exhausted, and drums and cannon fire were rumored to be heard in the distance, suggesting that reinforcements for the British would soon arrive.

The night was quiet but for the patter of rain on canvas and the murmur of men's voices wondering what morning would bring. At dawn, the beating of drums rolled from the French camp. In two columns, the French soldiers marched across the soggy meadow. In the British camp, the soldiers packed up all they could carry in their knapsacks. It was not much. All the horses for transport lay dead in the meadow, likewise the milk cows. Little food remained. By agreement, the British had to leave behind the nine swivel cannon, although Captain Villiers had allowed them in the terms of capitulation to take away one, in part to show "that we treat them as friends." Free to carry off their gunpowder, the British found their kegs too heavy. Washington and his officers ordered the kegs staved in and powder poured onto the ground to keep it out of French hands. Without horses or wagons, the British wounded would have to walk if they could or cling to their comrades' backs. For hostages, the British would leave behind with the French two young, unmarried officers, Captain van Braam and Captain Robert Stobo, the flamboyant Scottish merchant's son who had arrived in the Ohio wilderness with ten servants and a wagonload of Madeira.

Despite the "honors of war" accorded to the British in the articles of capitulation, it was not without its humiliations. The proud and outspoken Scots doctor, Adam Stephen, was covered in mud up to his thighs from the fight, without stockings, his face black with gunpowder. As he assembled his men, his personal servant ran up to him. "Major, a Frenchman has carried off your clothes."

Stephen saw his portmanteau—a traveling bag designed to go on horseback—disappearing on the shoulder of a Frenchman who ran into a crowd of milling soldiers. Chasing him down and grabbing the bag, Stephen kicked the Frenchman in the rear. Two French officers witnessed Stephen's kick and, thinking he was a common soldier in his mud-splattered clothes, warned him that this kind of behavior broke the terms of the capitulation, which meant that the French would not have to follow the terms, either.

Insulted, Stephen ordered his servant to open the portmanteau. He pulled out his blazing scarlet officer's uniform, with its lace trim, and put it on. Stephen later reported proudly that, seeing his officer's finery, the two French officers treated him with deference. Joking with him, the officers said that Stephen ought to make them hostages and take them back to Virginia because they had heard many "Belles Mademoiselles" lived there.

Washington, Mackay, Stephen, and the other British officers ordered their men into marching formation. Drums beat. The French formed two columns spread a short distance apart, their drums beating, too. The British columns with their beating drums marched between the French columns with their beating drums. One nation was subjugated by another but, in recognition of the losing side's valiant defense, honored nevertheless in the intricate choreography of capitulation observed in this remote wilderness where two empires collided. Washington had suffered his first, humiliating defeat. For the rest of his life, he would never forget that dark anniversary—July 3.

CHAPTER NINE

In mid-July, a fine dust floated up in the hot sun whenever a wagon or carriage rolled down broad, sandy Duke of Gloucester Street in Williamsburg. The temperature hovered in the high eighties, the humidity lay thick. Pedestrians clung to the shade. The town of one thousand felt drowsy, except that across the great lawn in the cool, shady confines of the Governor's Palace work proceeded at a flurry. The news of Colonel Washington's surrender of Fort Necessity had just appeared in hard print in the *Virginia Gazette,* between self-help advice on the evils of slander; testimony in the trial over the ship *Nightingale,* burned in an apparent insurance fraud; advertisements for large parcels of land, fine horses, and imported hair to make wigs; and notices of rewards for runaway slaves.

The *Gazette* article reprinted a joint report on the battle from Colonel Washington of the Virginia Regiment and Captain Mackay of the Carolina Independents. It enumerated the hardships, the lack of food, the killing by the French of "every living Thing they could, even to the very Dogs," and the pouring rain. But it never mentioned the words *surrender* or *retreat* or *capitulation*. Rather, Washington and Mackay's report stated that each side agreed to "retire"—the French to Fort Duquesne and the British to Wills Creek. It glossed over the fact that the battle was a crushing defeat for the British. Instead, it concluded on a valiant, upbeat note: "Our men behaved with singular Intrepidity, and we determined not to ask for Quarter but with our Bayonets screw'd, to sell our lives as dearly as possibly we could. From the Numbers of the Enemy, and our Situation, we could not hope for Victory; and from the Character of those we had to encounter, we expected no Mercy, but on Terms that we positively resolved not to submit to."

The report listed the British losses as thirty killed and seventy wounded—a huge number out of three hundred active men, the casualities amounting to about one-third of the British fighting force. Washington and Mackay could not account for the exact number of French casualties, they reported, but had heard from a Dutchman that it climbed upward of three hundred French and Indian dead and wounded. The two officers believed that figure accurate because the French were "busy all Night in the burying their dead. . . ."

This was a wild exaggeration of French losses. In his official report, Villiers listed only three dead—two French and one Indian—and seventeen wounded, in addition to slightly wounded men not requiring treatment. A powerful work of public relations spin, Washington and Mackay's report suggested, totally inaccurately, that the British had taken a huge toll on the French and Indians, that there was no capitulation or surrender by the British, and that the two parties had agreed to quit fighting and return to their respective outposts.

In reality, the British retreat from Fort Necessity was as humiliating as its surrender. The British had to abandon much of their baggage for lack of horses or wagons to carry it. A group of one hundred Indians, newly arrived at the French camp, pilfered the British belongings. Delaware and other Indians who had switched their allegiance and fought with the French greeted the retreating British officers familiarly. Not a single Indian had fought with the British.

Struggling along on foot, toting the wounded on litters or on their backs, the retreating British traveled only three exhausting miles before stopping. Low on food, they left their wounded at this camp. Seven of them died the first night, while their comrades hurried ahead to Wills Creek to summon wagons to bring food and evacuate the wounded. Upon arriving at Wills Creek, however, healthy men deserted. A group of sixteen Virginia volunteers approached Colonel Washington and stated they had joined up to settle the Ohio lands but had now—not surprisingly—

dropped the idea. They quit. He tried to persuade them to stay, but they quit anyway. Washington and Mackay hurried on horseback over the mountains to Williamsburg to inform Governor Dinwiddie personally of the loss of Fort Necessity. In their absence, another two or three men deserted from Wills Creek every day.

Villiers and his French forces, meanwhile, had headed on the rugged trails back to Fort Duquesne. En route, they knocked down Fort Necessity and set it aflame, paused at Gist's settlement to burn to ashes his houses and fence rails used as palisades, marched on to the mouth of Redstone Creek and the Ohio Company warehouse, with its squared-off logs and empty loopholes, and reduced it to a pile of embers, too.

No matter what young Washington claimed, and how adroitly his joint report with Mackay presented the battle and capitulation, the reality was, as Francis Parkman put it, "Not an English flag now waved beyond the Alleganies."

In the Governor's Palace, in the heat of mid-July, Governor Dinwiddie reacted to the news with anger at the loss, outrage at French pretensions in the Ohio, and a determination to strike back. Giving the governor personal accounts of the battle, young Colonel Washington and Captain Mackay emphasized that they could have defended themselves against the French force if Colonel Innes, leading reinforcements from New York, had arrived in time. Washington also convinced Governor Dinwiddie that he had not been the aggressor against Jumonville's party at the glen and that they were not a French diplomatic party but rather a pack of spies skulking about waiting to attack his camp. He managed to deflect blame for this incident, which triggered the cascade of events that ended so poorly at Fort Necessity, and for his role in advancing the Virginia troops too far too fast, thus making them vulnerable to the French response.

Governor Dinwiddie sprang for his quill and wrote angrily to Colonel Innes that the blame for the "Misfortune" lay entirely on him for his delay in bringing reinforcements. The governor ordered Innes to build a fort at Wills Creek, stock it with troops and provisions, and await the next British thrust into the Ohio, which he was determined to make. He then dispatched by the next ship a thick sheaf of letters to the London authorities to explain the events and to request all His Majesty's might to help repel the French "encroachment." (Exonerating himself in the Great Meadows debacle, Dinwiddie carefully noted to the Lords of Trade that he had requested Colonel Washington to wait for reinforcements before mounting any attack, but the French had fallen upon Washington too quickly for him to heed this advice.) Dinwiddie also wrote to the governors of nearby provinces asking for help, almost demanding it, and to his own Virginia House of Burgesses for funds to fight the French. Despite Dinwiddie's vehemence, no war against France had been declared. The two nations officially were at peace.

Two days after Washington arrived in Williamsburg to inform his boss of events in the Ohio wilderness, the *Virginia Gazette* editorialized that the troops of Virginia had performed bravely at Fort Necessity and would have won the battle if reinforcements from New York had not delayed. "Thus have a few brave Men been exposed, to be butchered, by the Negligence of those who, in Obedience to the Sovereign's Command ought to have been with them many months before. . . . [Had New York reinforcements arrived] our Camp would have been secure from the Insults of the French, and our brave Men still alive to serve their King and Country."

In the taverns of Williamsburg such as the Raleigh and its Apollo Room, or just down Duke of Gloucester Street in the new tavern opened by the widow Christiana Campbell, the patrons brandished tankards of ale and cups of rum punch by candle flame and firelight and no doubt weighed and dissected Colonel Washington's strategy and his Virginia troops' performance under fire. Many no doubt agreed with the *Gazette* that lack of reinforcements caused the loss. More expert commentators, however, spoke and wrote privately of young Washington's shortcomings. The Half King and his Indian warriors, along with their families, had disappeared from the Great Meadows even before the battle began, disgusted with Washington's inexperience, his arrogance, and his lack of preparedness.

In the battle's aftermath, the Half King enumerated to the Indian agent Conrad Weiser his complaints about young Washington's leadership: "He was a good-natured man, but had no experience," the Half King said. "[He] took upon him to command the Indians as his Slaves . . . [and] would by no means take Advice from the Indians; that he lay at one Place from one full Moon to the other and made no Fortifications at all, but that little thing upon the Meadow, where he thought the French would come up to him in open Field. . . ."

Expert Indian negotiator William Johnson, who worked closely with the Iroquois, took Washington to task for both inexperience and ambition for glory. "I wish Washington had acted with prudence and circumspection," he wrote three weeks after the debacle in a letter to an acquaintance, ". . . but I can't help saying he was very wrong in many respects and I [question] his being too ambitious of acquiring all the honor or as much as he could before the rest joined him. . . ."

As news of the defeat crossed the Atlantic, first in the *Virginia Gazette* article, it electrified the authorities in London. In letters and in conversations in drawing rooms, clubs, and wood-paneled offices, dukes and earls reacted with great displeasure—about both Washington's leadership and the position in which Britain now found itself in North America with regard to France, its longtime nemesis. The Earl of Albemarle warned the Duke of Newcastle that Mr. Washington of Virginia and other colonial officers did not measure up to the professional officer standards of the British Royal Army: "Washington and many such may have courage and resolution but they have no knowledge or experience in our profession; consequently there can be no dependence on them. Officers, and good ones, must be sent to discipline the militia and to lead them on as this nation; we may then (and not before) drive the French back to their settlements. . . ."

The Duke of Newcastle, as prime minister, pushed for an urgent response against the French to recover "our lost Possessions": "All North America will be lost if These Practices are tolerated; And no War can be worse to This Country than the Suffering of Such Insults as these."

The loss at Fort Necessity fell as a deep blow on British pride and mystified military authorities like the Duke of Cumberland, captain-general of the British Army, who simply could not comprehend how one hundred Indians could cause such harassment to three hundred retreating British colonial troops.

King George II himself commented cuttingly on Washington, after Washington's strutting letter to his brother Jack about the Jumonville incident somehow appeared in the London press, including the remark, "I can with truth assure you, I heard Bulletts whistle and believe me there was something charming in the sound." King George said about Washington's bravado: "He would not say so if he had been used to hear many."

Another controversy erupted over Washington's signing of the articles of capitulation with the inclusion of the word *l'assasinat*. By signing the French document, Colonel Washington and Captain Mackay had, in effect, admitted to the murder of Ensign Jumonville, because that is how the French word is translated. As details of the articles became known, Washington and his fellow officers came under heated criticism for this admission. Washington fiercely defended his signing of the document, as did his fellow Virginia Regiment officer Adam Stephen, the proud Scottish doctor who had served with the Royal Navy, who wrote an account of the Fort Necessity battle a few weeks later in a letter to the *Maryland Gazette*. Like Washington, who said he never would have signed a document admitting to the murder of Jumonville, Stephen blamed the Dutchman van Braam, who, either through "evil Intention or Negligence," mistranslated the French document by candlelight in the leaky fort that night, using only the word *death,* not *murder.*

"Let any of these brave Gentlemen, who fight so many successful Engagements over a Bottle," wrote Stephen, "imagine himself at the head of 300 Men, and laboring under all the Disadvantages above-mentioned, and would he not accept of worse Terms than Col. Washington agreed to?"

While some second-guessed his decisions, the local view in Virginia was that, generally, Washington and his Virginian troops had performed gallantly. Two months after the battle, the House of Burgesses voted an official thanks. But the farther one traveled from Virginia, the harsher the criticism of Washington and the difficult situation he had created.

INDIGNANT AND AFFRONTED by the French advances, ever mindful of his land investments with the Ohio Company in the wilderness, Governor Dinwiddie glared from mullioned office windows overlooking the luxurious lawns and gardens of the Governor's Palace in the warmth of July. He now ordered Washington to rebuild the Virginia Regiment to three hundred men, return as quickly as possible to Wills Creek and the Ohio wilderness, join up with Colonel Innes and his troops, and either seize Fort Duquesne at the Forks or build a British fort in a "proper Place" chosen by a council of war. All this must be accomplished before winter, as the French would surely reinforce their fort in the spring.

Washington was stunned. At such short notice, how could he possibly assemble a force large enough and healthy enough? Men had deserted in droves and remained sick and wounded. The Virginia Regiment lacked supplies and ammunition. Ordered back to the scene of his defeat with still-inadequate forces, he would surely face failure, defeat, and humiliation. But how could he refuse the orders of the governor, his commander? Washington, now based temporarily in Alexandria, looked instead to his patron, Colonel William Fairfax of Belvoir Manor, and wrote him a long screed on the subject of why returning to the Ohio wilderness at this point to attack the French was "morally impossible."

> *Consider, I pray you, Sir, under what unhappy circumstances the men at present are; and their numbers, compared with those of the Enemy, are so inconsiderable, that we should be harrassed and drove from place to place at their pleasure: and to what end would the building of a Fort be . . . I can not see, unless it be to secure a Retreat, which we should have no occassion for, were we to go out in proper force & properly provided, which I aver cannot be done this Fall. . . .*

Washington painted for Colonel Fairfax the brutal nature of winter in the mountains, having nearly perished in icy rivers and frozen woods during his long trek the previous winter. Neither troops nor horses could withstand it, he insisted. Despite a good tent, "the cold was so intense that it was scarcely supportable," and out of the five or six men in his party, three of them, despite good clothing, had been "rendered useless by the Frost" and were left behind. Likewise, the horses would slip on icy trails and die for want of forage or grain.

Washington complained to Fairfax that he currently lacked "Men, Arms, Tents, Kettles, Screws . . . Bayonets, Cartouch-Boxes, &c. &c. &c." Without money to pay them, how could he recruit more men and where find more clothing? The existing troops were "almost naked, and scarcely a man has either Shoes, Stockings or Hat. . . . [or] a Blanket to secure him from cold or wet." Nor did Washington possess the gifts needed to recruit Indian allies or the provisions to feed them. He wrote that earlier the Indians had asked him "if we meant to starve them as well as ourselves." He vented for pages about why he could not go back into the wilderness and fight the French this year.

Things fell apart. During August and September, Governor Dinwiddie's request to the Virginia House of Burgesses for £20,000 to raise troops to fight the French got tangled in old Virginia political feuds, with the equivalent of riders attached. Men deserted from the Alexandria camp. More fell sick, probably from unsanitary camp conditions. Angry and frustrated with his fellow politicians, Dinwiddie dismissed the House of Burgesses until mid-October.

With neither troops nor funding, the governor faced reality and suspended his plan to attack the French fort before winter. He also came to the hard realization that he could not rely on other provinces to provide troops or support for an attack. He ordered Washington to march to Wills Creek, at the edge of the wilderness, with his remaining Virginia troops, and join Colonel Innes's troops there while also sending a detachment of forty to protect frontier settlers against possible Indian attack.

Taking a break from matters of deserters and pay, Washington hired men and canoes and paddled the upper reaches of the Potomac. No doubt informed by veteran Virginia land speculators as well as his own geographic knowledge, Washington grasped that the Potomac could solve a great transport problem if it could connect ports of the Atlantic coast with the Ohio Valley and future settlements there. With both the Potomac and the Ohio rivers originating in the Appalachians, although flowing down opposite slopes of the mountain range, some easy undiscovered passage might link the two watersheds. A boat carrying freight would thus be able to travel from the Atlantic coast through the Appalachians and, via the Ohio and Mississippi, cut through the center of the continent all the way to New Orleans.

Washington took a certain pride in plunging into the wilds like this, the bold adventurer overcoming trials in distant, difficult lands to bring back knowledge that would benefit his own tribe, the Virginians. In this case, he did not find what he sought. He canoed the Potomac's upper stretches unhindered but then had to negotiate four waterfalls, one dubbed "the Spout," where rocky outcrops squeezed the river through a narrows and plunged it into a cauldron of rapids. Here his wooden canoe nearly swamped—filling with water and almost sinking.

The frustrations built as October wore on, for both Governor Dinwiddie and young Washington. The great Mingo Indian ally of the British, the Half King, suddenly fell ill. He died three days later, a victim, his followers believed, of French witchcraft. Matters of pride and rank bedeviled Washington. If he should return to Wills Creek, as Governor Dinwiddie had ordered, he would be subordinate to other officers posted there, including Colonel Innes and several captains with royal commissions. It was particularly affronting that he had led troops and survived combat with Jumonville's men and at Fort Necessity while they were far from the scene.

With the House of Burgesses about to hold another session in late October, Washington rode from Alexandria to Williamsburg to check in at the colonial capital and with his commander, the governor. He socialized at the Raleigh Tavern, a hub of the capital, which held weekly balls in its Apollo Room when the burgesses were in session. Meanwhile, Washington had ordered from London merchants many yards of gold lace, blue fabric, red velvet, gold buttons, and a gold shoulder knot, apparently planning to have tailored a resplendent uniform befitting a high officer of the Virginia Regiment and sure to look smart in Williamsburg.

Good news arrived for Governor Dinwiddie when London agreed to extend £20,000 to defend Virginia and also gave two thousand sets of arms, while the Virginia Burgesses, reconvening, agreed to kick in the same amount. The king also appointed Maryland's governor, Horatio Sharpe, a longtime Royal Army officer and combat veteran, to head all the colonial forces, including Virginia's. He traveled to Williamsburg to meet Governor Dinwiddie and Governor Arthur Dobbs of North Carolina to map a new strategy against the French.

With this new funding, Governor Dinwiddie could recruit many more troops for his Virginia Regiment, while Maryland and North Carolina would also raise troops. Tiring of squabbles about rank, of these "disputes between the regulars and the officers appointed by me," Dinwiddie decided to reorganize the Virginia Regiment into ten independent companies of one hundred men each under his command and all headed by captains. He would solve the issue of rank by asking London for the authority to give captain's commissions in the British Royal Army to those he chose to head the ten Virginia companies.

One of the captains would be George Washington. This came as an unhappy shock to the young colonel. In effect, Washington was being demoted—or at least that's the way he saw it—to serve on an equal basis with other captains and possibly take orders from officers whom he had previously commanded as colonel. Had Governor Dinwiddie done this consciously? Did he, after Washington's debacle at Fort Necessity, want to demote him and find it expeditious to go about it this way? Even with a Royal Army commission, Washington expected much more in terms of rank than captain on a par with nine other captains.

He quit.

It was already becoming a pattern with young Washington. He perceived a threat to his honor or pride or what he would later call his "reputation," so he quit, or threatened to, as if to say that a gentleman should not suffer such insults. Unsure of the appropriate force with which to respond, he reacted, in effect, to the inadvertent bump of a fellow pedestrian in the street with a knock-down blow. He had entered a world of hierarchy, posturing, and force while burdened with a personality sensitive to the smallest slight and lacking the skills to navigate the obstacles he believed others placed in his path. He craved respect and praise. Almost perversely, quitting offered one way to take measure of how others regarded him. Six months earlier he had threatened to quit. Now he actually did. Governor Horatio Sharpe of Maryland, the new overall commander, tried to persuade Washington to stay. He promised that Washington would not have to take orders from those who had served under him during the Fort Necessity affair and he could keep his colonel's commission. Governor Sharpe's offer was transmitted to the upset Washington by William Fitzhugh, who also appealed to him to stay. "[F]or my Part," Fitzhugh wrote, "I shall be Extreamly fond of your Continuing in the Service & wou'd Advise you by no means to Quit."

Washington had now left Williamsburg and was staying at Belvoir Manor. With its two-story brick estate house, sweeping view over the Potomac, its gardens, stables, coach house, rooms furnished with European and Chinese ceramics, spacious entry hall in which to invite neighboring plantation owners to balls, its polite and aristocratic inhabitants, Belvoir to Washington represented the pinnacle of elegance. He loved to visit, playing cards, talking, riding, dining with his mentor, Colonel Fairfax, and Fairfax's son and Washington's good friend George William. One pictures them discussing Governor Sharpe's offer to Washington in a drawing room, by the fire, on a chill November day. George William's charming wife, Sally, moves about, along with her mother-in-law, Deborah, overseeing servants and slaves, who have quarters in the house cellar, Washington acutely aware of Sally's comings and goings. The lightness of her bearing, her teasing nature and dark eyes, her "clever sprightliness"—as one description put it—contrast markedly with his big frame, weighty earnestness, and prickly sensitivity. Perhaps his blue-gray eyes under his heavy brow surreptitiously follow her as she moves across the room. Perhaps she sits and joins the conversation—"a woman of unusually fine mind," as an early biographer described her, "which had been enriched . . . with the best literature of the day." Outside the paned windows, and below the bluff, the broad Potomac flows by gray and ruffled and grim, a mile across, a dark strip of forest lining the far shore. The wilderness over the mountains no longer beckons Washington as it did a year earlier. The memory still lives fresh in his mind of that deep cold, the crackling, frozen fabric of the tent, the benumbed companions who could go no farther, the creeks and rivers silently locked in ice. So does the memory of the past summer, the mud and blood awash in the trenches, the men dying. The Fairfax life at Belvoir Manor offered a captivating domestic vision, a gracious alternative to the military and wilderness. One could have this instead.

Washington adamantly refused Governor Sharpe's offer. Petulant and proud, he responded with anger and indignation as well as defensiveness. He feared giving the impression of weakness far more than he feared the musket balls and scalping knives of the French and Indians. "[I]f you think me capable of holding a Commission that has neither rank or emolument annexted to it," he wrote back to Fitzhugh, "you must entertain a very contemptible opinion of my weakness, and believe me to be more empty than the Commission itself." He enumerated his complaints in a long letter to Fitzhugh. How the recent campaign had been exhausting. How it had left him in poor health. How every mere captain with a royal commission would rank above him. How he suspected that officers at Wills Creek, presumed rivals, had put Governor Dinwiddie up to this idea of reorganizing the Virginia Regiment into ten companies led by captains and in effect demoting Washington.

The letter swung between whininess and boasting, its tone revealing an author of two minds—one feeling injured and wronged and wanting to quit, the other proud of his service and wishing to pursue a military life. But it also displayed a remarkable caginess, a foretaste of the politic Washington of the decades far ahead. He appears to have given certain passages a great deal of thought, balancing and prioritizing his various self-interests to achieve maximum results. In the space of a few sentences he managed to boast of his bravery, suggest that he quit unwillingly except for his injured sense of honor, and leave open the possibility that he would consider returning, given the right offer.

I shall have the consolation . . . [t]hat I have hitherto stood the heat and brunt of the Day, and escaped untouched, in time of

> *extreme danger; and that I have the Thanks of my Country for the Services I have rendered it. . . . [A]ssure [Governor Sharpe], Sir, as you truly may, of my reluctance to quit the Service . . . [A]lso, inform him, that it was to obey the call of Honour, and the advice of my Friends, I declined it, and not to gratify any desire I had to leave the military line.*

He then added a last sentence that stood alone, the white space on the page drawing attention to it, as to a prophesy: "My inclinations are strongly bent to arms."

It was as if he had written, *If you want me badly enough, I will return.*

CHAPTER TEN

THE DUKE OF NEWCASTLE HAD A PROBLEM. IT WAS THE Duke of Cumberland. Known as the "Butcher of Culloden," ten years earlier Cumberland had bloodily put down a Scottish uprising to restore a Catholic king to Britain, routing the rebels at Culloden and ordering his soldiers to stab to death the wounded who lay on the battlefield. As the favorite son of King George II, the Duke of Cumberland exerted a great deal of influence in the royal court. The Duke of Newcastle, by contrast, preferred to negotiate rather than fight. Well educated, enormously wealthy, his concerns ran to matters such as interest rates and national debt. The Duke of Newcastle's problem was that he had to find a way to rein in the Duke of Cumberland.

When news of the British defeat at Fort Necessity reached London in the form of the *Virginia Gazette* article in late August or early September, subsequently reprinted in the London papers, Newcastle had instantly responded that these practices could not be tolerated or "all North America will be lost." Yet he did not want to go too far. He began to lay plans for a modest military action in North America that would force the French to negotiate the Ohio Valley dispute without triggering an all-out war that could explode across Europe, between Britain and France and their Continental allies.

A fragile peace in Europe prevailed due to the delicate web of alliances known as the "System." The Duke of Newcastle was a proponent of the System's balance of power. He believed it could remain intact if the military action in North America were properly executed. "The key to success short of war," historian Fred Anderson writes, "thus lay in moving swiftly, secretly, to strike a blow in North America before the French could ward it off."

The Duke of Cumberland pushed a much more forceful approach. Instead of concentrating on striking a blow against the French only in the Ohio Valley, he proposed to strike them in several strongholds across North America and push them deep into Canada. He agreed with Newcastle, however, in wanting to appoint a single British Royal Army commander to take charge of all forces in British North America.

In the struggle for influence in the royal court, it was the king's son, Cumberland, who won out. Cumberland's choice, Major General Edward Braddock, would take command of all British forces in North America. Instead of one campaign in the Ohio Valley, Cumberland's aggressive and hugely ambitious plan called for launching four different expeditions involving ten thousand men. The first of these four military thrusts, the Expedition to Virginia, would consist of a massive overland assault on Fort Duquesne, deep in the wilderness at the Forks of the Ohio.

General Braddock, serving as the king's representative, would lead these expeditions, thus holding enormous power over the colonies and their defense. Sixty years old, he was described as "rough & haughty" in one contemporary account. While lacking social skills, he nevertheless appears to have been a gracious host, who liked to head a good table and engendered loyalty from the men close to him. With long experience in the military although not in actual combat, he was in some ways the very embodiment of the British Empire and its Royal Army.

Braddock was descended from a family of musicians and military men, who established connections to the royal family through their talent and their ambition. His grandfather, also Edward, had a good voice and sang in a choir for the king. His uncle by marriage was a prominent baroque composer who served as personal musician to the king. The family entered the military when the king awarded the singer's son, Edward II, with a lieutenant's commission in the Coldstream Guards, the elite infantry charged with protecting the royal family, which had been founded a century earlier in the English Civil War. Edward Braddock II rose through the ranks and had a son, Edward III.

As historian David Preston points out, young Edward Braddock III grew up with a "visceral sense of regal power." This was embodied by events such as the coronation of Queen Anne in 1702, when he was seven. His father's Coldstream Guards participated as the thirty-seven-year-old queen was carried in an open chair to Westminster Abbey, draped in crimson velvet and a golden robe, a petticoat studded with diamonds, and a six-yard-long train. During her reign, British military power and national pride soared, with victories by the Duke of Marlborough such as the celebrated Battle of Blenheim in 1704. Marlborough marched three hundred miles through rugged forest and mountains from the Low Countries to the Danube River before meeting, and vanquishing, the French—a feat that may have awed the boy Braddock and served as inspiration and template for him fifty years later in North America.

As a boy, he surely heard the famous dispatch that the Duke of Marlborough wrote from the battlefield to his wife, Sarah, capturing the heroic moment: "I have no time to say more but to beg you will give my duty to the Queen, and let her know her army has had a glorious victory."

At the age of sixteen, young Edward III, with his father's help, acquired a position in the Coldstream Guards. His father, meanwhile, had risen to major general, an event marked with the awarding of a twelve-foot-long scarlet silk sash, embroidered with the year 1709, that the younger Braddock would inherit and one fateful day wear into battle. At the age of twenty-two, he purchased a lieutenant's commission—not an uncommon practice at the time—also in the Coldstream Guards. Both his parents died when he was in his early thirties, and most of his father's estate went to his two sisters. Scandal then touched the family. His younger sister, Fanny Braddock, "became a parable for young women of the perils of the age," as historian Preston puts it, when she fell in love with a rakish suitor and spent her inheritance to pay off both his gambling debts and her own. Destitute, she hanged herself in 1731.

As a young man, Braddock himself shared in the dissolute spirit of the time, according to Preston, and acquired a reputation for boorish and rakish behavior, including a duel with another officer in Hyde Park. As he grew older, however, he gained valuable military experience in the logistics of supporting a campaign, if not in combat. Often assigned back to London to guard the royal residences, he apparently never saw battle himself during the Scottish uprising of 1745, with its Battle of Culloden, and the War of the Austrian Succession. Braddock did, however, serve in the Coldstream Guards alongside the king's son, the Duke of Cumberland, and came to the attention of other powerful individuals. Colonel Braddock then won command of a regiment, the 14th Foot, stationed at Gibraltar. This great rock fortress guarded the Mediterranean's straits, anchoring part of the British Empire, and had a reputation as a difficult post.

Braddock performed well at Gibraltar, grew popular with locals and troops alike, and was named lieutenant governor of the outpost. In September of 1754, in the aftermath of the Fort Necessity disaster in the Ohio wilderness, the urgent need arose for a British Royal Army commander to take charge of all British forces in North America, colonial troops and regular Royal Army alike. Braddock had experience with difficult logistics and supplies at Gibraltar, in handling administrative duties, in dealing with the many foreign cultures who lived there—Jews, Moors, Spaniards, Portuguese. He had shown an ability to negotiate and an adeptness at analyzing military fortifications, such as Gibraltar's monstrous array of two hundred and twenty-five cannon and artillery. "Unlike modern conceptions that equate military effectiveness with combat experience," writes Preston, "in the eighteenth century experience was demonstrated by competent and lengthy service."

While he possessed administrative and logistical competency and long service, Braddock's personality was not an easy one. As Washington himself would describe Braddock years later: "His attachments were warm—his enmities strong—and having no disguise about him, both appeared in full force. He was generous and disinterested—but plain and blunt in his manner even to rudeness."

In early January of 1755, General Braddock set sail from Cork Harbor in southeast Ireland aboard the warship *Norwich*. A large fleet spreading out behind her crashed through the gray, whitecapping swells of January's North Atlantic—five other warships, thirteen troop transport ships, and three ships weighted low in the water with cannon, howitzers, and mortars, as well as sixteen enormous "King's wagons" for transporting goods, each weighing over a ton. In addition to a total of 1,664 men, the fleet carried one thousand barrels of salt beef, one hundred tons of Irish butter, 572 barrels of gunpowder, 6,450 cannonballs and shells, and 144,388 flints.

It was the equivalent of a small town heaving across the wintry North Atlantic—a very heavily armed one. Soon the flotilla would sail past the windows of the home on the Potomac of a proud young planter recently resigned from the Virginia military, George Washington.

In an ironic turn of events, Washington had just come into considerable land and had committed himself to a planter's life. Two years earlier, when his half-brother Lawrence had died, leaving his wealth of land and slaves to his widow, Anne, she had quickly remarried and moved to the plantation of her new husband, George Lee, starting another family. But now Lawrence and Anne's daughter, little Sarah, like so many small children of the era, had fallen ill and died. By terms of Lawrence's will, in the event of daughter Sarah's death, George Washington would receive a share of Sarah's wealth, including some of her slaves. The more valuable of her slaves, such as Pharrow, Judah, Jenny, and Nell, were listed as worth £40 each, while the slave children were identified not by how much they could bring but by their physical stature, such as "Prince at 3' 7''" and "Betty at 3' 1''."

This presented a tremendous opportunity for young Washington and his ambitions. Anne did not need Mount Vernon as a residence, having moved to her new husband's plantation, and Washington had just inherited slaves. A shrewd bargainer, always looking for a good opportunity, especially when it came to land, Washington negotiated with Anne to rent Mount Vernon for the remainder of her life for an annual payment of fifteen thousand pounds of tobacco, which included use of the plantation's eighteen resident slaves. On December 17, 1754, just as General

Braddock was making preparations to sail with his large fleet to America, George Washington and Anne signed a lease for the lifetime rental of Mount Vernon, formalizing his commitment to the life of a planter.

Washington spent that Christmas of 1754 at Belvoir Manor with the Fairfaxes and other Potomac planter families, and notably not at his mother's modest home at Ferry Farm on the Rappahannock, surely straining what was already becoming a difficult relationship. In the wealthy, aristocratic houses of Tidewater Virginia like Belvoir, Christmas and the New Year brought feasting and drinking—roast beef and turkeys turning on spits over big kitchen fires, bowls of wassail, flowing wine and beer, toddy, and cider—as well as dress balls and gambling at dice, cribbage, and the card games whist and all fours. As a strong, athletic horseman, Washington had a passion for the Tidewater foxhunts that chased over the landscape at the holiday season. He liked to dance at the balls, perhaps more comfortable with the young ladies while in a jig or reel or the intricate lifting steps and Z patterns of the minuet than in conversation. He also played many hours of cards. In his ledger book for that December of 1754, he recorded that his three-day winnings over Christmas were "14s, 3d."

One can see young Washington aspiring to his new life as a planter, sitting with the other men around a polished table designed for cards. The women, and perhaps Sally, joined or observed the game, while servants in the kitchen roasted meat and minced the fillings for pies. Colonel Fairfax's cousin, the eccentric Lord Fairfax, recently of Leeds Castle, England, and proprietor of the entire five-million-acre Northern Neck, may have ridden down from his new hunting lodge in the Shenandoah Valley for the holidays. Belvoir and the people who occupied it were, in a sense, Washington's real family in those years—or the one he wished he had.

CHRISTMAS OF 1754 MAY HAVE BEEN among Washington's happiest times. He had just quit the military life and all the problems and responsibilities of handling troops, deserters, supplies, and orders. He had gained something of a name for himself, at least locally, despite the humiliation of Fort Necessity. He had just leased Mount Vernon, a beautiful piece of land on a hill rolling down to the Potomac. It held a sound but modest wooden house built two decades earlier by George and Lawrence's father, Gus, standing a story and a half high, with four windows across the ground floor and three little dormers in the roof, a chimney at each end, and was equipped with the plantation's own complement of slaves. The land that came with it amounted to a sprawling twenty-three hundred acres. All possibility now lay before him. Life looked good.

What would have been his reaction, then, on opening the first *Maryland Gazette* of the New Year, dated January 2, 1755? Dispatches from London dominated virtually the entire front page: A large force was assembling in Britain under an unnamed commander, destined for America to confront the French. Two regiments in Ireland of a thousand men each under the command of Colonel Sir Peter Halkett and Colonel Thomas Dunbar were ordered to America for Virginia's defense. Another two regiments assembling under Colonels Pepperell and Shirley would augment the Irish regiments, a force totaling nearly four thousand men, in addition to existing British and colonial troops in America. The London newspapers' front pages listed British officers who should repair immediately to London and to Cork, Ireland, to await further orders. Meanwhile, the royal munitions works at Woolwich, on the south bank of the Thames, labored day and night to manufacture cartridges and firearms for the America-bound force, and workshops rushed to stitch hundreds and hundreds of small tents to house eight thousand men in the North American wilds. "Reviewing, Recruiting, and other Military Preparations go on here as if we were at the Eve of a War," came reports from London.

Upon reading all this in the *Gazette* immediately after those pleasant gaming holidays with the Fairfaxes and other Potomac families, would Washington have felt any responsibility for setting in motion the events that led to this precipice of all-out war? It would not be the natural inclination of a twenty-two-year-old to question himself. Far simpler to blame it all on French territorial aggression in the Ohio. Even ambitious young Washington could not have imagined how quickly this remote expanse of wilderness would become the main stage in a global battle for power.

Washington had flatly resigned only six weeks earlier, indignant and offended. He had served at the forefront of this action against the French in the Ohio wilderness (although he may not have recognized that he helped trigger it). Now it appeared that the commanders of the British Royal Army were ignoring him.

He busied himself with domestic duties such as buying slaves and furniture for his new home at Mount Vernon, attempting to embrace his planter's role and the graceful plantation life. But he found it difficult to stand by as the first ships of the great fleet arrived off the capes of Virginia on the night of February 19. The next day, the fifty-gun *Norwich,* one hundred and forty feet long, one thousand tons, and carrying General Braddock in its officers' cabins, heaved into Virginia's vast natural harbor called Hampton Roads. The anchor plummeted to the bay's bottom while sailors scrambled aloft up the ratlines—rope ladders—and climbed out on the masts' crosspieces—yardarms—to furl and tie up the big square sails. The other ships arrived, one after the other, the slower transport ships several weeks behind, having traced a more southerly route across the North Atlantic.

On February 22, the flagship of the great fleet, the *Centurion,* its foremast damaged by the North Atlantic winter gales, sailed into the harbor carrying thirty-year-old Commodore Augustus Keppel, sea-toughened by having served in the Royal Navy since the age of ten. The following day, a Sunday, the official British military contingent rode the forty miles from Hampton Roads to Williamsburg. This included General Braddock and Commodore Keppel, plus Captain Robert Orme, Braddock's principal aide-de-camp and personal secretary, and Captain William Shirley, son of the Massachusetts governor. The chill February night had nearly fallen when they rode up to the brick Governor's Palace and Governor Dinwiddie's hearty Scottish greetings.

"Under the drawing room candelabra," writes Braddock's biographer Lee McCardell,

> the lieutenant-governor and the major general could look each other over. Dinwiddie saw a shorter, stouter man than he had

> expected, but a smartly turned out British officer in red coat, crimson sash, and gold lace, exuding a faint odor of snuff. The general had a good face, regular features, and the unmistakable click and carriage of a guardsman. Braddock saw a blue-gray–eyed Scotsman of his own age, a little puffy from good living, possibly a little harassed, and maybe a little self-conscious of his embroidered buttonholes.

Governor Dinwiddie promised the colony of Virginia's support to General Braddock to accomplish the first of his hugely ambitious objectives—capturing Fort Duquesne at the Forks of the Ohio. In addition to the £20,000 voted by the Virginia Assembly to cover expenses, this assistance included contracts signed for eleven hundred head of cattle to be driven to Wills Creek, the jumping-off point to the wilderness, and enough supplies—pork, salt, and flour—to feed three thousand men for eight months. The governor had advertised for recruits, and claimed to have one thousand men. Other contracts called for delivery of twenty-five hundred horses and two hundred wagons to transport Braddock's army. In addition, the Virginia governor remained confident that great numbers of Indians would join the British expedition, especially after witnessing the might and splendor of the Royal Army, with its red coats and its cannon.

But less optimistic reports reached General Braddock about the readiness of other colonies. His deputy quartermaster general, Sir John St. Clair, a fierce Scottish baronet who was to oversee the expedition's supplies, logistics, and transportation, had arrived in America a month or so before the general and assessed the situation. He now reported to General Braddock that few of the American colonies had signed on to the expedition, that Pennsylvania's pacifist Quakers flatly refused to support a military undertaking, and that the logistics of the thrust to the French fort loomed far more daunting than first believed. The expedition would have to surmount a stretch of rugged mountains sixty or seventy miles in breadth instead of fifteen, as originally estimated, and negotiate long approaches through dense forest over roads that ranged from extremely bad to nonexistent. "The worst road I ever traveled," St. Clair pronounced of the eighty-five-mile stretch from Winchester to Wills Creek.

In his forties when he arrived in Virginia as deputy quartermaster general, Sir John had inherited his title as a youth, along with an estate in Scotland too small to support a baronet's life, when his father died in 1726. To earn both a living and further distinction, he had joined the Royal Army as a young man. He had served in combat in the 1740s during the War of the Austrian Succession under commander Baron von Browne of Austria, who had performed a rapid, difficult, and celebrated crossing of Italy's Apennine Mountains to seize Genoa. The bold and decisive Baron von Browne famously did not believe in half measures to accomplish such daunting tasks. Nor, as it turned out, did his volatile and hot-tempered Scottish protégé, Sir John St. Clair. "[A] mad sort of fool," one fellow British Royal Army officer called Sir John in a private letter to his brother.

St. Clair's experience in the Apennines, however, did not prepare him for the ruggedness of the American wilderness. A campaign here was not something that could be easily planned, or even conceived, from the chandeliered rooms of London's ministries. Once landed in America, St. Clair, with typical passion and certainty, had quickly developed a very dim view of the competence of American colonials and their military. "[They are] totally ignorant of Military Affairs," he wrote. "Their Sloth & Ignorance is not to be described." The titled deputy quartermaster general could see that a tremendous amount of logistical work lay ahead.

In Williamsburg, General Braddock received these reports from St. Clair while staying at Raleigh's Tavern, the clapboard inn on Duke of Gloucester Street with its Apollo Room dance floor around which whirled the little capital's social life, under the motto in gilt letters over its fireplace, *Hilaritus Sapentise et Bense Vitae Poriec*—"Jollity the offspring of wisdom and good living." From the good life at the Raleigh, General Braddock could walk or ride a few hundred yards down Duke of Gloucester Street, turn right across the broad lawn of the Governor's Palace, and confer with Governor Dinwiddie to lay the groundwork for the great expedition into the Ohio wilderness.

News of General Braddock's doings in Williamsburg soon reached young Washington, the new proprietor of Mount Vernon, about a hundred and twenty miles to the north. Having leased the plantation only two months earlier, he now had his hands full with the upcoming spring planting.

He ruminated. How could he not take part in this expedition, after his own forays into the Ohio wilderness? How could he watch these other young men—living and working alongside the general as secretaries and aides-de-camp—march through the wilderness to the Forks and not he? He had been there since the beginning, laid the groundwork for this massive campaign, trekked and paddled and froze and suffered and almost died. He had watched his Virginia troops fall. Now the honor and glory of the final blow to the French would shower down on these newcomers and fellow officers, soon to be heroes in the history of the great colony of Virginia. They would gain General Braddock's military wisdom and experience, receive his blessing, and reap the benefits of the general's connections with the uppermost echelons of the Royal Army establishment, with the Duke of Cumberland himself, the king's own warrior son. Amid this great and glorious campaign, Washington's early role in it would be forgotten and he left behind, stuck at Mount Vernon watching his tobacco plants grow.

It did not take long for this unhappy vision to eat at young Washington and spur him to action. Within days of General Braddock's landing on Virginia soil in late February, Washington wrote to him. While the letter's contents do not survive, Washington appears to have steered his quill in a deft path between flattering the general with congratulations on his appointment as North American commander of British forces, and promoting himself and his experience. Washington added, however, that concerns about rank and status between British regular and colonial officers had prevented him from joining the campaign.

General Braddock's main aide-de-camp, Captain Robert Orme, formerly of the Coldstream Guards, replied to Washington's letter almost immediately. He warmly extended the general's invitation to join his inner circle of officers—or "family," as he called it—where Washington need not concern himself about whether Royal Army officers outranked colonial officers: "[General Braddock] has ordered me to acquaint you that he will be very glad of your company in his family by which all inconveniences of that kind will be obviated."

It made sense militarily that General Braddock and Captain Orme should invite young Washington to join the general's inner circle. One of the few colonial British who knew the terrain, Washington had traveled twice to the Ohio wilderness, in both winter and summer, and possessed some familiarity with the Indian tribes there. Although the military expedition Washington had led to the Ohio had met with a humiliating defeat, the formidable General Braddock, and not an inexperienced twenty-two-year old, would head this campaign. Besides, they needed recruits of all kinds to fill out their ranks.

Washington wrote back to Orme the very next day. Expressing his appreciation for the offer to join the general's family, he admitted that, frankly, he had a selfish motive for wishing to accompany him: "To be plain, Sir, I wish for nothing more earnestly, than to attain a small degree of knowledge in the Military Art; and, believing a more favourable opportunity cannot be wishd, than serving under a Gentleman of his Excellencys known ability and experince, it does as you may reasonably imagine, not a little contribute to influence me in my choice."

Despite convoluted syntax and poor spelling betraying his lack of formal education, Washington nevertheless again showed an artful diplomacy in confessing to his own selfish interest—presumably a measure of his sincerity—while also flattering the general. (To his brother Augustine he was franker still, making it clear that he saw the campaign in considerable measure as a chance for personal advancement.) But Washington also had a conflict. He had just taken on the proprietorship of Mount Vernon. How could he manage the plantation and at the same time take part in General Braddock's campaign against the French over the mountains in the Ohio?

He asked the general for a little more time to consider the offer. He did not say exactly why, and did not refer to his plantation, keeping his reasons even vaguer and more circuitous than in his usual writing. He had to deal with "incoveniences" from "some proceedings" which would lead to a life of greater "retirement." He would not say more on the subject, he wrote to Orme, until he called on his excellency General Braddock and in the meantime very much looked forward to meeting Captain Orme.

While Washington juggled plantation and military careers, the armada of warships, transports, and ordnance ships weighed anchor from Hampton Roads and sailed up the Potomac to the port of Alexandria—near the river's head of navigation and as far as the troops could be carried by ship. Working their way up the broad, winding river estuary, they passed Tidewater plantations with their private wharves and large homes and saluted them by firing cannon, while the plantations raised flags and fired salutes in return. The passage of the great force upstream past his new property, sailing immediately under his mullioned windows, highlighted the dichotomy that would stay with Washington for the rest of his life—how to reconcile the adrenaline-charged pull and gold-braided flash of the military life with the long, slow rhythms and earthy duties of a planter.

His Majesty's warships and transports dropped anchor at Alexandria on March 22. Recently founded as a port on the Potomac at the urging of the Ohio Company, the small but fast-growing town offered a new entry portal to Virginia's western lands. With twenty drums rolling from each regiment and echoing out over the wide river, the 44th and 48th Foot Regiments in their black shoes and redcoats with contrasting lapels in yellow or buff marched from the wharves up Oronoko Street toward their tent camp at the edge of town, passing the town square with its courthouse, whipping post, and pillory. Alexandria offered few other amenities besides a single tavern. The townspeople gawked. They had never before seen the British regular army in their redcoats. The soldiers themselves had to refrain from gawking, too. Some of them had never seen a black person, and now they had entered a land of slavery.

Alexandria's most prosperous merchant, the Scotsman John Carlyle, recalled the optimism of the moment: "[The ships l]anded in high Spirits about 1600 men, besides a fine Train of Artillery 100 [artillerymen] &c & Seemed to be Afraid of nothing but that the French & Indians Would not Give them A Meeting, & try their Courage."

General Braddock arrived a few days later and took up quarters at

Carlyle's home, from which the merchant had helped supply the Virginia Regiment the previous year. Sitting on a low bluff about three blocks back from the river, Carlyle's two-story house dominated the little port town, designed in a simple but well-proportioned Georgian style and executed in sandstone. The enterprising and well-connected Carlyle, from an aristocratic Scottish family, was married to Sarah Fairfax, a daughter of Colonel William Fairfax and younger sister to Lawrence Washington's widow, Anne.

Carlyle would quickly grow tired of the demands of General Braddock and his fellow British Royal Army officers, and their dismissive attitude toward the colonials. "[B]y Sum means or another," Carlyle wrote,

> *[they] came In So prejudiced agains Us, our Country, &c that they used us Like an Enemys Country & Took every thing they wanted & paid Nothing, or Very little for it, & When complaints was to the Comdg Officers, they Curst the Country, & the Inhabitants, Calling us the Spawn of Convicts the Sweepings of the Gaols &c; whch made their Company very disagreeable.*

While General Braddock himself acted in a friendly enough manner toward Carlyle, he had "taken everything he wanted abused my house, & furniture & made me little or no satisfaction." Carlyle called him "Possitive" in his attitude—by which he seems to have meant overbearing—"& Very Indolent, Slave to his Passions, Women & Wine, As Great an Epicure as could be in his Eating."

In this dismissive attitude of the British Royal Army officers toward the American colonists, Braddock historian David Preston locates some of the "deep fault lines on which the Thirteen Colonies and the British Empire ultimately divided." Washington himself would meet similar treatment from British Royal Army officers in the months ahead. He ultimately would personify some of these very fault lines.

At the moment, however, Washington was struggling to come to terms with his mother's wishes. She had shown up at Mount Vernon from her regular home at Ferry Farm and made it clear that she wanted him to stay home and attend to the family's various plantations rather than go off to war in the wilds. This recalled her forbidding him to go to sea as a teenager. Washington had maintained an oddly formal relationship with her in the seven or eight years since then, almost as if avoiding her. One can view her as a demanding, overbearing mother who insisted her son stay nearby to help her, or as exercising the instincts of a protective parent, or both. It's not difficult to imagine her alarm at her eldest son going off to war and leaving behind the unmanaged family plantations, and telling him so in no uncertain terms.

"I find myself much embarrassd with my Affairs," he wrote to Orme, "having no person in whom I can confide, to entrust the management of them with." Washington asked Orme to ask General Braddock if he could meet up with the army at Wills Creek, just before it headed into the wilderness. This would allow him time to put his business affairs in order. He also asked for permission to leave if there should be a period of inaction, to return home for a visit. Before sending off the letter, Washington enclosed a map that he had sketched of the way into the Ohio wilderness, based on his earlier travels. The British had virtually no accurate maps of the Ohio region. He surely knew that sending one would boost his desirability in Braddock's eyes.

Through Orme, General Braddock responded the next day. The general took the "greatest satisfaction" in Washington joining his inner circle. He graciously granted all of Washington's requests: that he could meet Braddock's army at Wills Creek and that he could return home as he wished to attend to matters there. "[W]henever you find it necessary to return he begs you will look upon yourself as entire Master, and judge of what is proper to be done."

Young Washington could not have asked for a better response. All his requests had been granted, including the remarkable one that allowed him to leave at any time. According to his own wish, he would serve as a volunteer, without pay. Concerned with his reputation, he did not want people to think he joined General Braddock for the money. ". . . I can very truly say," he wrote to William Byrd, a wealthy older planter friend and Virginia burgess who had concerns about the French in the Ohio, "I have no expectation of reward but the hope of meriting the love of my country and friendly regard of my acquaintances. . . ."

Again and again, in these early years, one sees how Washington's sense of honor—or what he perceived as his sense of honor—determined the decisions he made. The concept of honor is as ancient as human societies and can mean many different things. It can refer to a code of behavior existing beyond the rules of law. It can encompass an individual's good name, reputation, or standing in the community or social hierarchy. A woman's honor has traditionally referred to her sexual purity. A stain on one's honor can mean an insult left unanswered. An "honor culture" also implies a culture of revenge, in which injured parties take it upon themselves to exact a punishment from the perpetrator, or, biblically, "an eye for an eye."

This sense of honor, anthropologists have observed, flourished in herding or seminomadic societies that were scattered thinly over broad landscapes and lacked a central law authority, whether in the ancient Mediterranean, the Scottish and Irish borderlands in centuries past, or today's mountains of Afghanistan and Central Asia. "Every man," went the saying, "is sheriff of his own hearth." This attitude found its way to the Scotch-Irish frontier of the American South.

Honor can also mean winning acclaim for a notable act, particularly an act of self-sacrifice for the common good, and especially on the battlefield. For the ancient Greeks, *tîmê,* or honor cult, immortalized through ritual the heroes who sacrificed themselves in battle. For the chivalric society in Europe's Middle Ages, honor meant, among other things, faithfulness to a knight. Taken up by the English aristocracy, it became an essential virtue to achieve social standing. "Mine honour is my life; both grow in one," Thomas Mowbray says to Shakespeare's King Richard II. "Take honour from me, and my life is done."

This attitude toward honor migrated with the English aristocracy and would-be aristocracy across the Atlantic. Some historians believe that the concept of honor, and other values such as belief in a strictly hierarchical society, were particularly imported to America by Royalists escaping the English Civil War in the mid-1600s. (This would include ancestral Washingtons.) In any case, Virginia's planter elite embraced a concept of honor having deep roots in English aristocracy. To young Washington, obsessed with honor, it meant, among other things, both his good name and reputation, and winning acclaim for his bravery and sacrifice on the battlefield, which were inextricably linked.

HIS PUBLIC IMAGE was of the utmost importance to young Washington, an image that was almost a product existing outside himself. When he signed on with Braddock's campaign in March of 1755, he hoped the terms by which he did so—without pay—would help to earn him respect. Starting in his early twenties, he appears to have begun sculpting this image for public consumption—a task at which he would succeed beyond his wildest dreams. From this point forward, having turned twenty-three the previous month and just joined the Braddock campaign, Washington would carefully retain his correspondence, as if for posterity as well as for the military record, as if aware that this was the point when he walked into history and wished to be seen in the best possible light. In his later years, he would go back and edit and revise this correspondence from decades before, crossing out words and phrases, inserting others. The changes generally cast him more favorably, less prickly and less forward.

As Washington at Mount Vernon fretted over his image and terms of service, General Braddock and his impassioned coordinator of supplies and logistics, Sir John St. Clair, wrestled with the problems of putting an army on the move in a continent new to them and in a rugged mountain wilderness. The Ohio would lack the amenities of the farm-and-village European theaters of war, where food and housing could be had for the taking, often leaving a destitute countryside in an army's wake.

April proved a difficult, maddening month. St. Clair could find neither wagons and horses to transport the great quantities of food and supplies Braddock's army needed to carry into the wilds nor the enormous loads of fodder needed to feed the workhorses that would haul the ponderous cannon. The unpredictable cold snaps and rain of early spring hindered the movement of officers and troops. The British Royal Army officers discovered the lack of training of the new recruits from the American colonies destined to fill out the two British Army regiments from Ireland, the

44th and 48th. The American provinces failed to contribute enough money, manpower, or transport to help in the campaign—especially the pacifist-leaning Assembly in Pennsylvania, although this was the province, along with Virginia, that stood to gain most if the French were expelled from the Ohio Valley. Pennsylvania was especially exasperating to the general and St. Clair.

To curtail the rowdiness of the bored troops of the 44th and 48th regiments now stalled in the sprawling tent camp outside Alexandria, General Braddock ordered strict discipline: deserters would be hanged even if they willingly returned, and disorderly soldiers would go without food and face the lash. Even so, the men managed to get drunk in the little port town and locate willing women, giving the general yet another reason to move his troops quickly to the interior.

The camp itself swelled with the hundreds of "camp followers" that supported (and at times burdened) armies on the move in the eighteenth century—wives and children of officers and soldiers, nurses, washerwomen, mistresses, prostitutes, sutlers (who sold consumer goods), and many others, including horses and pet dogs. Among the followers was a British nurse known as Widow Browne, sister of one of the British officers, who would keep a journal of her travels into the Ohio wilderness with Braddock's army.

In the first week of April, General Braddock traveled to Annapolis, Maryland, to meet the governors of other American colonies and rally and coordinate support for his campaign against the French. But an April snowstorm blocked the primitive roads from the northern provinces, postponing the meeting and forcing the general back to his Alexandria headquarters in merchant Carlyle's sandstone mansion.

The British campaign, however, could not delay, or it would permit more French troops to boat unopposed into the Ohio on the rising waters of the new spring. General Braddock ordered his first troops to march on April 10, and Sir Peter Halkett and his 44th regiment departed the tent camp beside the Potomac to begin the one-hundred-forty-mile overland journey to Wills Creek, the staging area for the expedition into the wilderness. Soldiers and carpenters labored on a new British fort, named after the king's strong-armed son, the Duke of Cumberland, on a low bluff overlooking the confluence of Wills Creek and the north branch of the Potomac. Behind it rose a steep green wall—the first of the tall mountain ridges between there and the Forks.

Two days after the 44th's departure, Colonel Dunbar and his 48th regiment marched from Alexandria, tracing an alternate route to Wills Creek and Fort Cumberland. General Braddock and Deputy Quartermaster St. Clair wanted to split up the two regiments to lessen the demands for food and transport on countryside farmers, although the route would eventually lead them through western Maryland's mountainous wilds. On April 15, Colonel Dunbar's men had marched sixteen miles through hot April weather on dusty roads when, that night as they slept, a spring blizzard swept over their camp, weighting down their sagging tent fabric with more than a foot of heavy, wet snow, which they had to knock off several times in the night to prevent collapse and broken tent poles.

On those same days the provincial governors finally gathered at Alexandria to meet General Braddock, after the earlier postponement. They jammed into the Blue Room of merchant Carlyle's house, a small downstairs parlor with high ceilings, paned windows deep set in the thick stone walls, and a blue marble fireplace that surely blazed as the cold front moved through. Out on the Potomac, the fleet of ships that had carried Braddock's army across the Atlantic rode at anchor. Alexandria had quieted down since the departure of the 44th and 48th and their huge array of camp followers, but not the Blue Room. General Braddock met with five colonial governors: Virginia's Robert Dinwiddie, Maryland's Horatio Sharpe, New York's James De Lancey, Massachusetts's William Shirley, and Pennsylvania's Robert Hunter Morris. Eventually known as the "Alexandria Congress," it was an unusual gathering of American colonial provinces that normally acted independently of each other, a coordination among the colonies that had been advocated by Philadelphia printer and scientist Ben Franklin.*

* The first generally acknowledged gathering of American colonies in the eighteenth century had occurred a year earlier, in June and July of 1754, at the so-called Albany Congress, to plan a mutual defense and negotiation with the Indians, although the

The governors told General Braddock that, despite his proposal to do so, they could not set up a common fund to administer defense of the colonies unless Parliament acted on it. (Some interpretations assert that the governors' refusal constituted an early act of colonial unity and rebellion against the king's representative.) The governors agreed to Braddock's proposal to launch a four-pronged attack against the strategic French forts, starting with Fort Duquesne. Of utmost importance, they agreed, was recruiting the Iroquois for the British cause, and that the best envoy for the task was Colonel William Johnson of New York, a fur trader, Indian negotiator, and honorary sachem of the Mohawk (one of the Six Nations). Braddock agreed to put up £2,000 to buy presents for the Indians, and requested that the Irish-born Johnson, fluent in Mohawk, write two speeches for the general to deliver to the Indians. Seeing an opportunity for self-advancement, young Washington rode the ten miles from Mount Vernon into Alexandria to meet some of the governors, reporting back to Colonel Fairfax at Belvoir that he was "perfectly charmd" by Massachusetts governor Shirley.

While the governors' meeting was wrapping up at Alexandria, across the mountain ridges an enraged Sir John St. Clair rode into the new Fort Cumberland. Having just traversed the road from Winchester to Wills Creek that was supposed to carry Braddock's army and finding it among the worst he had ever seen, he now discovered that flour and other provisions had not arrived at Wills Creek, nor had any wagons and horses been procured to move supplies, artillery, and all the other cumbersome baggage of a massive expedition.

When St. Clair found a contingent of representatives from Pennsylvania waiting for him at Wills Creek, reported George Croghan, he "stormed like a Lyon Rampant." Pennsylvania had ignored his order of months earlier to clear a road toward the Ohio and had not contributed provisions, wagons, or horses. The furious Sir John launched on a diatribe, recorded

---

participants also debated Franklin's plan of founding a unified colonial government. It was to encourage the colonies to join this gathering that Ben Franklin published his famous "Join or Die" cartoon depicting a snake cut into segments.

by eyewitness Croghan, about the consequences of Pennsylvania's utter lack of support for General Braddock's campaign:

> That instead of marching to the Ohio [St. Clair] would in nine days march his Army into Cumberland County to cut the Roads, press Horses, Wagons, & c; that he would not suffer a Soldier to handle an Axe, but by Fire and Sword oblige the Inhabitants to do it . . . ; that he would kill all kind of Cattle and carry away the Horses, burn the Houses, & c and that if the French defeated them by the Delays of this Province that he would with his Sword drawn pass thro' the Province and treat the Inhabitants as a Parcel of Traitors to his Master; . . . [that] he did not value any thing [Pennsylvania's governor and Assembly] did or resolved, seeing they were dilatory retarded the March of the Troops, and hang an arse (as he phrased it) on this occasion, and told Us to go to the General if We pleased, who wou'd give Us ten bad Words for one that he had given.

When Croghan tried to speak up, St. Clair cut him off. "To every sentence he solemnly swore," Croghan reported.

Word of St. Clair's rage soon traveled to Philadelphia and reached the ears of members of the Assembly. They decided to dispatch their all-purpose fixer, Benjamin Franklin, to calm the apoplectic Sir John, deal with General Braddock, who had called Pennsylvania's behavior "pusillanimous," and repair Pennsylvania's damaged reputation in the eyes of the king's representatives. Benjamin and his son William rode to the village of Frederick, Maryland, to meet Braddock, who had just moved his headquarters from Alexandria to the little settlement about fifty miles inland and northwest, as it lay where the rolling Piedmont rises to the first mountain ridge. In Frederick, newly founded at the junction of two great Indian trails that now served as crude wagon roads, Braddock took up quarters in a stone tavern atop a hill.

The Franklins arrived in Frederick under the guise of transacting postal business and found the last room at a nearby inn in the crowded little town. As deputy postmaster general of the North American British colonies, Franklin claimed he must meet with General Braddock to speed messages most efficiently during his Ohio campaign. At age forty-nine, Franklin was a well-known figure in Pennsylvania politics, and also a journalist, publisher, inventor, and natural scientist who had conducted internationally famous experiments, such as drawing electricity from a storm cloud using a kite. Clever and witty, curious and practical, Franklin held a passionate belief in political freedoms while also showing a marked practicality in dealing with problems of all sorts.

When the Franklins, father and son, showed up at the general's headquarters in the hilltop tavern, they delivered letters to Braddock's personal secretary, young William Shirley, son of the Massachusetts governor. The secretary received the letters and dismissed the father-son pair in cursory fashion, not realizing who they were. That night, perhaps back at their own inn, they happened to dine with Colonel Dunbar, commander of the 48th regiment, and a number of his officers, who cordially welcomed them. In the course of socializing, the Franklins managed to correct many of the "violent predjudices [the British officers] had entertained against Pennsylvania."

Colonel Dunbar informed General Braddock that Ben Franklin was not just an ordinary messenger. Braddock then invited Franklin and son to breakfast and dine with him each day at his hilltop headquarters. True to his hidden mission, over these cordial meals Franklin convinced the general that Pennsylvania was willing to help the expedition. Just before returning to Philadelphia, the Franklins were dining with Braddock on the evening of April 22 when reports came to the general of the dire shortage of wagons in Maryland and Virginia, both of which largely used boats for transport—only twenty-five wagons available, and some of those out of repair.

This put the room in an uproar. "The general and all the officers . . . declar'd the expedition was then at an end, being impossible," remembered Franklin in his *Autobiography,* "and exclaim'd against the [British] ministers for ignorantly landing them in a country destitute of the means of conveying their stores, baggage, etc., not less than one hundred and fifty waggons being necessary."

Sir John St. Clair, recently arrived from Wills Creek and his tumultuous meeting with George Croghan, was especially enraged. Sir John stormed about the tavern, a "most violent Creature," in Franklin's judgment. "[He] said the damned people of Pennsylvania could furnish them, if they would; but they were Traiterous Frenchmen in their hearts, and would do nothing; for of all the flour they had promised, but two Waggons-load were yet arrived &c. &c. and desired the General only to furnish him with a body of Troops, and he would scour the Country from one End to the other; and if they would not furnish Waggons and horses he would take them by force, and chastise the Resistors with Fire and Sword."

Amid the tumult, Franklin spoke up, diplomatically. "I thought it was a pity they had not been landed rather in Pennsylvania, as in that country almost every farmer had his waggon."

General Braddock turned to him. "Then you, sir, who are a man of interest there, can probably procure them for us; and I beg you will undertake it."

Toting £800 in gold and silver from General Braddock to hire wagons, teams, and packhorses from Pennsylvania farmers, Ben and William Franklin rode off. A few days of traveling brought them to Lancaster, in a fertile region of southeastern Pennsylvania recently settled by German or "Dutch" farmers. Franklin printed an ad to hire one hundred and fifty wagons and fifteen hundred horses to serve Braddock's expedition. He also published a warning that if his fellow Pennsylvanians did not offer up the wagons voluntarily for this good compensation, the British forces, in a bad temper, would most likely seize them, with no promise of payment. Worse, likely to be leading this rampage would be Sir John St. Clair—aka "Sir John the Hussar."

Franklin's threat was a clever one, as the German settlers knew of the terrifying Hussars in Europe—fierce Hungarian warriors on horseback, bedecked in uniforms of braid and gold trim, whose task it was to overtake and run down retreating enemy troops. It happened that Sir John St.

Clair owned a Hussar uniform that he was fond of wearing. For the purposes of recruiting wagons, Franklin thus branded Sir John as an actual Hussar whose wrath would sweep the countryside if wagons did not come forth from Pennsylvania's many German settlers. General Braddock, his aide Orme reported, laughed for an hour straight when he was shown Franklin's ad.

Nor had Franklin exhausted his bag of tricks. Colonel Dunbar had mentioned to Franklin over supper one night that his British junior officers lacked personal funds to purchase the small luxuries to which they were accustomed back home but that had proved expensive and scarce in America. They would especially enjoy these while on a wilderness campaign.

Franklin said nothing, but with the help of his son, who had served in the provincial military, drew up a list of the luxury items and wrote to a finance committee of the Pennsylvania Assembly, which allotted funds. He then arranged for the preparation of twenty gift parcels, each loaded on a horse and delivered to a British officer in Braddock's army, and containing loaf sugar, Gloucestershire cheese, a twenty-pound keg of butter, two dozen bottles of old Madeira, two well-cured hams, raisins, chocolate, coffee, and a smorgasbord of other treats.

Franklin's practical cleverness bought a great deal of goodwill for Pennsylvania from General Braddock and his officers, while the pacifist Quakers of the Assembly, with these gifts of food, did not have to trouble their consciences about spending funds on warmongering. But Franklin harbored some doubts about General Braddock, finding him cordial but overly confident. As Franklin later reported, the general told him how quickly he expected to progress: "'After taking Fort Duquesne,' says he, 'I am to proceed to Niagara; and, having taken that, to Frontenac, if the season will allow time, and I suppose it will, for Duquesne can hardly detain me above three or four days; and then I see nothing that can obstruct my march to Niagara.'"

Franklin wondered whether General Braddock could sweep through the French parts of North America so briskly. He knew that the general would have to cut a narrow road for a great distance through the Ohio wilderness before he reached Fort Duquesne. He told Braddock that he thought his army could make relatively short work of the fort if the general arrived there before the French could bring in reinforcements and if he brought with him a full complement of troops and artillery. But he forewarned Braddock on one account. "The only danger I apprehend of obstruction to your march is from ambuscades of Indians, who, by constant practice, are dexterous in laying and executing them."

General Braddock, however, dismissed these concerns with an arrogance that would come to characterize, at least in the American popular imagination, the British Royal Army's sense of superiority—and, indeed, Britain's—over all things American. The view from the seats of power in London was that both the American colonists and the Native Americans were primitive, savage, and unknowing.

Reported Franklin: "He smiled at my ignorance, and replied: 'These savages may, indeed, be a formidable enemy to your raw American militia, but upon the king's regular and disciplined troops, sir, it is impossible they should make any impression.'"

Ben Franklin would turn out to be prescient.

CHAPTER ELEVEN

AT THE END OF APRIL, AS FRANKLIN TRIED TO HIRE WAGons and mend Pennsylvania's reputation, Washington arranged his business and personal affairs at Mount Vernon before setting out to join the Braddock campaign. He wanted the world to know that he joined the campaign not for personal advancement but out of a sense of duty. He stated this so emphatically and repeatedly in letters to acquaintances that one suspects him of exactly that—joining Braddock to advance his career and his position in the world. "I have no expectation of reward but the hope of meriting the love of my country," he had written to the aristocratic burgess William Byrd.

He now wrote to his acquaintance Carter Burwell, a wealthy planter, Virginia burgess, and member of the military committee, requesting £50 to cover valuables he had lost to the French when he surrendered Fort Necessity—surveying equipment, books, clothes, and his servant (or slave), who had died of wounds a few days after the battle and was counted as another valuable piece of personal property. He closed with a statement of his noble intentions in joining Braddock's campaign: "I am just ready to embark a 2nd time in the Service of my Country; to merit whose regard & esteem, is the sole motive that enduces me to make this Campaigne; for I can very truly say I have no views, either of profitting by it or rising in the Service as I go a Volunteer witht Pay, & am certain it is not in Genl Braddocks power to give a Comn that I wd accpt. . . ."

Adding a last emphatic stroke about his altruism, he pointed out that, as his business affairs were in such disorder, far from gaining personally he would surely lose from the campaign. He wrote in a similar vein to John Robinson, Speaker of the House of Burgesses, asking him to support his £50 request for lost property and emphasizing that he was going as a volunteer without pay. "[T]he sole motive wch envites me to the Field, is, the laudable desire of servg my Country; and not for the gratification of any ambitious or lucrative ends. . . ."

When it came to denying that he had a great deal of personal ambition in joining Braddock, young George Washington, like the queen in Shakespeare's *Hamlet,* did "protest too much."

Washington bid his good-byes to those closest to him—presumably to his mother as well as to the Fairfax family, including his friend George William and his friend's wife, Sally, who so captivated Washington. To Colonel William Fairfax, his patron and the master of Belvoir Manor, away on business in Williamsburg, Washington wrote a good-bye letter and predicted a long, arduous journey into the wilderness. He expected progress would be slowed by the train of artillery and baggage that must accompany it, "which I am sorry to say, I think will be tedious in advancing—very tedious indeed—answerable to the expectation I have long conceivd, tho' few believ'd."

On April 23, business affairs completed, Washington left his Potomac house atop Mount Vernon and rode toward the mountains to join General Braddock's campaign. En route, along with other intermediary stops, he wanted to check in at his Bullskin Creek plantation in the Shenandoah Valley, which as a young surveyor he had purchased five years earlier. In his absence, he had decided to leave the management of both the Bullskin property and Mount Vernon to his younger brother Jack. Accompanied by his white personal servant, John Alton, and several spare horses for the upcoming wilderness campaign, it took him three or four days' hard riding from Mount Vernon to reach the Shenandoah, with one of his spent horses dying en route and the others beginning to fail. One imagines a reflective, brooding Washington in breeches and boots jostling along for hours on this long ride, pondering his fate during the upcoming campaign. He faced long periods of loneliness. He faced the possibility that he would never return. The sudden death of his horse may have churned up thoughts of his own mortality. Perhaps he thought of what truly mattered to him.

Traversing the Blue Ridge, he descended into the sparsely settled Shenandoah and, on reaching his Bullskin Plantation, sat down and wrote Sally Fairfax a peculiar and entreating letter. He asked to carry on a personal correspondence with her. He would earn her correspondence in return, he wrote, "by embracing the earliest, and every opportunity, of writing to you." He went on in his convoluted style: "It will be needless to expatiate on the pleasures that a communication of this kind will afford me, as it shall suffice to say—a corrispondance with my Friends is the greatest satisfaction I expect to enjoy, in the course of the Campaigne, and that none of my Friends are able to convey more real delight than you are. . . ."

Young Washington clearly cared very deeply for Sally Fairfax, married to his close friend and former traveling companion George William Fairfax, and she seems to have reciprocated on some level. How much of this correspondence was known to her husband is unclear, and whether it crossed the boundaries of propriety. But in some ways the correspondence made sense. She was a charming lady of a major Tidewater plantation family—a female pinnacle of the Virginia aristocracy and establishment. He was a young would-be hero heading off to war to defend that aristocracy and establishment. He clearly hungered for an audience to whom he could relate his exploits, a loved one who could embody all that he fought for and defended back home. Acutely aware of his public image, his honor, and his reputation, he longed to be cheered on, encouraged, and admired as he went off to war. Encouraging women had cheered on their men to the battlefield for countless centuries before him. Sally Fairfax could fulfill this role. She may have understood her part, too. He was defending Virginia and the Fairfax way of life, and perhaps a special correspondence with Sally Fairfax could be tolerated or even encouraged if it contributed to that defense.

ON MAY 1, Washington finally met up with General Braddock, still stuck at his hilltop tavern headquarters in Frederick, Maryland. Here the general and his staff had been waiting, frustrated and angry, for hundreds of wagons and thousands of horses to arrive to haul his massive supplies and artillery to Wills Creek and then over the high ridges into the Ohio wilderness. They had been led to believe that a road existed from Frederick to Wills Creek. When they learned that it did not, General Braddock's staff and the troops of Colonel Dunbar's 48th regiment changed plans. Instead of starting the march on the Maryland route, they diverted and marched fifty miles south to Winchester, in the Shenandoah. From there they would follow the Virginia route, over a very bad road, to Wills Creek.

When General Braddock arrived in Winchester, a Virginia frontier town of sixty houses along the ancient north-south Indian trail known as the Great Warriors Path, he expected to find many Cherokee and Catawba warriors from the Carolina region to the south who were eager to join his campaign, as Governor Dinwiddie had promised. There were none. No Indians, no wagons, no horses, no roads. General Braddock fumed. Ben Franklin, back in Pennsylvania, worked on drumming up hundreds of wagons and packhorses to transport corn, oats, and fodder into the wilderness. It pleased Washington that the Marylanders could not deliver on their promises. The general had a "good opportunity to see the absurdity of the [Maryland] route, and damning it very heartily," Washington wrote. It gave Washington "infinite satisfaction" that Colonel Dunbar's troops would have to abandon the Maryland route in favor of the Virginia one. This, however, would not end Washington's road jealousy.

Braddock paused at Winchester, regrouping, during the first hot week of May. The troops were on the move in separate regiments. Sir Peter Halkett's 44th Foot Regiment of seven hundred men, the British Redcoats, had already gone ahead toward Wills Creek and Fort Cumberland, while Colonel Dunbar's 48th Foot Regiment, about the same size, also British Redcoats, had diverted south from Frederick to Winchester and would soon follow. The general waited at Winchester for the massive artillery train that had been offloaded from the British warships at the port of Alexandria and now labored over the Blue Ridge, with the help of Royal Navy sailors.

Young Washington waited with him. He had now joined General Braddock's "family"—his inner circle of aides and officers. He dined with them but so far had been given no official duties. "I am very happy in the Generals Family," he wrote to his mother, "as I am treated with a complaisant Freedom which is quite agreeable; so that I have no reason to doubt the satisfaction I props'd, in making the Campaigne." This has the ring of defensiveness, of justifying his decision to join General Braddock despite his mother's objections.

During the pause at Winchester, Washington tried to solve his horse problems, having exhausted his mounts on the ride from Mount Vernon to the Shenandoah. He wrote to his brother Jack about a favorite horse named Gist (after the wilderness guide) and one named Countess, with instructions for their proper care. He also wrote to the British peer and vastly landed Northern Neck proprietor Lord Thomas Fairfax, cousin of Colonel William Fairfax, who now resided at his Shenandoah hunting lodge, to borrow £40 to buy more horses. A soldier marched on foot; an officer and a gentleman like Washington rode on a horse; and a general such as Braddock sat in a heavy coach brought by ship from England and accompanied by a mounted guard of light horses. This was not just a military campaign. This had the trappings of an empire on the move.

After four days at Winchester, General Braddock's entourage, and with it young Washington, moved out on May 7, heading for Wills Creek, some sixty miles away. Colonel Dunbar's 48th Foot Regiment had gone ahead, following Sir Peter Halkett's 44th. The route took them from the broad, partly settled Shenandoah Valley over the first corrugated ridges of the Appalachians, rising as if the landscape had folded, ridge after ridge, carpeted with forest, creased by streams, or runs. They crossed crude bridges that the troops had thrown across streams, load by load forded rivers on rafts, or floats, built for the purpose, skirted around hills, and dropped through notches, or gaps, in the mountain ridges.

But the road to Wills Creek, after leaving the fertile valley floor of the Shenandoah, reverted to a crude pathway through the woods, humping and twisting its way over roots, rocks, stumps, and ruts. It was so rough that the widowed nurse from Britain, Mrs. Browne, who rode in a wagon at the rear of the campaign and whose brother was serving up front, finally just got out and walked. "There is no describing the badness of the roads," she recorded in her journal. "I am almost disjointed."

The path crossed one stream twenty times in three miles and another nineteen times in two miles. Sloshing through the shallows, straining up the far bank, and then jouncing back down to the stream again, the troops had not yet reached the deep wilderness of the Ohio Valley itself, but, except for occasional tiny settlements, wilderness surrounded them. "There is nothing round us but trees, swamps and thickets," wrote one British officer in a letter home. Like British officers generally, he held a strict notion of what made for a proper battlefield. "I cannot conceive how war can be made in such a country."

Another letter-writing British officer sketched the crude homesteads sited in natural meadows or hand-chopped clearings, where the settlers had fenced off a garden plot and hoed tobacco and corn into mounds, the corn stalks shooting up like reeds eight or nine feet tall. Along with this Indian corn, eaten in cakes and mush, they subsisted on their herds of cattle—milk, curds, whey, cheese, butter, and lean, tough beef from old cows.

> *Their Cattle are near as wild as Deer; a Cow-pen generally consists of a very large Cottage or House in the Woods, with about four-score or one hundred Acres, inclosed with high Rails and divided, in a small inclosure they kept for Corn, for the Family, the rest is the Pasture; . . . they may perhaps have a Stock of four or five hundred to a thousand head of Cattle belonging to a Cow-Pen. . . . The Cow-Pen Men are hardy People, are almost continually on Horseback, being obliged to know the Haunts of their Cattle.*
>
> *You see, Sir, what a wild set of Creatures our English Men grow into when they lose Society.*

The British troops feared snakes in the swamps. But the mosquitos and chiggers were maddening. The tiny chiggers, or *Trombiculidae,* lay hidden in the thick, swampy grasses, in clods of mud, and low bushes. Resembling ticks but nearly invisible (less than one hundredth of an inch across in their larval stage), they wait for an animal host to come brushing through, then hop onto the host, crawl to an open patch of skin, and bite. Into the tiny wound they inject enzymes that dissolve the skin tissue. A minuscule tube forms of hardened skin cells. The chigger larva sucks the dissolved skin tissue up through this tube as nutrients to feed itself and grow to its next stage of development, the nymph. The skin-dissolving enzymes in the wound cause intense itching and irritation for twenty-four hours or so after the bite. "[T]he Wound itches and makes one ready to tear off the Flesh," wrote one British officer.

Itching as they went, with one soldier rumored to have had his leg amputated due to infections caused by chigger bites, and two reputedly dead of snakebites, the troops marched on. They crossed the upper Potomac on floats. Once on the north bank they found relatively easy marching. About ten miles short of Wills Creek and Fort Cumberland they camped for the night at the settlement of Thomas Cresap, the trader known to the Indians as "Big Spoon" for the kettle he kept over the fire. The Fairfax survey party had camped at Cresap's in 1748 and witnessed a dance by passing Indians that struck sixteen-year-old Washington as "comicle."

Yorkshire-born Cresap had built his settlement at a major crossroads of the wilds—where the north-south Great Warriors Path crossed the headwaters of the Potomac, as the crow flies about one hundred and ten miles northwest of Alexandria. The aggressive Cresap and his sharp-shooting wife, Hannah, had a history of settling on land not clearly their own. A skilled carpenter, Cresap had raised a wooden stockade around their current homestead and trading post to protect them from anyone who thought they did not belong there.

The "Seaman" was not impressed. An anonymous British Navy officer in General Braddock's army, he kept a detailed journal of the march. Along with thirty Royal Navy seamen, he had been assigned, by Commodore Keppel, to help Braddock's soldiers haul cannon over the mountains because the sailors knew how to work block and tackle—rope-and-pulley systems to lift heavy objects like yardarms and sails. Despite the steep-bluffed beauty of the surrounding mountains and the fertility of the valley where Cresap had settled, on the site of an old Shawnee Indian village, the Seaman showed nothing but contempt for Cresap himself. "Here lives one Colonel Cressop, a Rattle Snake Colonel, and a D____d Rascal; calls himself a Frontier man, as he thinks he is situated nearest the Ohio of any inhabitants of the country, and is one of the Ohio Company."

Catching up to the troops that had gone ahead, General Braddock and his personal entourage rolled into Cresap's trading post to spend the night of May 8. The troops held a horse race in his honor. After pausing at Cresap's for a day, due to rain, on May 10 most troops moved out early, marching toward Fort Cumberland along an easy trail on the Potomac riverbank. About noon, General Braddock and his entourage caught up to them and passed by in full military splendor—the British Empire displaying itself like a great red peacock emerging from the forest—with the drummers playing a march and the general's coach rumbling through accompanied by the hoofbeats of his light horse guard.

He disappeared down the crude wagon road. The Redcoats marched in unison, in linear order, in the sprawl of forest and mountain. Soon they heard the echoing booms of seventeen cannon. General Braddock and the might of the British Empire, with eager young Washington trailing in its wake, had arrived at Fort Cumberland—the jumping-off point to the Ohio wilderness and the French fort that lay deep within it.

ANOTHER PAUSE, again full of frustrations, descended on the army before the final push over the tallest mountain ridges into the Ohio. To move an empire demands a great deal of feed for horses and food for men. As the mass of troops assembled, neither enough wagons, nor horses, nor provisions had arrived at Fort Cumberland. Nor enough Indians. Instead of four hundred Cherokee and Catawba warriors, as promised by Governor Dinwiddie, only about one hundred Iroquois, including women and children, had encamped in bark lean-tos at the forest's edge near the fort. The promised flour had not arrived. Kegs of salted beef had jounced in on the crude road, but when twenty kegs were pried open for inspection, the meat, from Colonel Cresap, was not properly brined and had turned rancid. It had to be buried. This caused an especially noted hardship for the British Royal Army officers accustomed to the good life even on the march.

"Indeed the Officers are as ill off about Food as [the soldiers], the General himself, who understands good Eating as well as any Man, cannot find wherewithal to make a tolerable Dinner of, tho' he hath two good Cooks who could make an excellent Ragout out of a Pair of Boots . . . ," wrote one British officer in a letter to his brother. "[At Fort Cumberland] we can get nothing but Indian Corn, or mouldy Bisket . . . Bread we must bake in Holes in the Ground having no Ovens, so besides the Mustiness of the Flour, it is half Sand and Dirt."

General Braddock railed about the provincial governors who had promised him goods and had not delivered. He held a special contempt for Virginia, notably "the folly of Mr. Dinwiddie & the roguery of its assembly." A rumor circulated among the troops that the general would march his army back to the coast if, in two days' time, Virginia had not produced wagons and provisions. The disgruntled British Royal Army officer writing to his brother called this "the best News I ever heard in my Life. . . ."

Washington wrote to Sally Fairfax striking a brave-and-ready-for-battle note when he said that his dreaded concern before leaving home had been boredom, and now here they were, expecting to wait at Fort Cumberland an entire month for supplies and wagons. It was as if he wanted her to know that he was a seasoned veteran of combat and ready for more, as long as he did not have to wait around too long for the action to begin.

The fort itself consisted of a log stockade twelve feet high, over four hundred feet long, and more than one hundred feet wide, sitting at the foot of a tall, forested ridge. It held five officers' quarters, twenty-seven barracks, and corner fortifications housing a total of ten cannon. "Like a ship run aground on the eastern front of the Alleghenies," writes historian Preston, "Fort Cumberland symbolized the geographical limits of Britain's imperial expansion in mid-eighteenth-century America."

THE FORT'S BARRACKS could not house all the troops that had clustered at this final outpost for the push into the Ohio wilderness. Canvas tents marched in rows on the open ground outside the fort, in addition to bark Indian lean-tos at the forest's edge. From his Fort Cumberland headquarters tent, General Braddock wrote letters and orders to rally provisions and horses while his troops drilled and practiced musket firing, amazing the Indians with their linear marching order. Among many other instructions, the general officially named young Washington an aide-de-camp, one of several. Self-satisfied with this appointment and hyperconsciously measuring his place in the hierarchy, Washington boasted to his younger brother Jack that he had only the general's commands to follow and that orders issued by Washington that had originated with the general must be "implicity obey'd" by everyone else. In other words, only General Braddock could tell him what to do. To Jack, he also noted the career-advancing opportunities of getting to know the general. "I have now a good oppertunity, and shall not neglect it, of forming an acquaintance which may be serviceable hereafter, if I can find it worth while pushing my Fortune in the Military way."

Ever conscious of an officer's appearance, he asked Jack to send him a "neat" pair of boots—meaning well-made but not ostentatious—as his had worn out, and "wearing Boots is quite the Mode."

Before Washington even sent off the letter to Jack, however, General Braddock dispatched him on his first mission—a hasty ride all the way back to Virginia's Tidewater to fetch badly needed funds—silver—to pay farmers and contractors for hundreds of wagons, horses, and supplies expected to arrive at Fort Cumberland soon. Washington rode hard over the mountains, struck by the lack of rain. He avoided a visit to his mother, though passing close by her house at Ferry Farm, but pointedly stopped at Belvoir Manor. Needing a fresh mount, he borrowed Sally Fairfax's personal horse and rushed southward to Williamsburg to collect £4,000 sterling from agents of the British Royal Army for the general's urgent needs. On the return trip, he paused briefly at Port Royal on the Rappahannock River to buy a slave named Harry for £45 and rode on to Winchester. Here he impatiently waited two days for mounted guards to accompany him and the money after he left the settled country, writing to Jack that the frontier town was a "vile hole" after the gracious pleasures he had just enjoyed at Belvoir with the Fairfaxes, and advising his younger brother to stay in the Fairfaxes' good graces and visit often because the family could be very helpful to "young Beginners" like the two brothers.

On May 30, he rushed into Fort Cumberland laden with silver specie for General Braddock. The full army had now assembled there, twenty-five hundred strong—seven hundred Redcoats in each of Sir Peter Halkett's 44th Foot and Colonel Dunbar's 48th Foot, three independent companies with one hundred men in each, a fifty-man company from Maryland and one from North Carolina, and nine companies from Virginia of fifty men each. In addition, there were sixty British regulars to attend the artillery train, plus the thirty Royal Navy sailors with block and tackle to winch the big cannon over the Appalachian Mountains. The last company to march into Fort Cumberland, arriving May 30, had been summoned from North Carolina. It numbered among its men a twenty-year-old wagon driver named Daniel Boone and another named Daniel Morgan, the latter to play a key role in the revolution that lay twenty years in the future, and the former one day to open the Kentucky country to settlement.

Along with the army arrived wagons carrying wives, children, nurses, washerwomen, mistresses, and prostitutes. Doctors checked the latter on May 23 to make sure they were "clean" before the march into the wilds. Dogs trotted alongside, horses neighed, and cows ambled, all descending on Wills Creek and Fort Cumberland. The general enforced discipline in the camp by issuing one thousand lashes each to two men who stole money and attempted to desert, parading them in front of the men to beating drums with halters around their necks, a meting out of justice not lost on young Washington.

He eyed the future in other ways, too, asking his brother Jack to subtly feel out—"fish out," as he expressed it—whether Colonel Fairfax or other prominent local planters had any intention of running for a Fairfax County seat in the House of Burgesses that soon would open up. If not, Jack should delicately find out whether they thought Washington might make a good candidate, but without letting them know he was interested. If they preferred some other candidate, Jack should quietly let the matter drop, advised the elder brother, trusting the younger brother's "good sense" and discretion. "[C]onduct the whole, till you are satisfied of the Sentimts of those I have mention'd, with an air of Indifference & unconcern," he instructed.

It was an early and very visible example of young Washington beginning to mask his hard-driving ambition with a feigned indifference, a strategy he would employ to great effect in the decades ahead.

General Braddock was still waiting for the Cherokee and Catawba to come up the Great Warriors Path from the south, as Governor Dinwiddie had promised. With none arriving, he dispatched Indian trader George Croghan to summon the chiefs of other tribes to Fort Cumberland, such as the powerful leader of the Delaware, Chief Shingas. The general ordered his officers and troops to treat any potential Indian allies with special hospitality: gifts, food, and "let them want for nothing." When Chief Shingas finally met with General Braddock, however, gifts did not sway him. Rather, he focused on the fate of the great hunting grounds of the Ohio and just who had access to them. Chief Shingas would later recount the conversation he had with General Braddock on this occasion, as Shingas understood it. While recent historians have questioned whether Braddock actually was so dismissive toward Indians and actually spoke the words quoted by Chief Shingas, the chief and his Delaware warriors nevertheless appear to have viewed Braddock in this light.

If the general succeeded in driving the French and their Indian allies away, Shingas asked, what did Braddock intend to do with the lands of the Ohio?

"The English Should Inhabit & Inherit the Land," replied General Braddock.

But could the Indians who were friends and allies to the English live and trade and have hunting grounds in the Ohio?

"No Savage Should Inherit the Land," General Braddock replied.

Shingas and the other chiefs responded that if they were not free to live on the land, they would not help the British fight for the Ohio. "To wch Genl Braddock answered that he did not need their Help," Shingas recounted, "and had No doubt of driveing the French and their Indians away."

CHAPTER TWELVE

General Braddock's plunge into the American wilderness began on May 29, 1755. The first obstacle nearly brought it to a halt. Braddock had ordered Sir John St. Clair to "open & prepare a Road over the Allegany Mountains, towards the great meadows." From the assembled British Redcoats and colonial troops that had massed at Fort Cumberland, St. Clair had chosen six hundred and fifty men. At dawn that morning they marched to the forest's edge and started to chop a road through the wilderness toward Fort Duquesne, some one hundred and twenty-five miles away.

Just to the west of Fort Cumberland rose a steep, forested ridge fourteen hundred feet high, known as Haystack Mountain. A path of sorts existed over the crest. The road builders followed it, widening it as they went, accompanied by fifty wagons, Ben Franklin's efforts finally having delivered an additional one hundred and fifty wagons. With their teams of horses straining in taut harnesses up the steep slopes, some of the fifty crested the top, but the precipitous descent nearly destroyed them, as wagons broke loose and bounced and crashed down among trees and rocks. "The ascent and descent were almost a perpendicular rock; three waggons were entirely destroyed, which were replaced from the camp; and many more were extremely shattered," recorded Washington's fellow aide-de-camp, Captain Orme.

General Braddock, on horseback, inspected the route over this first obstacle and ordered more might and manpower thrown at it, as if sheer will and discipline alone could subjugate the American wilds. It was then that a young British naval officer named Spendelowe had a different idea. Commanding the detachment of thirty sailors from the Royal Navy, Lieutenant Splendelowe also showed a talent for reading terrain. He noticed a deep valley, or gap, just to the north of Haystack Mountain. Exploring it, he found that it cut right through the great ridge, and had inexplicably escaped the attention of St. Clair and others laying out the road. The general immediately ordered a detachment of one hundred men and an engineer to cut a new road through the gap. After two days of hard work, they had a passable route around Haystack Mountain.

While troops cleared the road, young Washington wrote a packet of final letters before stepping into the wilds. To his mother he wrote a brief, almost curt, letter, dutiful but not affectionate, apologizing that he could not provide a Dutch manservant or the butter she had requested. She undoubtedy thought that, as he was near Pennsylvania, where the Dutch (actually German) farmers lived in numbers, these two items would be readily available. He also apologized for not visiting her on his recent journey back to Williamsburg for the general's cash. All in all, it read like the letter of a son trying to avoid a demanding mother.

He wrote to Sally Fairfax in far more affectionate terms and with the convoluted syntax he associated with polite society, noting that she had asked to be informed of his safe arrival at Fort Cumberland. He pointed out that she had requested that he communicate this in a letter to some mutual acquaintance. He had not heard a word from her. Did this mean she did not want to correspond with him, or were his fears about her silence unfounded? He wanted reassurance. "[H]ow easy it is to remove my suspicions, enliven my Spirits, and make me happier than the Day is long, by honouring me with a corrispondance which you did once partly promise." About to enter the wilderness and face an unknown fate, he could not let go of his thoughts and feelings for Sally and a half promise she had supposedly made. His fears seemed to center less on the possibility of death than on the dreaded possibility that she did not care for him as he cared for her.

To his brother Jack he gave an update on his arrival at Fort Cumberland and their continued lack of intelligence about the French except that

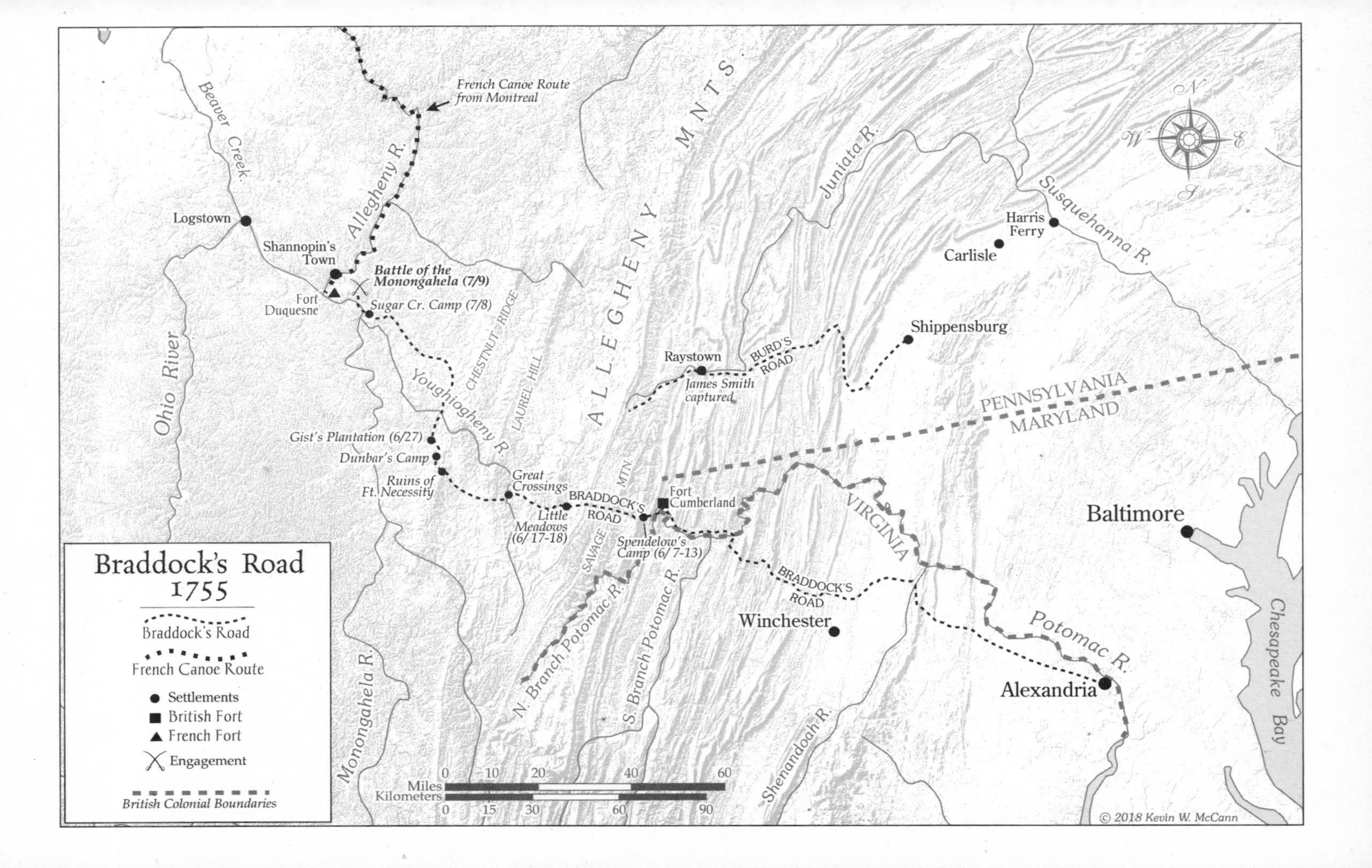
Braddock's Road
1755
Braddock's Road
French Canoe Route
Settlements
British Fort
French Fort
Engagement
British Colonial Boundaries
Miles
Kilometers
0 10 20 40 60
0 15 30 60 90
French Canoe Route from Montreal
Beaver Creek.
Allegheny R.
Logstown
Shannopin's Town
Battle of the Monongahela (7/9)
Fort Duquesne
Sugar Cr. Camp (7/8)
Ohio River
CHESTNUT RIDGE
LAUREL HILL
Youghiogheny R.
Gist's Plantation (6/27)
Dunbar's Camp
Ruins of Ft. Necessity
Great Crossings
BRADDOCK'S ROAD
Little Meadows (6/ 17-18)
MTN.
SAVAGE
Fort Cumberland
Spendelow's Camp (6/ 7-13)
N. Branch Potomac R.
S. Branch Potomac R.
Monongahela R.
ALLEGHENY MNTS
Raystown
James Smith captured
BURD'S ROAD
Juniata R.
Shippensburg
Carlisle
Harris Ferry
Susquehanna R.
PENNSYLVANIA
MARYLAND
VIRGINIA
BRADDOCK'S ROAD
Winchester
Shenandoah R.
Potomac R.
Alexandria
Baltimore
Chesapeake Bay
N
E
S
W
© 2018 Kevin W. McCann

French reinforcements were expected anytime at Fort Duquesne. Also, that the "bloody flux"—dysentery—had hit the camp, with several soldiers dying and many others sick. Washington himself would soon fall victim, too.

In these final days before the march into the wilds, he dined with Braddock and the other close officers in the general's family, attended to by the general's two cooks and presumably off the complete silver service that the general carried in his luggage. As one of the few Americans at the general's table, Washington took the side of the American colonists against the dismissive attitude of General Braddock and his British staff and found himself in debate with a heated general he thought at times beyond reason. As Washington wrote to his patron, Colonel William Fairfax:

> *The General, by frequent breaches of Contracts, has lost all degree of Patience . . . [and] will, I fear, represent us home [to England] in a light we little deserve; for instead of blaming the Individuals as he ought, he charges all his Disappointments to a publick Supineness; and looks upon the [colonies], I believe, as void of both Honour and Honesty; we have frequentes disputes on this head, which are maintaind with warmth on both sides especially on his, who is incapable of Arguing witht; or giving up any point he asserts, be it ever so incompatable with Reason.*

In these heated discussions at General Braddock's well-laden dinner table, young Washington, in part due to Braddock's stubbornness and sense of superiority, had planted within him the growing seeds of an American identity separate from the British Empire.

The main body of troops marched into the wilds on June 7, when Sir Peter Halkett and his 44th Foot Regiment, having sailed from Ireland and marched from Alexandria, strode out of Fort Cumberland on the new road skirting Haystack Mountain to the symphonic accompaniment of thunder and lightning. The next day, June 8, the independent companies from New York and South Carolina and the six Virginia companies of rangers marched forth, along with a company each of rangers from Maryland and North Carolina, as reported by aide-de-camp Orme. On June 10, Dunbar's 48th Foot Regiment left the fort's protection for the wilderness edge. The first climb through the winding mountain gap exhausted the horses, straining to pull wagons far overloaded for the steep grades. As Halkett and his regiment waited at the first overnight stop, Spendelowe's Camp, just beyond the gap and only five miles from Fort Cumberland, the other troops and wagon trains inched forward, some taking two days to reach it. When General Braddock himself arrived at Spendelowe's Camp, he ordered a council of war among his closest officers to determine the next step.

Among them were Colonels Halkett and Dunbar, both of the British Royal Army, but not young aide-de-camp Washington of Virginia. He knew this wilderness terrain, unlike the British officers. Unlike them, he had fought the French and Indians in the wilds. But he held neither high enough rank nor an officer's commission in His Majesty's Royal Army. Another attendee at the general's war council was thirty-five-year-old Lieutenant Colonel Thomas Gage. (Two decades in the future, Gage, as military governor of the Colony of Massachusetts Bay, would stand at the center of Britain's attempts to thwart the American Revolution.) The second son of a British viscount, Gage had joined the Royal Army and fought in Europe during the War of the Austrian Succession but, on his return to England, suffered heartbreak when a wealthy and aristocratic young woman broke off their engagement. Soon after, Gage, as well as his father, both lost their bids for seats in Parliament. He hoped for better fortune with General Braddock in the wilds of America.

With advice from his war council, General Braddock decided to lighten the load by sending back the heaviest baggage and wagons and spreading out the rest. The general was now experiencing on a far greater scale what Washington had sampled on his winter journey to the French commandant—the cruel logistical calculus of wilderness travel. The king's wagons had to go. These sixteen large and heavily reinforced wagons had been brought from England on the flotilla across the Atlantic and hauled tons of gunpowder for the expedition's cannon and muskets.

The men transferred the gunpowder kegs to regular farm wagons, pulled by four horses. Each piece of the heaviest artillery required seven horses to draw it. To further lighten the load, the general sent back two of the six-pounder cannon—wheeled cannon weighing about fourteen hundred pounds and shooting a ball that weighed six pounds—and several four-inch mortars. He still had plenty of artillery, hauling twelve cannon, four howitzers, fifteen mortars, and three hundred rounds for each, plus three months' provisions for two thousand men.

The general asked the officers to kick in, too, from the mounds of extra baggage they carried. Officers had brought their own large tents, portmanteaus with fancy uniforms, luxury items such as Ben Franklin's hams and Madeira, and other "necessaries." They each possessed several horses to haul all this, but the general made it known that it would not be forgotten which officers gave up their special tents for regular ones and contributed some of their horses to help carry the train's baggage. His inner circle stepped forward with twenty horses. This included young Washington's contribution of his best horse, although he traveled with four as well as his personal servant. He also reduced his personal belongings to half of what fit in his portmanteau. Other officers contributed another eighty horses, for a total of one hundred additional horses to help carry the expedition's essential baggage. With his experience in the wilderness, Washington had advocated using pack animals to carry baggage over the rugged mountains rather than hauling it in wagons. Now General Braddock was discovering the value of that advice.

To reduce the weight of passengers and provisions, Braddock ordered that he would provide food for only two women per company. The rest—wives, washerwomen, mistresses—should return to Fort Cumberland. It was further decided to reduce the load of each wagon from a little more than two thousand pounds to about fourteen hundred pounds. All this helped. The expedition's load lightened, and with an extra sixty thousand pounds, or thirty tons, of carrying capacity, the great column marched on.

Even then, they faced a Herculean struggle. Imagine, first, road engineers under Sir John St. Clair studiously selecting a route through thick forest and mountain ridges, attempting to avoid streams and swamps and to contour around, rather than climb, the steepest slopes and rock outcroppings. With axes and long military blades called "fascine knives" they hacked blazes on tree trunks to mark the way for the work party. Next came guards bearing muskets to defend the unarmed laborers against Indian or French ambush from the woods. Under the guards' watchful eyes, the army of axmen swarmed over the landscape, whacking down trembling saplings with one or two quick blows, or biting resoundingly into the solid bole of a thick oak, hickory, or maple. *Thunk . . . thunk . . . thunk,* the forest resounds with hundreds of ax blows. Cries echo through the woods to get out of the way as a toppling tree smashes through the forest canopy and booms onto the earth, shaking the ground, dead leaves and bark rising in a dark dust cloud as it hits the forest floor. A team now springs forward to chop and saw the tree into sections to drag or roll away. Other teams of two men each set to work on the largest trees with one of the twelve whipsaws, backs bending and arms pumping as they smoothly draw back and forth, the jagged teeth cutting green wood, sweat dripping in the hot, closed-in forest, the smell of fresh sawdust in the nostrils as the saw slices through the ancient tree and it, too, crashes down, its great crown leaving a hole in the leafy forest canopy open to the warm blue sky. A patch of sunlight shines onto a littering of wilting leaves and wood chips.

Stumps are sawn or chopped level with the earth to permit a wagon to pass over them, and logs dragged or rolled to the side. The diggers come in, dozens upon dozens of them. They dig with shovels and picks to level a road twelve feet wide to permit the passage of the largest wagons and red-coated columns of troops. The clang of metal on earth and stone rings through the forest. Dirt clods fly. Rocks rumble down hillsides. A team of former miners, adept at prying, digging, and levering, goes to work to "spring" the largest boulders. Some won't move. They pound away with hammer and chisel. Some are simply too large. The miners call the blasters. They stoop low to lay a charge of black powder under the rounded boulder, quickly pull back. The boom resonates across the wooded ridges. Rock shards fly. A column of white smoke erupts. The diggers move in again to clear the sharp-edged fragments of the shattered boulder, level the ground, fill it in, make a road.

They knew how to make a road, these British engineers, and they did—up, down, and across the hills, bridging streams, spanning swamps. The boggy ground they crossed with corduroy—logs laid down next to each other, like so many pencils in a box, to form a rough log road. Or with their long knives they chopped twigs and branches and bound them into tight bundles, called "fascines," to place in shallow streams or swamps, allowing water to pass freely while giving purchase to shoes and hooves. The scouts and engineers always chose to ford streams just downstream of incoming tributaries, whose discharge formed gravel bars and shallow riffles that offered an easy place to cross.

But the road was rough, everywhere.

Then the two thousand soldiers marched through, in their red columns, packs on their backs, keeping rhythmic step over the uneven ground. Then hundreds of horses, each carrying one hundred pounds in sacks of flour and loads of bacon, and officers sitting erect on their mounts, keeping with their soldiers. Then followed the enormous parade of wagons and artillery. They needed to carry feed—so much feed—for the horses, plus the tents, the powder, and all the accoutrements. Two hundred wagons spaced fifty feet apart would create a wagon train two miles long, one wagon after the other, wheels sinking into moist earth, rumbling over broken rock, slamming over low stumps. The uphills were excruciating, the teams dragging nearly a ton up the steep inclines, up the slanted rocks, hooves scraping, nostrils flaring, whips cracking. They could not manage. Men were called. They leaned into the rear of the wagons or tugged on ropes, dozens of them, hefting the slowly jouncing wagons up the mountainsides.

In some ways, down was worse. How to hold a loaded wagon or a cannon of half a ton or more from rolling down a steep mountainside? The wagoners locked the brakes. This still could not prevent a skid to disaster. The horses could not hold them. Men held back the wagons with ropes, planting their feet and leaning uphill, coarse rope fibers cutting against their palms. Finally, they tied big logs to the rear of the wagons to act as anchors on the steep descents, or brought in the Royal Navy team of sailors to rig the block-and-tackle systems to lower them slowly down the steepest pitches.

It proved a monumental task, pushing this linear road, this projection of an empire's power, into the jumbled mountain wilderness. They accomplished it one tree, one rock at a time, multiplied by hundreds and thousands, all working at once, like a giant organism eating a tunnel into the forest. The forest itself spooked them. Dark and heavy, it held dangers they could not see. Having departed Fort Cumberland and skirted Haystack Mountain, they had left even scattered British settlements behind. There were none ahead. Indians roamed these forests, or so they feared.

The old soldiers worried that the French would attack while the British column marched through wooded gullies, one British officer reported in a letter home. But he added that General Braddock wisely reassured them that the thick forest would hinder an attack by the French against the British in the same way it would an attack by the British against the French. General Braddock had gained his military tactics from the battlefields of Europe, however, and the great rock fortress of Gibraltar, with its two hundred cannon aimed at enemy fleets on the sea. The Indians had a very different kind of battlefield experience, and had taught the French well.

Veterans of Indian combat and wilderness guides warned General Braddock that "formal attacks & Platoon-firing would never answer against Savages and Canadeans," as Adam Stephen, who had survived the French and Indian attacks at Fort Necessity, wrote in a letter to a friend at the time. Washington, writing many years later, said he repeatedly advised General Braddock to beware of unusual forest-and-mountain warfare techniques. "His Excellency," apparently, did not listen.

On June 15, Braddock's army crossed a major barrier—the Allegheny, or Big Savage, Mountain, the last rise of land before the Eastern Continental Divide. Streams on the eastern side of this divide eventually flow into the Atlantic Ocean, and streams on the western side make their way to the Mississippi River and thence the Gulf of Mexico. "We this day passed the Aligany Mountain," reported aide-de-camp Orme, "which is a rocky ascent of more than two miles, in many places extremely steep; its descent is very rugged and almost perpendicular. . . ."

Beyond the mountain, they passed through a gloomy, low-lying pine forest, draped in moss and ridden with bogs. In these pre-Romantic times of the mid-eighteenth century, when wildness still connoted biblical evil instead of something pristine or beautiful, General Braddock's men, many of them from the soft, green, sheep-dotted hills of England or Ireland, or the airy Scottish Highlands, found this dense, mossy forest "dismal" and called it "Shades of Death." They saw this raw wildness captured in the flesh when one of the hunters brought in a bear he had killed, having also shot a wolf and tracked a cougar for six miles. The next day he hauled in two elk, and the men dined on bear meat and rattlesnake.

On June 16, they camped at the Little Meadows, a grassy forest opening similar to the Great Meadows, which still lay many miles ahead. At Little Meadows, General Braddock called a second council of war. Progress remained achingly slow. The baggage train stretched out three or four miles, leaving it thinly guarded and vulnerable to Indian attack. The horses weakened daily, pulling the laden wagons over steep, rough terrain. Washington complained in a letter to his brother Jack that the general stopped to level every mole hill and bridge every little brook. Much more had to be done to speed their progress.

Before holding the council, General Braddock consulted with young Washington, the aide-de-camp who knew the route into this forest wilderness. Washington advocated using all possible haste to reach Fort Duquesne before French reinforcements arrived there, even if it meant charging ahead with a smaller, lighter party. "I urgd it in the warmest terms I was master off, to push on," he wrote to his brother Jack, "if we even did it with a chosn Detacht for that purpose, with the Artillery and such other things as were absolutely necessary, leavg the heavy Artilly Baggage &ca with the Remainder of the Army, to follw by slow and regular Marches. . . ."

While intelligence reports from the Ohio indicated a small and weak French presence at Fort Duquesne, more might arrive at any time. Other reports gathered by Indian agent William Johnson to the north and delivered to Braddock via Mohawk messengers told of French troops moving from Montreal up the Great Lakes and destined for the Ohio. Some had stalled due to low water on the Rivière Le Boeuf. Johnson urged the general to make haste. Washington doubted that the reinforcements would reach Fort Duquesne while the drought held, speculating that canoes and bateaux could not float on the Rivière Le Boeuf. General Braddock considered these intelligence reports. He listened to what Washington had to say about a smaller, faster party. He met with his senior officers to hear their opinions. He decided. He would send ahead toward Fort Duquesne a smaller advance party under the fiery Sir John St. Clair, whom some fellow officers parodied as "the Knight." He would place other troops and supplies in a slower column led by Colonel Dunbar of the 48th.

The "flying column" under Sir John consisted of about twelve hundred chosen men, whom Braddock regarded as the cream of his army. Most of them were British regular army veterans of the 44th and 48th regiments rather than the less well-trained American provincial soldiers. General Braddock selected artillery, including four eight-inch howitzers, each pulled by a team of nine horses, and four twelve-pounders each pulled by a team of seven horses, but leaving some of the pieces behind. About thirty-five wagons hauled goods and ammunition for this advance party, as well as road-building tools and gifts for the Indians they hoped to win over.

Some men assigned to Colonel Dunbar's slower column worried that they would miss their chance for glory and petitioned General Braddock to name them to the flying column. He granted the request to a few British regular soldiers, but denied officers who wished to move forward. One embittered officer left behind described the departure of the flying column from Little Meadows: "[It] marched out the Knight swearing in the Van, The Genl Curseing, & bullying in the Center & their Whores bringing up the Rear."

Fever and bloody flux had run rampant in the low-lying Little Meadows camp, the result of bad water, badly sited privies, and poor sanitation. Bloody flux—the inflammation of the intestines from the *Shigella* bacteria or an amoeba that is spread through feces, causing severe, bloody diarrhea, fever, cramping, and dehydration and known today as dysentery—apparently had struck Washington, too. He had weakened so much coming down Big Savage Mountain that on the approach to the Little Meadows camp he had climbed off his horse and crawled into a wagon. He had hoped to join the flying column, but General Braddock ordered him to stay behind in a rear camp until he was strong enough to follow. The general committed him to the care of a physician traveling with Colonel Dunbar and also ordered him to take a powerful dose of Dr. James's Powders.

Washington received what amounted to a get-well card inscribed with orders, delivered by fellow aide-de-camp Roger Morris. "It is the Desire of [everyone] in the Family, & the Generals positive Commands to you, not to stirr, but by the Advice of the Person under whose Care you are, till you are better, which we all hope will be very soon. . . ."

Over the corrugated ridges of the Alleghenies the flying column traveled, through forests and across bogs. The half-mile-long column paused for two days while Sir John St. Clair and his forward party hewed a switchbacking road up a particularly steep, wooded ridge. Progress lagged; it took four days to cover twelve miles. Through knee-deep water, they waded eighty yards across a branch of the Youghiogheny River, flowing north and west off the Eastern Continental Divide. They had now entered the enormous Ohio Valley watershed, almost equal in size to France itself, and coveted by both empires. The swift current surged against the legs of men and horses and the wheels of wagons, until they heaved themselves, dripping, up the far bank. Here they encountered the first signs of the French and Indians.

Most of the one hundred Iroquois previously encamped at Fort Cumberland had gone off, but eight Iroquois sachems and warriors remained as scouts with General Braddock's flying column. These included Monacatoocha, who had been with Washington and the Half King at Jumonville Glen, as well as his son. Scouting beyond the branch of the Youghiogheny, now in the Ohio watershed, territory that the French considered theirs,

Monacatoocha was captured by a roaming party of French and Indians. His son found him tied to a tree, unharmed, after the French had interrogated him. A few days later, on June 24, the flying column came on a deserted French and Indian camp, with lean-to huts and the dead ashes of campfires. The Iroquois scouts with Braddock estimated the size of the force that had camped there at about one hundred and seventy. The French and Indians had left messages for the British, reported Captain Orme. "They had stripped and painted some trees, upon which they and the French had written many threats and bravados with all kinds of scurrilous language."

On June 25, early in the morning, a wagoner went out in the woods to gather his grazing horses but was discovered by a party of Indians who shot him four times in the belly. He stumbled into camp, to die a few days later. That same day, four other men from the flying column also went out to look for horses, and were scalped. The Indians and French stalked somewhere close about, unseen. Nerves tightened.

The next day, June 26, the column traversed a difficult mountain and camped below a rocky hill, where a party of French and Indians had slept the previous night. Here, they had drawn on trees in red paint symbols depicting the scalps they had taken, and French officers had written their names—Rochefort, Chauraudray, and Picauday.

The British column marched through the Great Meadows and past the charred ruins of Fort Necessity, its stockade logs pulled down and torched the previous summer by French troops after Washington's surrender. The scattered human bones of the dead surrounded it, dispersed by wild animals. Smashed swivel guns lay in the ditch. Beyond the Great Meadows, the column pushed up the steep, wooded slopes of Chestnut Ridge and traced the ridgeline northward several miles, passing the rocky hollow where Washington had opened his dawn volleys on Jumonville's breakfasting party. Dropping off the ridge's far slope, they arrived at the green valley that cupped the ruins of Gist's settlement. The French troops had burned that homestead, too, leaving blackened splotches on the green meadow.

Messengers arrived bearing news from Fort Cumberland back at

Wills Creek. Indian parties had scalped two families of frontier settlers within only a few miles of the fort. Rescuers had discovered a child of seven standing in a creek, scalped but alive, crying. The child died a week later under medical care at the fort. General Braddock worried that more Indian attacks might cut the mountainous supply road leading from Fort Cumberland to his forces. He wrote to Governor Morris of Pennsylvania urging quick completion of a road from Pennsylvania to the Forks as an alternative supply route to the one from Fort Cumberland.

On June 30, the flying column approached the main branch of the Youghiogheny River, a considerable two hundred yards across. They braced themselves for an Indian ambush in the vulnerable midst of the crossing, setting guards as they stumbled over the rocky bottom and plowed through water three feet deep. Nothing happened.

They marched onward. Still no attack, nor shootings nor scalpings. The size and strength of the flying column perhaps intimidated the French troops and Indian warriors, though not its ponderous progress. The confidence of General Braddock, his officers, and his men surged. Some fretted that the French, knowing they were overpowered, would blow up Fort Duquesne and flee before they arrived. They pushed on with haste, before the French and their Indian allies could run from the might of the British Empire bearing down on the Forks of the Ohio.

As June turned to July, young Washington, weak but recovering from the bloody flux after his powerful dose of Dr. James's Powders, and no longer requiring a nurse, still waited at a camp some twenty-five miles to the rear, stuck with the plodding Colonel Dunbar's group. He could not bear to miss the action and the glory. Although so weak he could barely finish the short letter, he wrote to his friend and fellow aide-de-camp Orme, asking where to meet the flying column before it reached the French fort. It was an opportunity he would not miss, he wrote to Orme, for £500.

CHAPTER THIRTEEN

AT THE SAME TIME, ABOUT SIXTY MILES TO THE NORTH-east, another company of three hundred frontier woodsmen chopped a forest road from Shippensburg, in Pennsylvania colony, over the wild Appalachians to join Braddock's Road near the Youghiogheny. This would serve as another supply route—essential if Indian attacks severed the Virginia–Fort Cumberland route. Among those swinging an ax at the solid oaks was an eighteen-year-old named James Smith, whose settler family had pioneered a homestead in the Conococheague Creek region, a fertile valley at the foot of the Allegheny ridges that marked Pennsylvania's leading edge of settlement. Smith's brother-in-law, a local gristmill owner, served as one of the road commissioners overseeing the cutting of the Pennsylvania supply route to Braddock's army.

The teenage James Smith, as he put it years later in his autobiography, had "fallen violently in love with a young lady, who was possessed with a large share of both beauty and virtue." Despite his infatuation, the excitement of Braddock's army crossing the Alleghenies to conquer the French and their wilderness fort proved a powerful draw. "[B]eing born between Venus and Mars, I concluded I must also leave my dear fair one, and go out with this company of road-cutters, to see the event of this campaign; but still expecting that some time in the course of this summer, I would again return to the arms of my beloved."

As June turned to July, with Washington ill in a rear camp and Braddock's flying column crossing the main branch of the Youghiogheny, sixty miles to the north the crew of three hundred Pennsylvania woodsmen that included James Smith had pushed the road deep into the mountains, largely tracing an ancient Indian trail that led from the Atlantic to the Ohio. On July 3, the road bosses dispatched Smith and a companion, Arnold Vigoras, to the rear on horseback to tell the provision wagons to hurry. The pair of messengers found the supply wagons moving as quickly as they could down the newly hewn track, and they swung their horses around to ride back to the front party of axmen.

Unbeknownst to them, three Indians had planted branches in the earth fifteen yards off the road to make a blind. As Smith and Vigoras rode past, they jumped out and fired. Vigoras fell dead from his horse. No bullets touched James Smith, but his horse bucked at the gunfire, throwing Smith to the ground, uninjured. The three Indians instantly seized Smith, ripped Vigoras's scalp from his skull, and fled into the forest with their captive.

Two of them were Delaware—one of whom spoke some English—and one a Canastauga. That first day they ran about fifteen miles through the woods and slept on Allegheny Mountain without a fire in order to avoid detection. After a breakfast of moldy biscuit and roasted groundhog, they ran westward again, through rocky terrain and laurel thickets, toward Fort Duquesne. In another fifty or so miles of fast travel they reached Loyalhanna Creek, about thirty-five miles short of Fort Duquesne, where they met another returning Indian raiding party. They shared roasted turkey and venison, which, despite the absence of salt and bread, Smith savored after his long run. They slept again, met another party, and, as one large group, kept on. In four days, Smith's Indian captors traveled one hundred miles through forest and mountains—a distance that took Braddock's army several weeks.

As they approached Fort Duquesne, the returning Indians erupted in the "scalp hallo" that announced each captured scalp, followed by shouts of joy. The fort responded with a deafening chorus of celebratory gunfire and the boom of cannon, such a cacophony that Smith believed the French had recruited thousands of Indians.

Scores of Indians ran from the fort to meet them, stripped naked except for breechcloths and painted in "the most hideous manner" in vermillion and black, brown and blue. They made two long rows about thirty feet apart. The Delaware who spoke English instructed young Smith that he must run between these two rows while the Indians flogged him. He advised the captive to run as fast as he could because they would stop beating him if he could reach the far end.

"There appeared to be a general rejoicing around me," Smith reported,

> yet I could find nothing like joy in my breast; but I started to the race with all the resolution and vigor I was capable of exerting, and found that it was as I had been told, for I was flogged the whole way. When I had got near the end of the lines, I was struck with something that appeared to me to be a flick, or the handle of a tommahawk, which caused me to fall to the ground. On my recovering my senses, I endeavored to renew my race; but as I arose, some one cast sand in my eyes, which blinded me so, that I could not see where to run. They continued beating me most intolerably, until I was at length insensible; but before I lost my senses, I remember my wishing them to strike the fatal blow, for I thought they intended killing me, but apprehended they were too long about it.

When he regained consciousness, Smith found himself inside Fort Duquesne and attended by a French doctor. The physician had sliced open an artery in Smith's left arm to bleed him and had washed his wounds in brandy. As Smith lay there, Indian allies of the French arrived to interrogate him, threatening him with torture and death if he did not tell the truth. When they demanded to know the size and arms of the woodcutters' force, he truthfully gave the size—three hundred—but lied that they were well armed when in fact they possessed only about thirty muskets. He hoped this might prevent the Indians from making what would otherwise be an easy attack.

After the interrogation, he received a visit from the Delaware Indian who had captured him and spoke English. Smith asked what he had done wrong to receive such a severe beating. He learned that the beating did not constitute some form of punishment but rather was an old Indian custom—their version, as the Delaware put it, of "how do you do." In the future, the Indian said, Smith would receive good treatment.

"What news from Braddock's army?" Smith asked the Delaware. The Delaware told him that every day the Indians spied on Braddock. He proceeded to sketch out with a stick on the ground exactly how the army was arrayed as it marched through the forest. When it came time to attack the column, the Delaware said, the Indians would hide behind trees.

"Shoot um down all one pigeon," the Delaware confided, as if Braddock's army were a giant bird sitting dumbly in the forest.

* * *

By July 8 General Braddock's and Sir John St. Clair's axmen had chopped to within fifteen miles of Fort Duquesne. Braddock had held another council of war: Should they wait for Colonel Dunbar's slowly advancing rear column to join them before attacking the fort? No, it was decided. It would be too long a wait, provisions would run low, and the French might blow up the fort and flee, leaving the British without their great victory.

"The British Gentlemen," wrote Adam Stephen in the aftermath, "were confident they would never be attacked, and would have laid any Odds that they never should, until they came before the Fort, yea, some went further, and were of the opinion, that We should hear the Explosion of the French Fort blown up and deserted, before We approached it."

Yet others advised caution, especially at river crossings. Two days earlier, on July 6, Indians ambushed three soldiers and a woman herding the column's cows, killing and scalping the victims. In the confusion that followed, Braddock's soldiers fired on what they believed were enemy Indians but turned out to be their own scouts. This volley of friendly fire killed the son of Monocatoocha, the Iroquois sachem who had staunchly allied himself with the British. While the father grieved, shattered that his warrior son had died at the hands of his own allies rather than valiantly in battle, the son received a British military burial with volleys fired over the grave.

As they drew nearer the fort at the Forks, they encountered tricky jumbled terrain around the confluence of the two rivers, the Monongahela and the Allegheny. To avoid a loop of the Monongahela called the Narrows, which would squeeze his column between the riverbank on his left and a steep mountainside on his right, General Braddock, with the advice of his officers, decided to ford the Monongahela twice within the distance of about a mile. The drought had lowered the river and rendered the fords shallow and relatively easy to cross, and he could send a party ahead to the opposite bank to guard the fords against attack. His scouts reported that once across the second ford of the Monongahela and up a bluff of land, the terrain would allow a relatively easy march to Fort Duquesne.

A torrential rain soaked much of the flying column's flour, and General Braddock sent back to Colonel Dunbar's rear column to request one hundred packhorses loaded with flour sacks. Dunbar had struggled with a lack of strong horses to pull his heavy baggage wagons, having only half the number needed, but he followed the general's orders and sent the flour-laden horses. In the wake of the pack train jounced a covered wagon that carried George Washington, having recovered somewhat in the rear camp but still too weak to ride. The heroic path to glory he had envisioned for himself did not include bouncing weakly along as a passenger in a flour wagon. Brooding during his long illness, Washington had plenty of time to wonder why no friends had written to him. He wrote with whiny bitterness to his brother: "You may thank my Fds for the Letters I have recd; wch has not been one from any Mortal since I left Fairfax, except yourself and Mr. Dalton."

He wrote to Colonel James Innes, commander back at Fort Cumberland, that there had to be some mistake and his mail was not forwarded properly, as if he could not believe that people back home, perhaps including Sally Fairfax, were not thinking of him as much as he would like: ". . . I have been greatly surprisd at not receiveing any Letter's from my Friends

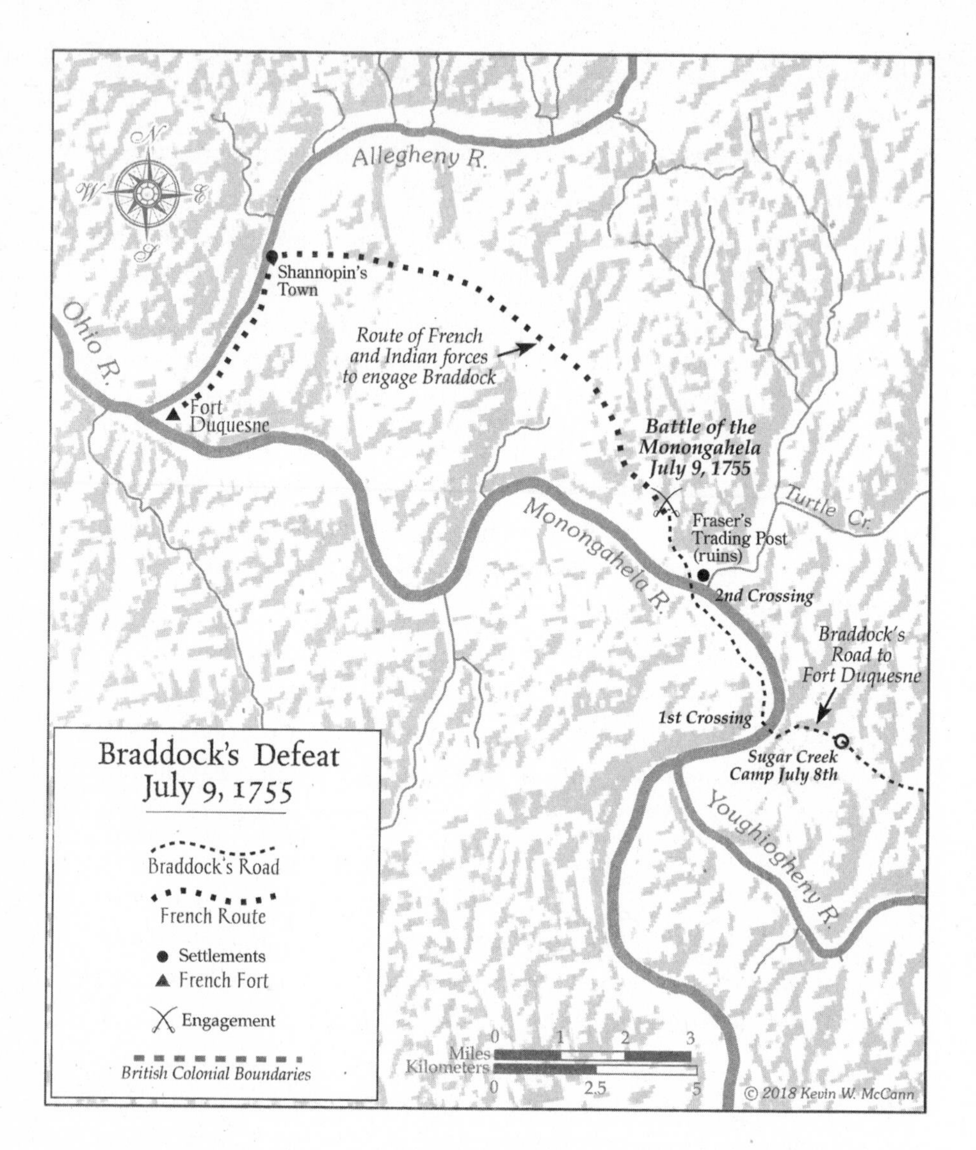

since I came out; and must impute it to miscarriage, somewhere, for I am certain it cannot be owing to their not writing."

He arrived at General Braddock's camp, situated on a wooded plateau above the Monongahela River, on the evening of July 8, climbing down from the covered wagon. One pictures dozens of campfires flickering in the evening light and scores and scores of white tents, all in tight lines as if the tents themselves marched in formation, confined within a space of three hundred yards along the stump-studded road opening. Soldiers move about in the woodsmoke preparing their dinner of bread and salted beef. The axmen have cut a large oval around the camp, a secure perimeter, where pickets stand at regular intervals, and another cleared perimeter beyond forms an enormous rectangle that contains the oval within it. The soldiers secure wagons and cannon for the night. It is a well-ordered camp, a grid of linearity and preparedness in the jumble of forest. Horses grind away on feed brought by wagons. Cows moo. The general and his family dine well at his tented quarters, with his ham and Madeira and Gloucestershire cheeses. He welcomes young Washington back to camp. The mood is upbeat, confident. All the might of the British Empire has been transported deep into the wilderness and stands at the ready. General Braddock dictates orders for the next day. The column will proceed in three parties, leaving consecutively, cross the Monongahela River twice, and establish another camp within a few miles of Fort Duquesne. From there the siege of the French fort will begin.

The first party roused at 2 A.M. after a few hours' sleep. These were grenadiers chosen from both the 44th and 48th regiments of British Redcoats, elite soldiers named after the grenades they sometimes threw, who traditionally led assaults and sieges and distinguished themselves by their size, fearlessness, and tall pointed hats. Under Lieutenant Colonel Thomas Gage, and accompanied by other soldiers who brought the party to about three hundred men, the grenadiers marched forth in the predawn hours with their flags flying, yellow silk regimental banners, the king's colors, and hauling two six-pounder cannon to protect the river crossings. Christopher Gist, George Croghan, and other frontier guides would lead them down to the first Monongahela ford.

At 4 A.M., the two hundred axmen and shovelers—known as the "carpenters and pioneers"—assembled under Sir John St. Clair and followed the grenadiers, clearing and enlarging the forest path down toward the river as they went, flushing out a party of Indians, who fled. The previous evening, "the Knight"—Sir John—had proposed that he lead a lightning night raid on Fort Duquesne, but more cautious officers immediately shot it down.

At 5 A.M. the main party came to readiness—nearly a thousand soldiers, plus more cannon, mounted officers, wagons, and packhorses. If an attack were to come, most likely it would strike them at the second ford of the Monongahela, less than ten miles from Fort Duquesne.

They marched downward in a long, winding column as dawn's gathering light lifted the forest shadows from heavy black to soft gray to leafy green. Still weak from his bout with fever and dysentery, young Washington insisted on traveling with the general as one of his aides-de-camp but had to tie cushions to his saddle to pad himself against the hard jolts of the ride.

At around 8 A.M., the main body of the great procession reached the first ford of the Monongahela at a long bend of the river bounded by steep bluffs on the near bank and a flat on the far. General Braddock arrayed troops and wagons cautiously for maximum protection during the crossing. He ordered a large guard of soldiers to the high ground above the river. They watched at the ready as beneath them passed, first a vanguard of one hundred and fifty troops, then half the wagons, then another guard of one hundred and fifty soldiers, followed by horses, cattle, artillery, the rest of the baggage, and finally all the troops that had guarded the crossing from the surrounding heights. It was well planned, orderly, defense-minded. A messenger arrived from the first party. The grenadiers had already crossed the second ford, had taken positions on higher ground with their two six-pounder cannon, and had secured the second crossing from possible attack.

The main party marched along the flat, wide bank the two miles from the first to the second crossing and halted at the river's beach. Here they were delayed until 1 P.M. while, across the river, St. Clair's party prepared the steep far bank for wagons, packhorses, and heavy artillery to surmount it. While they waited they had a chance to look around. The river here was about two or three hundred yards across and a little over knee deep, dancing over shallow riffles that sparkled in a warm July sun. The river had undercut and eroded the far bank to create a steep wall of sandy clay about twelve feet high. The axmen and diggers swung to excavate a ramp, an impressive cut in the wall-like riverbank to create an incline perhaps fifty or one hundred feet long sloping up from water's edge. Beyond, the land flattened briefly in a narrow bench, a shelf-like piece of land between river's edge and the rising bluff, where amid old Indian fields sat the remains of a trading post built by John Fraser but burned by the French. From the bench, the land lifted in a giant step, carpeted with leafy hickory forest, to the higher plateau some four hundred feet above and the blue skies of summer. Posted on this bank where the diggers labored, the grenadiers, wearing their tall wizard-like hats, called "mitres," meant to intimidate the enemy and lacking brims to permit free arm movement for grenade throwing, guarded the crossing, their two cannon at the ready. Suspecting that the French and Indians spied on the crossing, General Braddock had ordered the troops to ford the river in greatest order and in their finest regalia.

This moment embodied all the power and the glory of His Royal Majesty's empire. A Delaware Indian spy secreted on the wooded plateau across the river, perhaps high in an oak's branches, would have looked down on a great blood-red swath in the flat green bottomland, interspersed with rectangles of the provincials' blue uniforms, wagon trains like beads on a string, and bright, distant sparks where sunlight glinted off cannon barrels. Wafting up on the gentle summer breeze, delayed by distance for a few heartbeats, the first ripping *rat-a-tat-tats* would have reached the Delaware spy with the ululating whistling of fifes as the head of the long red swath crawled down the riverbank.

They marched to the music and words of "The British Grenadiers." The lyrics compared the British grenadiers favorably to the warrior heroes and demigods of classical antiquity, names that meant nothing to the Delaware in the wilds of the Ohio:

> *Some talk of Alexander, and some of Hercules*
> *Of Hector and Lysander, and such great names as these.*
> *But of all the world's great heroes, there's none that can compare*
> *With a tow, row, row, row, row, row, to the British Grenadiers.*

Horses' hooves splashed into the shallows. Hundreds of men marched in lockstep into the river—red coats with white lace and long tails, yellow or buff lapels, tight red knee breeches, white gaiters over shoes and calves, muskets over shoulders, black tricorn hats—flags flew, the yellow silk banners of the 44th, the red cross of England and white X of Scotland joined in their strength to make the flag of Great Britain. The column marched across the shallow river, on and on, stretched out over a mile, drums beating, fifes playing the grenadiers' march, muskets shining, their hundreds of sharpened bayonets glistening, wagon wheels churning over the riverbed, cannon lumbering down the bank, officers poised atop their horses in gold-embroidered uniforms pulled from portmanteaus, monitoring, directing—and it kept coming, row after row, wagon after wagon, all in order, all in a line, all geometry and squareness, as if something as inconsequential as a river and a vast forest wilderness could not stop the might of the British Empire.

It was glorious. It was spectacular. It was the most beautiful sight, some later recalled, that they ever saw.

The soaring words of "The Campaign," Joseph Addison's famous poem about the Duke of Marlborough's march through swamp, forest, and river to meet the enemy at the Battle of Blenheim, perhaps rang in the minds of officers and men, and General Braddock himself, who surely learned them as a boy.

> Nor fens nor floods can stop Britannia's bands,
> When her proud foe rang'd on their borders stands.

They heaved up the incline cut into the clay bank, column after column, cannon after cannon, wagon after wagon, flag after flag, beating drum after beating drum, and, dripping in the midday sun, reassembled on the far side in the open fields of the bench. The crossings had gone without a hitch. General Braddock and his army had passed through this most vulnerable of moments. The power of Braddock's army marching through the wilderness and across rivers had intimidated the French. If they had not attacked by now, it was unlikely that they would.

By 2 P.M., the whole procession had completed the second crossing, reassembled on the far bank in the old fields near Fraser's burned post, and paused to await further orders. General Braddock ordered the advance party and the fearsome grenadiers under Colonel Gage to continue marching until 3 P.M. With George Croghan and the scouts just ahead, and British engineer Gordon making blazes on trees to mark where to cut the road, they started up the Indian trail that slanted up the bluffs, the grenadiers' tall mitre hats bobbing. Sir John St. Clair's woodsmen were to follow Gage's advance party of grenadiers, then the main column, wagons, artillery, packhorses, and finally the rear guard. As usual flanking parties, these of twenty men plus an officer, would range out into the forest a hundred yards or so to the sides of the column to guard against side attack. These flanking parties had just taken position and the main column was about to march when, from somewhere up front, the sound of musket firing erupted in the woods. Many muskets. It did not stop.

CAPTAIN BEAUJEU HAD BEFORE HIM a seven-hundred-mile journey and a race against the turn of the seasons to reach Fort Duquesne—the equivalent of paddling a canoe from New York City to Savannah, Georgia, in eight weeks' time. The last of winter's ice still lingered when he departed from Montreal on April 23, 1755, with forty bateaux laden with provisions and two hundred and forty men. Soon to follow were three more convoys, each consisting of ten bateaux and one hundred men. Captain Beaujeu well knew that as winter's snow melted into spring runoff, river levels would rise, but as spring warmed to summer, rivers would drop, possibly grounding his craft in rocky shallows en route to Fort Duquesne. Montreal, where Captain Beaujeu's men began their seven-hundred-mile paddle, sits on the Saint Lawrence River roughly halfway between the Atlantic Ocean and the Great Lakes. The Saint Lawrence, in effect, forms the outlet drain of this enormous water basin in the North American interior, opening like a funnel in reverse as it spills from the narrow outlet of Lake Ontario and eventually empties into the Gulf of Saint Lawrence and the Atlantic. Captain Beaujeu and his convoys would have to paddle up it.

He had made this journey into the wild interior of the continent many times, as had his father before him. His father had served as commandant of Michilimackinac, a key French fur and military post on the island of that name far up the Great Lakes where Lake Huron and Lake Michigan join. He had won respect and status and gained an appointment as mayor of Quebec. As a teenager, his son, Daniel-Hyacinthe-Marie Liénard de Beaujeu, joined the French Marines in Canada, known as the Troupes de la Marine. The younger Beaujeu earned a reputation for bravery and leadership in the 1740s during bitter winter attacks against the British in Acadia. As he later put it, he idealized the kind of leader who, instead of turning back at danger, "goes toward the musket shots." Beaujeu commanded the French posts at Detroit and then at Niagara, where he improved the difficult portage around the big falls and mastered good relations with the Indian tribes. When the Marquis Duquesne, the governor of New France, resolved to send reinforcements to the Ohio Valley in the spring of 1755, the forty-four-year-old Captain Beaujeu—brave, strong, a natural leader, and much admired by the Indians—was chosen to lead a large body of troops from Montreal.

His route followed one that French voyageurs had traced for decades, and Indians for thousands of years before them. From Montreal, they wrestled their bateaux up the rapids of the Saint Lawrence. The boats typically measured about thirty-six feet long and six feet wide and were constructed of oak and fir, with a flat bottom designed to take heavy loads through shallow water or rough conditions. Usually powered by oars or a small sail, they could carry several tons. At the rapids of the Saint Lawrence above Montreal, the men had to drag the laden boats by ropes from the riverbank or pole them up the rapids, leaning their bodies into the long poles to push against the river bottom, sometimes wading deep into the cold current to muscle them around projecting points of land, careful not to get swept away themselves.

Above the rapids, Captain Beaujeu's party rowed and sailed up a one-hundred-thirty-mile stretch of the Saint Lawrence, wove through the fifty-mile-long archipelago of the Thousand Islands, and in early May entered Lake Ontario, the most eastern of the five Great Lakes. One can picture both to the left and right the low, dark-green shorelines of the big lake widening apart, sheathed with forest, fringed with cobble rocks or sandy beach, uninhabited by Europeans, stretching into the hazy distance until each dimming band thins to nothing and fades into the horizon. Ahead lies open water, large as an ocean seemingly, reaching to meet the sky. A few unseen Indian villages cling to the forest's edge. Fleecy clouds drift over ruffled blue waters until a squall arises, a patch of fierce gray sweeping across the surface, whipping it into a frenzy of whitecaps, when the bateaux with oars stroking in unison make for the safety of the shore, and two small ships carrying troops lower their sails.

Captain Beaujeu's convoy hugged the south shore of the huge lake, riding out its good weather and bad, its squalls and combing waves. They skirted around the one spot of European habitation along the two hundred miles of shoreline, the British post at Fort Oswego, although British officers scanning the lake spotted the dark dots of the French convoy. By messenger, they forwarded the intelligence southward to their superiors. About three weeks out of Montreal, in mid-May, Captain Beaujeu and his bateaux reached the western end of Lake Ontario and entered the Niagara River. If they continued twelve miles upstream, the low riverbanks would lift into rock walls and they would come face-to-face with the thundering white curtain of Niagara Falls. En route to the Atlantic, the waters draining from the upper Great Lakes here leaped over a shelf of bedrock hundreds of miles long known as the Niagara Escarpment. Captain Beaujeu knew this area well from his two years as commandant at Fort Niagara. He also knew the difficulty of the eight-mile portage. It took nearly two weeks in the latter part of May to haul the convoy's provisions and bateaux around the falls, time lost against the dropping water levels of the rivers that lay ahead as summer approached.

Entering the big waters of Lake Erie, they again hugged the wild south shore. In another eighty miles, on June 8, they reached the sandy hook and natural harbor at Presque Isle, where the previous year French troops under Marin and Le Mercier had built a substantial fort of squared chestnut logs. Here they faced the worst portage of all, a sixteen-mile trek that took them from the shores of Lake Erie southward, up rising slopes to the subcontinenal divide that separates the waters of the Great Lakes from the waters of the Mississippi and Ohio valleys. Up and down ravines and across swamps bridged by log corduroy roads they had to haul tons upon tons of goods. French-Canadian voyageurs typically hauled on their backs two ninety-pound parcels of goods, for a combined load of a crushing one hundred and eighty pounds, resting every half mile or so. The mounds of baggage that arrived with Beaujeu's convoys far overtaxed the available porters and carts. In a two-month period, the porters would haul 2,378 barrels of food across the portage, and plentiful casks of wine and brandy.

At the portage's south end stood Fort Le Boeuf, on the headwaters of the Rivière Le Boeuf, where, a year and a half earlier, at the start of this clash of empires in the wilderness, young Washington had delivered the letter to the French commandant. Launching hundreds of canoes, both birch-bark and dugout, in the winding stream that looped past the fort, Captain Beaujeu's convoys then had to negotiate the Rivière Le Boeuf. Spring's high water had already drained away and a summer drought had set in, dropping river levels so low that his canoes could barely float and his bateaux crunched their hulls against the gravelly shallows. Paddling, poling, and hauling their way through its countless bends, they emerged, seventy miles later, onto the broad waters of the Allegheny River.

Here stood the Delaware Indian town of Venango, where the French had earlier seized another former trading post of John Fraser's and reinforced it into a small military camp. By now it was mid-June. Captain Beaujeu delayed at Venango in the hope that his rear troop convoys would catch up. While at Venango, however, a messenger brought a letter from Contrecoeur, commandant of Fort Duquesne, about four or five days' voyage down the Allegheny, who was in poor health and pining for his wife and children after years of wilderness expeditions. With alarm, Contrecoeur reported that an English deserter had arrived at the Forks and told of a force of three thousand English soldiers closing in to besiege Fort Duquesne with eighteen-pounder brass cannon, mortars, bombs, "and many grenades."

"Since the English are coming on fast," Contrecoeur wrote to Beaujeu, "I beg you, my dear friend, to come with as many men as possible, as soon as you receive my letter, and bring all the food you can."

Unable to wait any longer for his following convoys, Captain Beaujeu swung his vessels out into La Belle Rivière—the Beautiful River, the broad Allegheny—and started paddling and rowing down its swift waters looping between high, forested bluffs.

On June 27, as General Braddock's flying column traversed the Great Meadows and the charred ruins of Fort Necessity, and while young Washington recuperated at a camp to the rear, the first of Captain Beaujeu's craft landed at Fort Duquesne. All his first convoy had arrived by July 2, although the rear convoys were still negotiating the long portage at Fort Le Boeuf and the shallow Rivière Le Boeuf. While many French troops remained en route, another ally awaited Captain Beaujeu at the Forks—six to seven hundred Indian warriors camped in bark wigwams outside Fort Duquesne.

This remarkable gathering of warriors had assembled from tribes throughout a wide swath of North America. Some had come from nations nearby or to the east. A great many had paddled long distances from the region around Lakes Superior and Michigan, known to French fur traders as the "*pays d'en haut,*" or "upper country." Others had journeyed still farther, even from beyond the Mississippi River. There were Abenaki, Iroquois, Nipissing, Shawnee, Miami, Hurons, Ojibwa, Potawatomi, Sauc, and Fox. And there were Delaware, or Lenape. This latter tribe had good reason to fight the British. They had already lost much of their land in eastern Pennsylvania to incursions of British settlers and crooked land deals such as the Walking Purchase.

As historian David Preston points out, it is unclear just how so many Indians from so many different tribes came to gather at Fort Duquesne in the summer of 1755. Contrecoeur summoned some from places like the French post at Detroit, where Indians had settled. He may have called many others from deeper in the continent. In any case, they were there—loyal to the French and armed with muskets, tomahawks, scalping knives, war clubs, and special cords for binding captives. In addition, there were the French regulars, the Independent Companies of the Marine, and the

Canadian militia. The officers, like Beaujeu, were of an exceptional caliber, veterans of combat, knowledgeable in techniques of guerrilla and small-scale warfare, and on good terms with the Indians, knowing their customs and speaking their languages.

Now it remained to rally the Indians to the French design. On June 16, before Beaujeu had arrived, Contrecoeur had urged the warriors to "take up the hatchet" against the British, calling the British the common enemy of both Indians and French. The French had maintained that the border between England and France in the Ohio Valley ran along the crest of the Appalachians. Contrecoeur reminded the Indian warriors that they—or at least some of them—had pledged to fight the British if they ventured beyond that crest into the Ohio Valley itself. Braddock's army had just done so, leaving Wills Creek and crossing beyond Big Savage Mountain. "They have scorned your words," Contrecoeur told the warriors. "Now it is up to you to show them your resentments."

On June 21, the ailing Contrecoeur put his thoughts about his military strategy in a letter to a colleague. To defend the fort, he would employ to his best advantage the tremendous woodcraft and fighting skills of the Indians by sending them out of the fort and into the forest: "[I]f we are attacked, I plan to put all our savages outside. They will have the woods to themselves, with the French to guide them, in order to strike behind [Braddock's] army day and night."

When Captain Beaujeu paddled into the Forks—young, fit, warrior-like, and sure of his command in battle—he, too, rallied the Indians. He spoke confidently in the name of Onontio, the metaphorical French "father" who looked over his Indian "children" and who had been invoked for decades by Jesuit priests, French traders, and military officers to embody the French Crown. "Children, now is the Time to come of which I often told," one Indian who was present recalled a French officer saying, most likely Beaujeu himself. "Such an Army is coming against you, to take your Lands from you and make Slaves of you—You know the Virginians;—they all come with him.—If you will stand your Ground I will fight with you for your Land, and I don't doubt we will conquer them."

Beaujeu wanted to attack the British army before it reached the fort.

The Indians at first resisted this plan. They held their own council of war on July 8, with the bold and battle-hardened Captain Beaujeu in attendance.

"Do you want to die, my father," they asked him, "and sacrifice us besides?"

"I am determined to meet the English," Beaujeu replied. "What! will you let your father go alone?"

Captain Beaujeu then sang their war song. It was both a pledge of solidarity and a fierce, impassioned challenge. With their warriors' pride inflamed, hundreds of Indians at the fort rose up and rallied beside him. The Potawatomi, however, wished to wait until the next day before going to fight. Scouts' reports told of the approach of the British army to within six leagues of the fort. Beaujeu decided to ambush the British the next day, July 9, at their second river crossing.

One can sketch the two camps that evening of Tuesday, July 8, based on participant accounts: One forms the great oval perimeter hacked into the forest and bisected by the crude road where the fifteen hundred British of the flying column have erected hundreds of tents in a row as if the tents themselves marched forward into the wilderness. At the center of the French camp, by contrast, stand the heavy walls of a star-shaped fort protected by ramparts of earth and timber twelve feet thick. Inside, officers and men eat, drink, and prepare for the next day's battle with letters home, diary entries, cleaning of weapons, prayers, and confessions, while in the distance they can hear muffled drums and rhythmic cries. Around the fort extends a forest clearing and at its edge are scores of bark lean-tos, arrayed not in the straight lines or geometric stars of European thinking but scattered organically over the contours of forest and landscape, where Indian campfires burn into the night, flames jumping and shadows flickering in the trees, as they dance, and sing, and drum, and prepare for war, with the bright stars above.

Early the next morning, Wednesday, July 9, Captain Beaujeu attended confession and prayer in the fort's chapel, planning to leave at dawn at the head of his Indian and French force. The Indian warriors, however, were still making their preparations. He delayed.

Captive James Smith was present, too, that morning at Fort Duquesne. He had run the gauntlet a few days before and was now recovering from his wounds. ". . . I heard a great stir in the fort," he later remembered. "As I could then walk with a staff in my hand, I went out of the door which was just by the wall of the fort, and stood upon the wall and viewed the Indians in a huddle before the gate, where were barrels of powder, bullets, flints &c, and every one taking what suited. . . ."

The frenzy of preparation was under way. In solidarity with their Indian comrades, Captain Beaujeu and several other French officers stripped to the waist, wearing on their naked torsos only their silver gorgets, the ornamental band of armor hanging below the throat that dated to the time of medieval knights. It represented the European version of the bear claw's necklace of Indian warriors. At 8 A.M. Captain Beaujeu set off with a total force numbering nearly one thousand, most of them—six hundred and fifty—Indian warriors summoned from tribes across a vast sweep of North America. The French and Canadians in Beaujeu's forces numbered about two hundred and fifty—regulars of the Troupe de la Marine, including officers and officers-in-training, as well as members of the Canadian militia. Painted with fierce designs in red, black, brown, and blue, the bare-chested Indian warriors did not march in metronomic European-style order but moved to rhythms set by the uneven forest terrain.

Captain Beaujeu's force followed a path a few miles north to Shannopin's Town, the old Indian village on the Allegheny where, eighteen months earlier, young Washington and Christopher Gist had thawed out after their raft disaster. From Shannopin's Town, Captain Beaujeu swung his forces eastward on another Indian path. This traversed the forested plateau that separated the Allegheny from the Monongahela and led to the second ford of the Monongahela, about nine miles away, where he hoped to attack Braddock's army during the vulnerability of their river crossing. Although he hurried troops and warriors after the late start, he did not reach the far edge of the wooded plateau until early afternoon.

Here the plateau fell away steeply to the Monongahela River, which lay three or four hundred feet below. The river ran through a broad gorge created by the wooded plateau on which Beaujeu stood and a similar plateau on the river's opposite side. At the foot of each plateau, running along each riverbank, lay a flat bench of low and partly open ground, where Fraser had sited his now-destroyed trading post. Small streams cut straight down the hillsides to the river, creating a series of wooded ravines like the corrugations of a roof. As Captain Beaujeu looked down on the sparkling Monongahela from the plateau's edge, the Indian path led off to his left, angling downward across the face of the hillside and the creases made by its ravines to the ford, perhaps a mile and a half or two miles distant.

Beaujeu's Indian scouts reported. He had arrived too late to ambush the British at the Monongahela ford below. Braddock's massive army had waded and ridden and rolled across the river and now reassembled on the flat bench on this near bank, at the foot of the steep hillside of the gorge, near Fraser's burned post. As Braddock's force assembled, Captain Beaujeu's scouts had carefully noted the formation of the enormous column. A mile long, it was spearheaded by an advance guard of grenadiers with their tall mitre hats and red coats marching in a column fifteen wide backed by two horse-drawn six-pounder cannon. A guard of troops on horseback followed immediately thereafter, and then a team of woodsmen numbering in the hundreds and wielding axes, and behind them the main train of troops in red and blue bearing muskets and flags, dozens of wagons, cannon, lowing cattle, and a rearguard of light horsemen. The column possessed eyes and ears of a sort—guides and scouts roamed a few hundred yards out in front, while the column's sides were protected by perhaps a score of flanking parties, each led by a sergeant or subaltern and consisting of ten men, working through the forest about one hundred yards off to the sides, alert for ambush aimed at the column's vulnerable flanks.

Assessing the layout of Braddock's army from his scouts' reports and taking stock of his own strengths, Captain Beaujeu gave orders. His French-Canadian Marines and militia would form a column fifteen abreast and descend the Indian path that angled across the hillside and its stream ravines. This would meet the British column, also marching fifteen abreast but angling up the hillside, head-on. He directed his six hundred and fifty Indian warriors to split into two groups, one to slip through the forest to the path's left and the other to the right. He further divided these two groups into numerous small bands overseen by French officers who spoke the various tribal languages. The Indians were not to reveal themselves or to shoot until he gave the signal.

CHAPTER FOURTEEN

NEAR THE HEAD OF THE BRITISH ADVANCE PARTY THAT afternoon was the "batman" of Captain Robert Cholmley—the personal servant responsible for Captain Cholmley's horses, baggage, and other comforts due an officer, especially one from an aristocratic and landed British family such as the twenty-nine-year-old Cholmley. After fording the Monongahela once early in the morning and the second time at midmorning, Braddock's advance party, led by Colonel Thomas Gage and the fearsome grenadiers, paused on the flat, relatively open bench below the steep hillside. They had eaten nothing since leaving camp hours before dawn. While waiting for the long procession to complete the second crossing in all its drum-beating, flag-waving glory, Cholmley's batman, who tended his master's seven horses and a cow, took out some ham and Gloucestershire cheese for his master, milked the cow, and mixed in rum from the captain's baggage, making a milk punch. Captain Cholmley ate and drank a little punch, recorded the batman in his journal, noting that most of the soldiers had nothing to eat. As the main train forded the Monongahela and assembled on the bench, spirits soared. If the French and Indians had not attacked at the crossing, it was unlikely that they would. They envisioned the French fort lying just ahead, ready for the taking.

"Every one . . . hugg'd themselves with joy at our Good Luck in having surmounted our greatest Difficultys, & . . . Concluded the Enemy never wou'd dare to Oppose us," recorded Harry Gordon, the British engineer marking the road.

Colonel Gage's advance party now received orders from General Braddock, delivered by aide-de-camp Morris, to march forward until 3 P.M., slanting on the Indian path from the river up the long, wooded hillside to the plateau. Leading the axmen, Sir John St. Clair asked the leader of the advance party, Colonel Gage, if he wished to take the two six-pounder cannon in front with him again, as he had on the morning's march.

"No, Sir I think not," Colonel Gage replied, the batman recorded, "for I do not think we Shall have much Occation for them and they being troublesome to get forwards before the Roads are Cut."

Colonel Gage's advance party and the grenadiers started up the slanting Indian path, St. Clair's two hundred axmen and shovelers just behind them. The trees here grew widely spaced—far enough apart, one participant remembered, that one could easily drive a wagon between them. Wearing his father's long silk sash, General Braddock rode on horseback farther back in the column, letting the advance guard forge the way. With him rode his aides-de-camp, including Washington, seated painfully on his cushioned saddle and still weak from his bout with fever, delirium, and bloody flux, but with his silver-handled sword at his left side. Despite his illness and whatever warnings he may have given General Braddock, Washington shared in the jubilation of the crossing.

How could he not get caught up in the drum-thumping rhythm, scarlet-uniformed discipline, and cannon-wielding might of the British Royal Army marching into the American wilderness? He, twenty-three-year-old George Washington, lately a middling Virginia planter and freelance surveyor, rode amid the highest ranks of one of the most brilliant military expeditions ever launched by the British Crown. And now with the second dangerous crossing completed, its glorious victory—and his—was almost assured.

Those heroes of antiquity ne'er saw a cannon ball,
Or knew the force of powder to slay their foes withal.
But our brave boys do know it, and banish all their fears,
With a tow, row, row, row, row, row, for the British Grenadiers.

Having received orders from General Braddock to march until 3 P.M., Colonel Gage's advance guard moved forward and with it Captain Cholmley, on horseback, who led an advance battalion of one hundred foot soldiers. As his batman recorded it:

> *So we began our March again, Beating the grannadiers March all the way, Never Seasing. There Never was an Army in the World in more spirits then we where, thinking of Reaching Fort de Cain the day following as we was then only five miles from it. But we had not got above a mile and a half before three of our guides in the front of me above ten yards spyed the Indiens lay'd down Before us. He Immediately dischared his piece, turned Round his horse [and] Cried, the Indiens was upon us.*

Harry Gordon, the British road engineer, had ridden ahead of the grenadiers to choose a route, working his way through the large, well-spaced trees and fallen logs that lay rotting on the forest floor. To his right, the forested hillside rose toward the wooded plateau, while to his left the terrain, more open but brushy, sloped down to the flat bench skirting the river. The guides, about two hundred yards ahead, suddenly rushed back, shouting out the warning. Gordon spotted movement ahead. French troops and Indian warriors. Many of them. Three hundred, he estimated, not more, coming through the forest, crossing down through a ravine that ran down the hillside. George Croghan, in front with the guides, estimated about the same number. What they saw, however, was not the half of it.

The French now spotted the British movement. Three French officers pulled their hats from their heads and waved signals to their men. One officer was stripped to the waist like the Indians and wore a silver gorget around his neck.

Another wave of the hat. The French troops in the center take firing position. A billow of white smoke erupts from their muskets. The blast shakes the foliage. But the shots fall short of the British, the musket balls thudding into the earth and dead leaves.

Across the ravine, British officers on horseback draw swords and wave them in the air, shouting commands to the grenadiers.

Make ready!

The scarlet front rank of British soldiers drops to one knee.

Present!

The front rank raises musket barrels and levels them, aimed slightly down toward the ravine.

Fire!

Another cascade rattles the forest, another white billow, this from the British side. Planted above the ravine, the British have the advantage of higher ground relative to the French troops down in the ravine, although to their right the wooded slope rises above the British troops and to their left drops toward the river. Officers on both sides shout out orders, keeping battle formation, trying to prevent chaos.

From farther back in the advance guard, Colonel Gage rides to the column's front, bringing forward the two-horse-drawn six-pounder cannon. Hidden behind the mass of British soldiers and grenadiers in their orderly rows, the cannon are swung into position, aimed. An officer shouts. The rows of British troops suddenly part, a curtain drawn back to reveal the two heavy barrels staring at the French troops down in the ravine and across the far side.

Boom!

Grapeshot—a swarm of heavy metal pellets—sprays at a velocity of 1,000 feet per second toward the French troops and Indian warriors. Tree trunks splinter, leaves flutter gently to the earth, severed by the demonic power of these European machines of war reaching into the leafy quiet of the Indian forest. A few men stagger. The French troops hunch behind trees. The officer with the naked torso and silver gorget stands in the open, unprotected, waving directions with his hat.

The scarlet curtain of British troops opens again. The forest shakes. Leaves fly.

Boom!

Across the ravine, one hundred French-Canadian militia turn and run, overwhelmed by the whistling cloud of metal balls.

"Every man for himself!" someone shouts.

The tremendous blast terrifies the Indian warriors in the center with the main body of French soldiers. They run back into the forest, too. The French officer with the naked torso and silver gorget urges his troops forward, but the ferocious roar and whistling grapeshot stun those who remain, dropping back ten paces at each blast.

This emboldens the British troops across the ravine, ramming charges down their musket barrels, shoving grapeshot into cannon mouths.

"God save the King!" they shout.

And then a third British volley.

Boom!

The French officer in the silver gorget drops limply, naked torso ripped by grapeshot or musket balls, tumbling into the bottom of the ravine. It is the fearless Captain Beaujeu. The French troops shudder. His next-in-command, Dumas, steps forth, attempts to rally the French troops. They hang back.

Sensing the French reluctance, the British prepare to advance in their orderly red rows, oblivious to movement that stirs on the wooded hillside above their right flank. The crack of musket fire suddenly reverberates from above—not in volleys but individual shots in rapid sequence.

Slipping from tree to tree, hundreds of Indian warriors have crept unseen across the wooded hillside, while hundreds more have infiltrated the brushy slope below the British left flank. On the rising slope to the right, a British flanking party walks into hidden Indian warriors. Musket fire intensifies.

"Fix your bayonets!" commands Colonel Gage, directing the action in the column, sword drawn in one hand, reins gripped in the other.

He orders his grenadiers to prepare to take the hillside above with steely bayonets mounted on musket barrels. They face about, shaping into phalanxes to march up the wooded slope, orderly rows ready to mow down with musket balls or impale with bayonets any living creature who resists their advance. But before they start, an unearthly noise erupts from up in the forest. It is a shrill, inhuman scream, like the shriek of a lone panther in the echoing forest—full of bloodlust, darkness, and violence. It is the death cry of the Indian warriors, jubilant and impassioned, as their scalping knives inscribe the skulls of the flanking party's fallen soldiers. Turning a soldier facedown, the warrior plants a knee in his back, grabs a forelock of the soldier's hair, and in a single motion jerks it upward. Victims who were still alive, scalping survivors have reported, would hear a sound resembling the rumble of distant thunder as their flesh was ripped from their skull.

It's the British turn to shudder. On horseback, sword in hand, Colonel Gage gives orders that his platoons with their mounted bayonets face up the wooded hill and advance to take the high ground. They remain rooted in place. His officers implore the men, furiously trying to set platoons in motion up the wooded hill. They cling to the crude road and the original line of march as if they find safety there, on familiar ground and clustered in a herd amid the other troops. Instead of advancing up the hill, they begin to fire blindly at the woods, not in coordinated volleys as they've been trained, not aiming, simply firing up at the forested hillside. They carry twenty-four shots each in their cartridge boxes. Gage and his officers shout to them, "Don't throw away your fire!" but they keep firing, even though they can't see more than two Indians at once, hidden as they are. Instead, their shots drop their own flanking soldiers who are fleeing through the trees.

Extraordinarily well-aimed musket balls zip down from the hillside and thump into the red masses of British troops in the road. Soldiers groan and fall away, chipped off from the square of red.

Colonel Gage gives the order to fall back fifty or sixty paces, out of the Indian range of fire from the hillside. White smoke puffs from behind tree trunks, mossy fallen logs, little rises of ground. They aim for the perfect targets offered by the British officers sitting tall on their horses and wearing bright scarlet coats—the officers that constitute the nerve system that directs the marching, wheeling, and firing of the great red organism. The officers tumble, one after the other, from horses that themselves crumple along the crude road. Within ten minutes of the first shot that is fired, fifteen of the eighteen officers of the advance guard are killed or wounded, among them the batman's master, battalion commander Captain Cholmley.

* * *

Several hundred yards behind Colonel Gage's advance guard and Sir John St. Clair's axmen, General Braddock, riding with the main body of troops, listens with alarm to the heavy and rapid firing that breaks out at about 2 P.M., the ripping volleys, the boom of cannon. He calls a halt. He orders Colonel Burton to move forward with a large detachment of about five hundred men taken from the main column, keeping another four hundred behind to guard the baggage train—the wagons, cannon, packhorses. He also sends one of his aides-de-camp to the front, riding quickly, to report back. The firing does not stop. General Braddock decides to move forward himself with his staff, leaving Sir Peter Halkett in charge of the four hundred troops guarding the long baggage train.

Colonel Burton's body of five hundred troops separates from the main column and marches four abreast up the Indian path that the axmen and shovelers have just widened into a crude road, angling from the riverbank up the forested hillside toward the gunfire. With this detachment rides General Braddock, saber at his side and accompanied by his staff and aides-de-camp, probably including young Washington, astraddle his cushioned horse. The adrenaline of battle floods through him, despite his weakness. *The French and Indians are upon us!* He knows that at General Braddock's side, he rides near the center of this great historic moment, surrounded by the power and glory of this great empire, its brilliance shining on him.

The drummers again beat out the grenadier's march as Colonel Burton's column proceeds up the hillside trail, banners flying, towing three big twelve-pounder cannon. The cacophony of musket fire ahead grows louder. They see no enemy as they enter the trees. A few men fall. Colonel Burton's troops crowd closer, jamming the orderly rows. The colonel calls a halt. He orders his men to form and turn to their right. They will make an assault on the hillside sloping above them.

"[I]f we saw of them five or six at one time [it] was a great sight,"

Captain Cholmley's batman recorded, "and they Either on their Bellies or Behind trees or Runing from one tree to another almost by the ground."

Ahead in the crude road, the advance guard of Colonel Thomas Gage have abandoned their six-pounder cannon and are retreating, pushed back by Indian fire. The rumor spreads that the main baggage train is being attacked—their ammunition, provisions, personnel, their entire lifeline in the wilds. While Colonel Gage's advance troops and grenadiers are retreating down the trail, Colonel Burton's five hundred men from the main column are marching up the trail to bolster them. These two bodies of troops under Colonel Gage and Colonel Burton collide in the narrow road, piling up like an accordion squeezed together, with Sir John St. Clair's two hundred axmen and shovelers mixed among them. The orderly columns disintegrate into a mass of milling bodies in red and blue uniforms, beating drums and waving silken banners, while the Indian warriors, making a half-moon shape around them, tightening closer, fire from hiding places in the brushy downslope and wooded upslope. Animal fear leaps from man to man.

"The whole now were got together in great confusion," recorded aide-de-camp Orme.

"Nothing afterwards was to Be Seen Amongst the Men But Confusion and Panick," wrote road engineer Gordon.

"The men from what storys they had heard of the Indians, in regard to their scalping; and [tomahawking], were so pannick struck, that their officers had little or no comand over them . . . ," recorded a British officer.

"The yell of the Indians . . . ," wrote another officer, "will haunt me until the hour of my dissolution."

Through the confusion, officers bark commands to ready all the big twelve-pounder cannon for firing. Other officers attempt to form the men into marching order to make an assault on the hillside, or form platoons to fire volleys. But the army surges in a mass in the crude road, no longer four abreast but piled atop each other.

"[T]he whole army was very soon mixed together, twelve or fourteen deep, firing away at nothing but trees . . . ," remembered Colonel Gage.

The forest shakes with the roar of the twelve-pounders, finally opening fire into the brushy ground sloping down on the left toward the Monongahela. Nearly surrounded by the hidden Indians, the British platoons that do manage to form must swivel back and forth to fire on Indians whose musket shots whistle in from right, left, and center. Soldiers burrow for safety into the thick of the milling bodies.

"Men dropped like leaves in *Autumn,*" another British officer wrote.

Sir John St. Clair races on horseback to the front of the column to see what is happening, then back along the confused mass to seek out General Braddock, trying to bring order. Blood soaks through the back and chest of St. Clair's uniform. He has been shot through the shoulder, breaking bone, the bullet penetrating his left breast.

"[I] beg'd of him for God-Sake to gain the rising ground to our Right to prevent us from being totally surrounded," St. Clair later remembered.

Speaking in Italian so the troops will not understand, Sir John tells General Braddock that as he expects to live only a few minutes more he has no reason to lie to the general. He is defeated, Sir John tells General Braddock, and all is ruin and he should retreat. Then he falls, unconscious. His servant ties the slumping St. Clair to his saddle and evacuates him down toward the Monongahela River ford.

A half mile back in the column, Sir Peter Halkett, left by General Braddock in charge of four hundred troops to guard the baggage train, rides boldly back to the column's rear to summon forward a twelve-pounder cannon to defend the train's wagons. Directing soldiers and Royal Navy gunners, he manages to bring the horse-drawn cannon forward to the baggage train and swing into position. As he does so, a musket ball topples him, dead, from his horse. His son, Lieutenant James Halkett, sees his father tumble, and when the younger Halkett goes to help him, he, too, is shot, according to some accounts, falling dead across his father's body.

Ahead, where troops have collided in the column's confused front, General Braddock charges about on horseback with sword in hand and wearing his father's red silk sash, ordering the two regimental colors separated—the yellow silk banner of the 44th to one end of the column and the banner of the 48th to the other—with the idea that this will bring order to the troops. It doesn't. Some of the men, especially the "irregulars," the American colonial soldiers, take to the woods themselves and hide behind trees in order to protect themselves while they shoot, like the Indians, while others run into the ranks, according to one British officer, which puts "the whole in confusion." Some of the American colonial officers try to organize advances up the hill, moving tree to tree, but the British officers disparage this in favor of their orderly marches and firing platoons. The colonial troops who take to the trees also make themselves targets for their own British comrades milling down in the road.

"If any [in the road] got a shot at [an Indian], the fire immediately ran through the whole Line, tho' they saw nothing but Trees," wrote a British officer. "The whole Detachment was frequently divided into several parties, and then they were sure to fire on one another, the greatest part of the men who were behind trees, were either kill'd or wounded by our own People, even one or two officers were killed by their own Platoon."

The Indians, meanwhile, slip through the forest toward the baggage train in the rear and begin to attack it. The wagon drivers, some of them frontier settlers, guess the tilt of the battle. They know what the Indians, if victorious, will do. They cut loose their teams of horses from the wagon hitches, leap onto their backs, and gallop toward the Monongahela ford. Among the fleeing drivers are young Daniel Boone and his cousin, nineteen-year-old Daniel Morgan, future heroes of the frontier and the Revolution.

About a quarter mile ahead, in the middle of the melee in the road, is the general. Bullets zing past. One horse is shot out from under him, then another, and another. Captain Cholmley's batman gives his dead master's horse to the general, cautioning him that a musket ball has lodged in the animal's shoulder and to secure another mount before this one fails. General Braddock charges about on this horse, trying to force order, trying to organize troops to advance up the hill in battle order, shouting, swearing, wielding his sword. He chases down men who move into the trees to fight in the Indian style, striking them with the flat of his sword.

"Cowards!" he shouts.

Imagine that you were General Braddock, and your whole life has led up to this moment. You've been steeped in the British military tradition since youth, with your father in the elite Coldstream Guards, a military career offering your ticket to a good life. You've been brought up on the legends of Marlborough, the duke who, when you were a boy, crossed rivers and forests to confront the enemy and became the hero of epic poems. You've never seen combat although you're sixty years old, and you've been chosen to lead in this moment. You've had visions since your youth of what this moment's heroism would look like, mounted on your horse, waving your sword, wearing your father's red sash from Marlborough, leading the men into battle. And here it is, there is no visible enemy to fight, no enemy army across a battlefield at whom you could order volleys fired, cannon aimed, no commanders to outwit in the strategy of moving men and artillery like pieces on a game board. Instead, trees, forest, wild shrieks, and now the cries of the wounded, the screams of pain, the mushroom puffs of smoke from the woods. Now your men hide behind trees. It's all coming unraveled.

"Cowards!" He beats them forward with the flat of his sword.

Now imagine that you are a soldier, one of the American recruits from the frontier. You have a frontier homestead and a wife and children. You've joined the military for a brief stint to defend your home against the invasion of the French, who, you've heard, threaten to take the Ohio Valley and, who knows, perhaps lands beyond, including your own. You know something about Indians. You know how they attack, with stealth, with surprise, with quick thrusts at their enemy rather than stand-alone battles, and then withdraw. You know how they scalp and torture the survivors. But now the general forces you to stand in the middle of a crude road, unmoving, formed into regular rows to shoot volleys up a hillside at an unseen enemy. They shoot directly at you from behind the thick tree trunks and moss-covered logs. You're nothing but a target. You disobey the general and sprint from the road, ducking, to hide behind a tree, shooting up the hillside. You huddle in a posture that perfectly suits the woods but to the general violates all the manliness and honor of His Royal Majesty's throne.

"Cowards!"

Or you're a Redcoat, a British soldier from England's rolling green pastures, for whom these woods are new and foreign and strange, this endless tangled growth of the New World suddenly crawling with these fierce, shrieking creatures, these savages, these Indians. You've come to fight the French—the incompetent French, whose moral decay and strategic weaknesses on the battlefields of Europe you've heard about—and instead you're caught in a trap by these invisible, skulking creatures of the forest. Men are dropping and screaming and dying all around you. Smoke wreaths the nightmarish scene. There is no way out. Surrender is not an option. The Indians will torture you to death if you surrender. The puffs of smoke shoot from the woods. This will be your fate, to die tortured and screaming and then left to rot in the tangled growth of the New World, where the wild animals will pick your scattered bones.

The general shouts, and commands, and strikes the huddling men with the flat of his sword. If aide-de-camp Washington is at his side in this moment, does he try to restrain the general from striking, to let the men fight from behind trees? Does he offer to lead the men in an Indian-style attack against the Indians, as he remembers years later that he did? He finds himself pulled by conflicting impulses—he knows, or believes he knows, a way to counter this fierce onslaught of Indians. But how can he defy His Majesty's commander, having joined the general's staff specifically to acquire the general's military experience and help in advancing his own career? Challenging his commander's battle tactics might have been a hard point for the young Washington to push with General Braddock.

By now it's 4 P.M. The battle has been joined for over two hours. The artillery is in trouble. The Indians target the trained experts, Royal Army artillerymen and Royal Navy gunners, who rapidly load and fire all the twelve-pounder cannon. General Braddock orders a detachment to defend the gunners. Aides must deliver the order. But most of the general's aides have been wounded or killed—young Shirley dead with a bullet to the head, Orme incapacitated with a bullet above the knee, Roger Morris with a bullet through the cheek. It's left to Washington, still unharmed, to deliver the general's orders, riding about on his cushioned saddle amid the fighting, weak and ill, bullets piercing through folds in his clothing without touching him. He now finds nothing charming in their sound.

As the general sends a detachment to protect the twelve-pounders on the left, he simultaneously orders Colonel Burton to take the wooded hill on the right. Colonel Burton, who has led the detachment of five hundred men forward from the main column, manages at last to form out of the milling, panicky masses about one hundred and fifty grenadiers of the 48th Regiment into battle order to advance up the hill. He gives the command, and they start to march in formation up through the big, widely spaced trees. Two or three of the grenadiers are quickly picked off by hidden Indians above. Colonel Burton himself takes a musket ball in the hip. The entire body suddenly turns of its own accord, while officers shout to advance, and hastens back downhill to the main body milling in the road.

These are the general's last orders. The general's horse goes down again, the fourth horse shot out from under him. He calls for another horse and begins to remount. He suddenly flinches. His left hand grips his right arm and chest. He's been shot. No one knows where the shot originated. He slips from his horse to the ground. Lying there, he remains conscious. His officers charge forward, carrying out his orders to take the hillside and protect the twelve-pounders. They manage to form a body of men again and try to press it forward with purpose, but it collapses back into the milling mass. The puffs of smoke now come faster from the wooded hillside, as if the forest itself senses the column's writhing vulnerability. Officers stagger back from the column's front, wounded. Others don't stagger back at all but lie in a heap in the rough clearing of the road.

Fear surges in the surviving troops. Few officers stand to lead the men. Those who do shout, swear, scream to the men to hold fast. But the milling column starts to move back, like a creature with a thousand legs that doesn't know which way to move, but move it must, as if the heat to the front and to the sides has grown too intense. The horrible shrieks of the Indians intensify on the right hillside. The twelve-pounder cannon have fallen silent, the powerful booms giving way before the shrieks. The creature needs to move. It can't move forward, it can't move sideways, and so it moves back, slowly at first, and then all of a sudden it breaks, and the thousand legs move, the creature breaking into pieces, and those pieces running, and breaking into still smaller pieces.

"[M]any of the officers called out Halt, Halt," reported guide George Croghan.

One segment of the red organism, about two hundred men, holds for a time in good order before they, too, are driven back. Two American companies of South Carolina regulars and Virginia rangers who are guarding the column's rear, behind the baggage train, abandon the orderly, platoon-style formations and take to the trees. They include veterans of the Fort Necessity battle who have witnessed Indian forest fighting and now, like the Indians, shoot from cover at French soldiers pushing down the road and Indians ducking through the woods trying to surround the column entirely—certain to end in the gruesome deaths of all.

The British troops abandon the fallen General Braddock, bleeding into the earth. Washington, his only unwounded aide, stays by his side, along with the general's servants. They unwrap from his torso his twelve-foot-long red silken sash from the Duke of Marlborough and Coldstream Guards, place his body on it, and, with Washington directing, lift him into a small, covered cart, laying him alongside the wounded aides-de-camp Orme and Morris. Hastening past the long line of abandoned wagons, they pull the cart toward the ford and, on reaching the riverbank, quickly lift the general out and with the help of other officers carry him across, either on horseback or on foot, perhaps cradling him in the red silk sash embroidered with the year 1709.

Up the road, the two hundred men retreating in good order, shooting as they go, now fall back to the abandoned wagons. There is a pause. No one fires. The rest of the army has fled toward the ford, including the wagon drivers and the women who travel with the baggage train. The cannon have fallen silent. The Indians remain hidden in the trees. The two hundred gather around the abandoned train of wagons, heaped with supplies, goods, and ammunition.

The silence suddenly breaks with ferocious screaming.

"Our Indians," reported a French officer, "seeing them withdraw toward the tail of the column after having fired, took this movement for a flight, and came out of the woods, war clubs in hand, and made their horrible cries, and fell upon them with inexpressible fury."

Unable to reload quickly enough before the Indians pounce, the soldiers can only thrust with their bayonet-tipped muskets as they are swarmed by Indian warriors. The Indian war clubs, designed for close, high-speed combat, swing through the air and smash into soldiers' skulls, felling them limply. The Indians jerk off scalps with glorious, blood-curdling screams of triumph, holding up the bloody trophies of their bravery. Soldiers run.

"The war cries of the Indians, echoing through the forest, struck terror in the heart of the enemy," wrote another French officer. "The rout was complete. . . ."

"The officers used all possible endeavours to stop the men, and to prevail upon them to rally," recalled Orme, "but a great number of them threw away their arms and ammunition, and even their cloaths, to escape the faster."

Everyone now is running for the ford. Many of the wounded are abandoned, their screams mingling with those of the Indians, as the warriors with their clubs fall on the stragglers. If the soldiers can cross the river, some measure of safety might await them on the far bank. They sprint into the shallows, spray flying in the bright summer afternoon, into water ankle-deep on their gaiters, then calf-deep, and knee-deep, bogging down in the current, tripping over the uneven bottom, the officers' wives, mistresses, and washerwomen dragging their long, billowing dresses. The Indians pause on the high riverbank to aim for the fleeing soldiers taking to the water, the soldiers stumbling at the roar of the muskets, flopping into the current, flailing, sinking, gently swept downstream as their blood dyes the summer-warm water with swirling clouds of red. Those who are able scramble up the far bank and run down the newly cut road two miles toward the earlier crossing of the Monongahela. Others lie gasping on the bank.

Back on the battle side of the Monongahela, fleeing soldiers and personnel clog the ramp sloping down to the water's edge that the shovelers had cut in the twelve-foot-high sandy bank just that morning. Road engineer Harry Gordon, riding a second horse after his first had been shot and with his right forearm dangling after a musket ball had shattered the bone, arrives in the midst of the melee at the "choked" ramp. He sees he cannot get down it. He rides to the brink of the high riverbank and prepares to leap off it, when a chunk of the bank under his horse suddenly collapses, carrying him and his horse down with it to the water's edge. His horse manages to keep its feet and Gordon manages to stay in the saddle. As he and his horse make their splashing escape across the three-hundred-yard-wide river, he hears Indian screams. When he's forty yards out, he pauses to look back and sees behind him Indians tomahawking women in their long skirts and shooting soldiers in the water. Another musket ball now rips through his right shoulder. He spins and urges his horse onward to the far bank, where, in "Utmost pain," he finally reaches the wounded General Braddock and his officers and wounded aides-de-camp, and probably Washington, too, the last of General Braddock's aides-de-camp who remains unharmed and alive. The general needs him to deliver urgent orders. Among them also is Colonel Burton, shot through the hip, who climbs a small knoll and tries to convince the troops to make a stand on this far side of the river.

"Coll: Burton thho' very much Wounded attempted to Rally on the Other Side," wrote engineer Gordon, "& made a Speach to the Men to Beg them to get into some Order, But Nothing would Do, & we found that Every man wou'd Desert us; therefore we were oblig'd to go along."

The morning's glorious river crossing with flags flying, fifes playing, drums beating, muskets and cannon and swords glistening in the early sun, all in perfect order and regimental rows, had now reversed, as if playing itself backward, into chaos and blood.

PART THREE

CHAPTER FIFTEEN

James Smith waited anxiously at Fort Duquesne all that afternoon. The battered eighteen-year-old captive hoped to hear that General Braddock's army had decimated the French and Indian forces. The news was not good for the captive Smith, however. A runner arrived at the fort that afternoon with the message that the Indians, hiding behind trees and in gullies, had surrounded General Braddock and kept up a constant fire. The English were "falling in heaps."

As Smith recorded,

> Some time after this, I heard a number of scalp halloo's and saw a company of Indians and French coming in. I observed they had a great many bloody scalps, grenadiers' caps, British canteens, bayonets &c. with them. They brought the news that Braddock was defeated. After that another company came in which appeared to be about one hundred, and chiefly Indians, and it seemed to me that almost every one of this company was carrying scalps; after this came another company with a number of waggon-horses, and also a great many scalps. Those that were coming in, and those that had arrived, kept a constant firing of small arms, and also the great guns in the fort, which were accompanied with the most hideous shouts and yells from all quarters; so that it appeared to me as if the infernal regions had broke loose.

Some of the victorious Indian warriors, Smith noted, donned full British officers' dress, including half-moon gorgets and laced hats. Some surely wore as trophies the massacred grenadiers' mitre caps—tall and conical, covered in back with a rich red fabric and shimmering gold in the front, emblazoned with the swirling initials GR for Georgis Rex, or King George, and the emblem of a galloping steed that symbolized his House of Hanover, along with the motto Nec Aspera Terrent—Difficulties daunt us not.

While the Indians at Fort Duquesne celebrated wildly, one British survivor of the slaughter, Duncan Cameron, still lay on the battlefield. After taking a musket ball early in the action, he had fallen stunned on the hillside where the battle began and was left for dead. When the British retreated toward the Monongahela River ford in the face of overwhelming fire and tomahawk attacks, Cameron came to consciousness. The woods around him appeared deserted except for the corpses of the British soldiers, the dead horses, the abandoned cannon. He stumbled away from the sprawled bodies and into the forest, where he found a large hollow tree, probably a giant sycamore, which often grow with hollow centers large enough to easily shelter a human. He crawled into the hole and hid.

> . . . I had not been there long before these ravenous Hell-hounds came yelping and screaming like so many Davils, and fell to work [scalping]. . . . About a Foot above the Entrance into the Tree I was in there was a good Foot-hold for me to stand on . . . and against my Face there was a small Knot-hole facing the Field of Battle, by Means of which I had a fair Prospect of their Cruelty: But Oh! what a Pannic was I in, when I saw one of those Savages look directly at the Knot-hole, as I aprehended, gave a Scream, and came directly up to the Tree? But what Inducement he had for doing it I cannot tell, for he went off again without shewing any Signs of his discovering me. The whole Army of the Enemy fell to plundering, &c. Though I must do the

> French Commandant this Justice, that as soon as possible he could, he put a stop to the Indians scalping those that were not yet quite dead, and ordered those Wounded to be taken Care of.

Amid the plunder on the battlefield, the Indian warriors and French soldiers discovered two hogsheads of rum and promptly broke them open. Drunk, with night falling, they abandoned the battlefield and returned to the fort, while young private Cameron "when the Coast was clear . . . got me out of my Hiding-place."

Some accounts stated that General Braddock, lying badly wounded, asked to be left to die on the battlefield alongside the rest of his fallen troops. His request was ignored. After the general was carried across the Monongahela ford, he lay on the far bank with a bullet hole through his breast and calmly issued orders. With the general's other aides-de-camp Orme and Morris both wounded, these were delivered to his subofficers by his only aide-de-camp who remained standing, George Washington.

Bearing the general's orders, Washington galloped off two miles along the flats to the far ford of the Monongahela and delivered the message to officers there: Halt the retreat. He then galloped back to General Braddock, whom he met partway across the two-mile flats, borne on a litter carried by several men. With evening falling, the general now dispatched Washington with another set of orders—these for Colonel Dunbar, commanding the rest of Braddock's army at a camp forty miles to the rear, near Gist's settlement and the Great Meadows. The general ordered Colonel Dunbar to rush forward with provisions, with troops to help defend the flying column's retreat, and with wagons to evacuate the many wounded.

Accompanied by two guides on horseback, Washington rode off into the night. No moonlight illuminated the way—the old moon had waned and the new moon's sliver had yet to appear—but where the forest opened into clearings, or above the river, bright clusters and ribbons of stars littered the skies. In the wooded hollows, however, darkness pooled so thickly that Washington's two guides had to dismount and feel for the road with their hands. Their saddles creaked as they rode through the forest's silence, their horses huffed, the hooves clopped. Across swampy ground, insects trilled and frogs croaked and fireflies blinked yellow in the warm July night. Then mournful human cries would erupt in the darkness where the wounded staggered and crawled and expired along the trail.

"The shocking Scenes which presented themselves in this Nights March are not to be described—The dead, the dying—the groans—lamentations—and crys along the Road of the wounded for help . . . were enough to pierce a heart of adamant," Washington remembered many years later.

With his two guides, Washington rode sixty miles through the night, past the dead and the dying, the wounded and the fleeing. At midmorning the next day, he reached Dunbar's camp pitched atop Chestnut Ridge, midway between Gist's settlement and the Great Meadows. This was as far as Colonel Dunbar's ponderous baggage train had inched forward in General Braddock's push to the Forks. The first battle survivors had already arrived at camp—wagon drivers fleeing on horses they cut free from their hitches—and breathlessly told of the slaughter and the ferocious force of Indians. No one knew if the Indians would pursue the fleeing British, nor how far. Panic rippled through Dunbar's camp. Deserters began to stream out.

Finally resting briefly after the battle and all-night ride, Washington could assess his narrow escape. Ill at the start with fever and the bloody flux, weak and riding on a cushioned saddle, he had rallied his strength and found himself in the thick of the battle, where he had delivered orders for General Braddock under a shower of musket balls before finally helping remove the wounded general from the field. Now, in its aftermath, still weak and ill, he discovered that four musket balls had pierced his coat without even grazing his flesh. Two horses had been shot out from under him. Carnage had surrounded him, but he had escaped, miraculously, unscathed. He thanked Providence for delivering him safely. Like his escape from harm at Jumonville Glen, and again during the intense battle at Fort Necessity, this confirmed for young Washington that he was somehow under divine protection. "[T]he miraculous care of Providence . . . protected me beyond all human expectation . . . ," he wrote to his brother Jack after the battle.

Much would turn on this revelation of Washington's. In the years and decades ahead, he would see this divine protection as a sign that he was chosen for some greater purpose. This conviction helped to make Washington, unflinchingly brave to start, utterly fearless in the face of physical danger and unconcerned about his personal welfare in battle. It was as if he had discovered a special power that, as he understood it, could allow him to accomplish great things.

General Braddock fared not nearly so well. Six men carried him on a makeshift litter first to Gist's plantation and then up the rocky trail of Chestnut Ridge to Dunbar's camp in a flat spot near the crest. Gravely wounded with a musket ball through the lung, he quietly issued further orders for the protection of his troops and the recovery of the wounded, including instructions to leave food along the roadside so the wounded would have sustenance as they struggled to retreat the sixty miles from the battlefield to Dunbar's camp. In defeat, he cared deeply about the welfare of his men. Yet unsubstantiated rumors would persist for decades that one of his own men deliberately fired the bullet that passed through General Braddock's lung.*

At Dunbar's camp, the general could see the hopelessness of the situation, both for himself and his army. In order to make space in Colo-

* This was first reported in the immediate aftermath of the battle by a letter writer who was not present, describing the battle to his son. One Thomas Fausett of Pennsylvania would later claim to have shot Braddock intentionally. According to Fausett's version of the story, General Braddock, wielding his sword, had stabbed Fausett's brother, who was hiding behind a tree, and Thomas Fausett then fired on Braddock in revenge and to save the army from being slaughtered by Braddock's mistaken tactics of fighting in the open. Historians have questioned this story ever since Winthrop Sargent looked into it in the mid-1800s. That the story arose and persisted for decades is nevertheless testimony to the American troops' questioning of Braddock's tactics and his intentions.

nel Dunbar's wagons to pick up the wounded along the trail, the troops first had to empty them of mortar shells, cannonballs, and barrels of gunpowder. The weakening commander wished to leave nothing that would help the enemy. He ordered the ammunition destroyed, mortars and cannon disabled or buried, spare wagons burned, to destroy everything that the troops could not themselves carry in haste. As the general faded, the fires roared, a hundred wagons burning at once. Everything had to go. "[T]he Confusion, hurry and Conflagration attending all this, Cannot be describ'd," wrote one officer of the chaotic scene at Dunbar's camp on Chestnut Ridge.

With the wagons ablaze for lack of horses to pull them, barrels of gunpowder broken open and poured into a spring, and provisions discarded as too heavy to carry, the men now had more food than they could possibly use. They ate nothing but the luxurious hams hauled into the wilderness at such difficulty and expense. "I my self got six or Eight [hams] it beeing as many as I Could well Carry on my Horse," reported Cholmley's batman.

Amid the chaos, amid this collapse of a powerful army in the wilderness, saving only its wounded and its hams from the flames, General Braddock remained a calm point as the foot-borne procession left Dunbar's camp for Fort Cumberland far to the east. Some accounts say he uttered few words beyond issuing orders. George Croghan said that at one point Braddock asked for his pistols, in order to "die as an old Roman"—an honorable suicide after a crushing defeat. According to Captain Robert Orme, Braddock's wounded aide-de-camp, who recounted the events to Ben Franklin, the general was largely silent the first day after the battle. The second day he spoke, reflectively.

"Who would have thought it?" he asked.

He fell silent. The next day he spoke again.

"We shall better know how to deal with them another time."

Then General Braddock died. It was 8 P.M. on July 13, 1755. They had traveled only a few miles east from Dunbar's camp.

With so many of Braddock's officers, and all of his aides except Washington, dead or wounded, it fell on young Washington to conduct

The Washington Family Runs Aground in British Virginia. In the mid-1600s, George's great-grandfather joined an English trading venture and arrived among these estuaries of Chesapeake Bay. He married a wealthy tobacco planter's daughter and became a planter himself.

George's Revered Older Brother Lawrence. The teenaged George Washington greatly admired his older half brother Lawrence, who had sailed aboard a British Royal Navy fleet to lay siege to a Spanish fortress in South America.

George Washington's Boyhood Home at Ferry Farm. By George's birth in 1732, the Washingtons were solid citizens and slave-owning planters of a middling level in the hierarchy of Virginia plantation society, living in this modest homestead. (Artist's rendering based on archaeological evidence; further research by the George Washington Foundation has since modified the house's facade.)

Rules of Civility & Decent Behaviour In Company and Conversation

Washington's Youthful Writings on Self-Improvement. Despite limited formal education, Washington pursued his own self-improvement, laboriously copying out "Rules of Civility and Decent Behaviour," such as Rule #9: "Spit not in the Fire."

Land Surveyor Instead of Sailor. After his father had died and George was fourteen, his older brother Lawrence suggested he go to sea. His mother emphatically rejected the idea, and George learned land surveying instead.

On Barbados, Washington Glimpses a Wider World. Accompanying Lawrence on a sea voyage to Barbados to treat the older brother's consumption, nineteen-year-old George found himself at a Caribbean crossroads of the British Empire, meeting a glittering array of commodores, judges, governors, and merchants.

Longhouses of Delaware Indians. On his winter journey into the wilderness, young Washington encountered Indian longhouses like these and cultural complexities opaque to an outsider like himself.

The Hardships of a Winter Journey into the Wilds. Washington's frontier guides and Indian scouts saved his life several times as they struggled along frozen rivers, chopping and hauling canoes and rafts over shoals of ice.

A Tidewater Virginia Plantation in the Eighteenth Century. Virginia's big tobacco plantations along the tidal river estuaries resembled discrete villages with their own wharves and a "great house" surrounded by kitchens, dairies, workshops, and slave quarters.

Young Washington's First Commander, Governor Robert Dinwiddie. A former Scottish merchant lacking military experience, Virginia governor Robert Dinwiddie aggressively pushed back against France when it claimed the Ohio wilderness as its own.

Fort Necessity, Young Washington's Makeshift Redoubt. Washington's rashness in pushing ahead too fast and too far forced him to hole up in this ramshackle structure, as Indian warriors and French soldiers bore down.

The Surrender of Fort Necessity and Admission of Assassination. Signed by Captain Mackay of South Carolina and Major Washington of Virginia, the document in French admits to the assassination (*"l'assissin"*—see second paragraph, third line, fifth word) of a French officer and surrenders the besieged fort.

Virginia's Fanciful Claims to North America. The British venturers who founded the Virginia colony claimed that its western boundary extended all the way across the uncharted continent to the Pacific Ocean.

Sally Fairfax, Young Washington's Infatuation. Charming, well-read, and beautiful, Sally Cary Fairfax, the wife of Washington's friend George William Fairfax, utterly captivated young Washington.

Apollo Room of Raleigh Tavern. When the House of Burgesses was in session in the Virginia capital of Williamsburg, the Apollo Room served as social hub for balls, dinners, and other events.

Belvoir Manor, Estate of the Fairfaxes. By ancient English title, the Fairfaxes possessed some five million acres of Virginia lands. Lawrence married a Fairfax daughter, and his plantation on the Potomac, named Mount Vernon, neighbored their elegant Belvoir.

General Edward Braddock, Commander of the British Wilderness Assault. After Fort Necessity's surrender to the French, outraged London authorities dispatched General Braddock of the British Royal Army to seize a French fortress in the Ohio wilderness. The ambitious young Washington volunteered as the general's unpaid aide-de-camp.

Sir John St. Clair, "A Mad Sort of Fool." A volatile and unpredictable Scottish baronet, Sir John, responsible for the logistics of General Braddock's two-thousand-person column, threatened to seize farmers' wagons "by Fire and Sword" if they refused to volunteer them.

British Grenadiers, Elite Strike Force of the Royal Army. Selected for their fearlessness and famous for their tall mitre hats, designed to intimidate the enemy, the grenadiers led British troops in assaults on enemy strongholds.

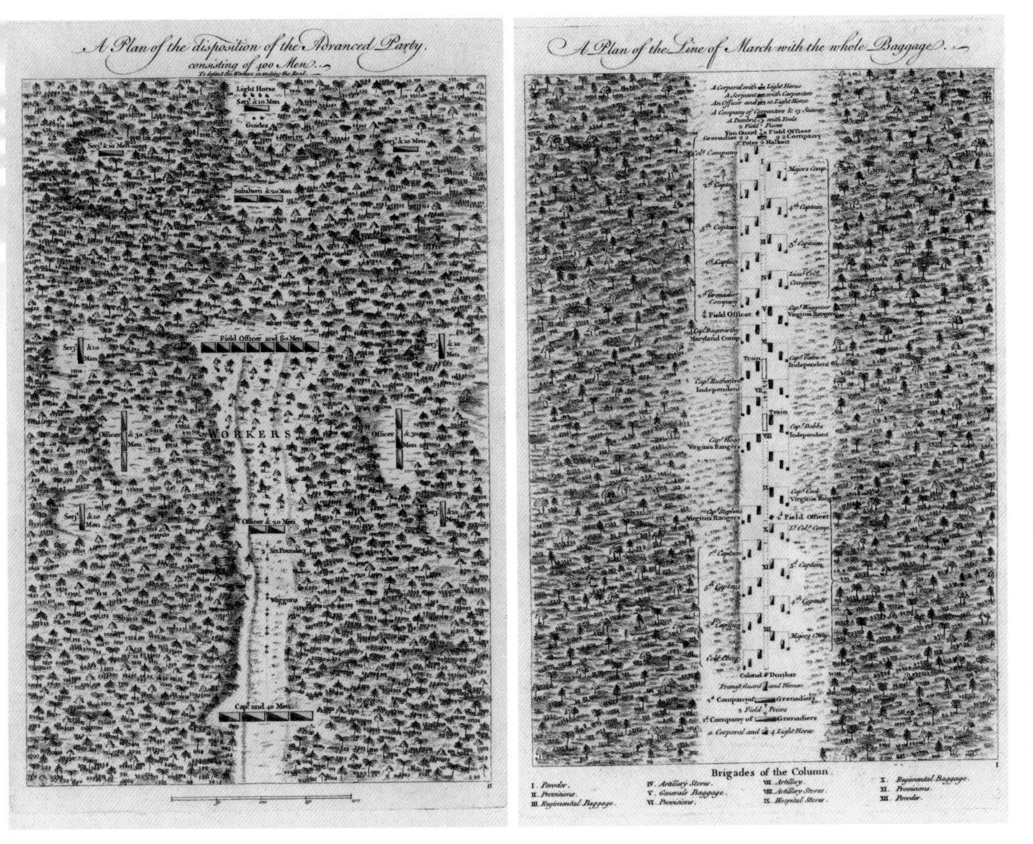

Left: **General Braddock's Advance Column.** General Braddock's column bored into the North American wilderness, guarding itself against surprise attack with bodies of troops flanking the mass of two hundred "carpenters" who chopped the road.

Right: **The Might of the British Empire Marches into the Wilds.** In May 1755, General Braddock's full column marched into the North American forest in orderly fashion—British army Redcoats along with provincial troops and a contingent of women at the rear.

Indian Scouts Watch Braddock Crossing. Indian warriors allied with the French tracked Braddock's army as it hacked its way toward French Fort Duquesne and forded the Monongahela River.

The Wounding of General Braddock. The general had five horses shot from under him as he lashed out with his sword and furiously tried to keep his troops in battle order, while Indian warriors sneaked from tree to tree, surrounding him. He was finally dropped by a musket ball through the chest.

Indian War Club. Ferocious Indian death yells resounded through the forest as warriors fell upon Braddock's flanking parties, dispatching soldiers with musket shots, tomahawks, and crushing blows to the head with traditional war clubs like these.

Indian Warrior Scalping a Victim. Terror engulfed the British troops, who suddenly broke and ran for the river, some of them shedding muskets and uniforms as they went, knowing that this fate awaited stragglers.

General Braddock's Red Silk Sash. Using Braddock's twelve-foot-long silk sash as an emergency stretcher, young aide-de-camp Washington and others hurried their commander from the field and across the river ford.

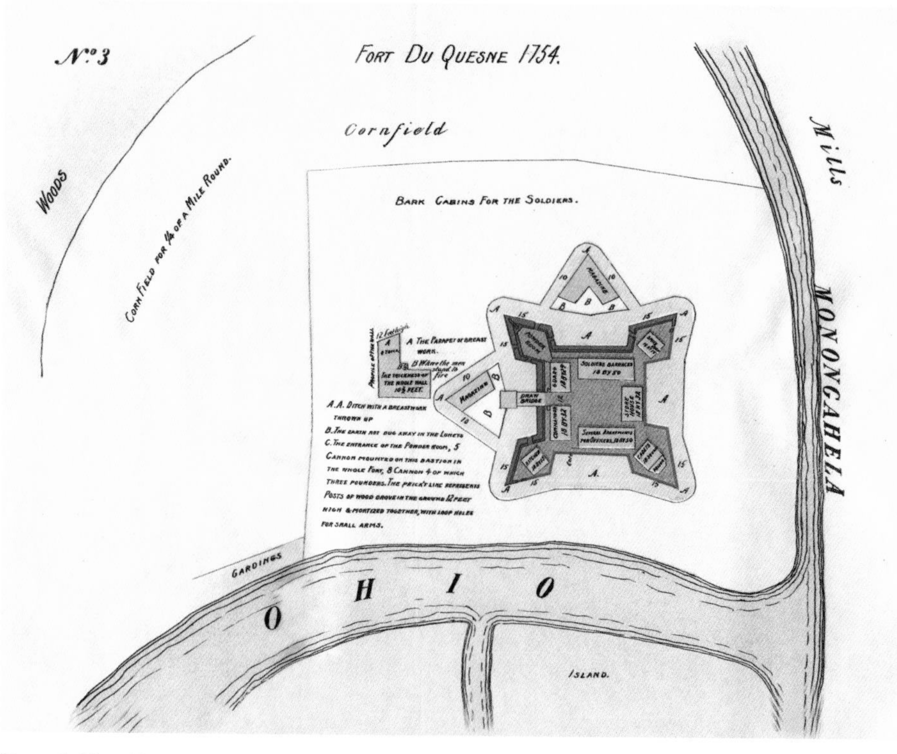

French Fort Duquesne. After Braddock failed to reach it, wave after wave of Indians and French emerged from the wilderness stronghold of Fort Duquesne to massacre and kidnap British settlers who had homesteaded along the frontier and in the wilds.

Washington's Frontier Post at Winchester and Fort Loudoun. As the new commander of the Virginia Regiment, young Colonel Washington made his headquarters in Winchester, in the Shenandoah Valley, and oversaw construction of Fort Loudoun and a string of smaller forts to protect Virginia's long frontier.

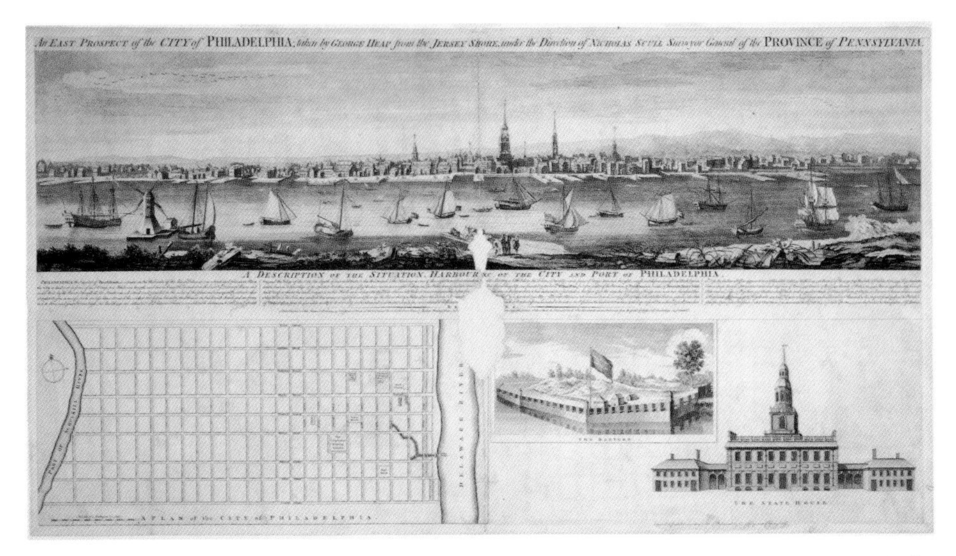

A Journey to Philadelphia. Frustrated by feuds with a fellow officer over rank and his lack of an officer's commission from the British Royal Army, the ambitious Colonel Washington rode to Philadelphia and beyond to petition in person to British Royal Army commanders.

Disciplining the Troops. Beset by desertions, young Washington turned to draconian methods of discipline common in the British military at the time. Deserters could be sentenced to up to fifteen hundred lashes or death by hanging or by firing squad.

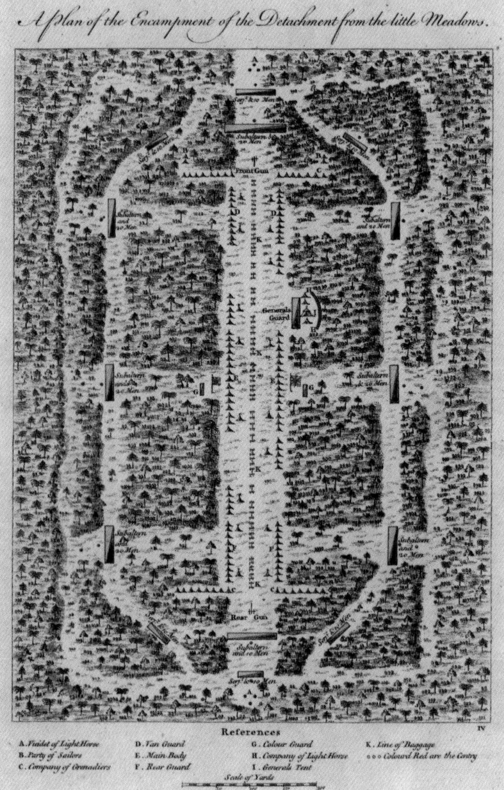

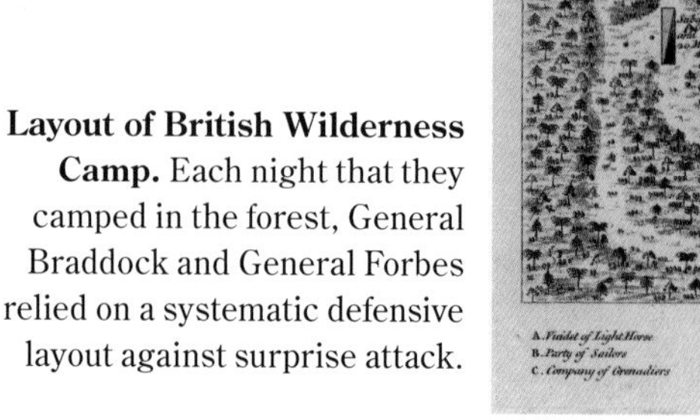

Layout of British Wilderness Camp. Each night that they camped in the forest, General Braddock and General Forbes relied on a systematic defensive layout against surprise attack.

Power, whose Influence he feels and must ever Submit to. I feel the force of her amiable beauties in the recollection of a thousand tender passages that I coud wish to obliterate, till I am bid to revive them.—but experience alas! sadly reminds me how Impossible this is.—and evinces an Opinion which I have long entertaind, that there is a Destiny, which has the Sovereign controul of our Actions—not to be resisted by the strongest efforts of Human Nature.—

You have drawn me my dear Madam, or rather have I drawn myself, into an honest confession of a Simple Fact—misconstrue not my meaning—'tis obvious—doubt it not, nor expose it,—the World has no business to know the object of my Love, declared in this manner to—you when I want to conceal it—One thing, above all things in this World I wish to know, and only one person of your Acquaintance can solve me that, or guess my meaning.—but adieu to this, till happier times, if I ever shall see them.—the hours at present are melancholy dull.—neither the rugged Toils of War, nor the gentler conflict of A—— B——s is in my choice.—I dare believe you are as happy as you say—I wish I was happy also—Mirth, good Humour, ease of Mind and.—what else? cannot fail to render you so, and consummate your Wishes.—

If one agreable Lady coud almost wish herself a fine Gentleman for the sake of another; I apprehend, that many fine

Gentlemen

Washington Confesses His Love for Sally Fairfax. Heading into the wilderness for a decisive battle, Washington made this remarkable confession to the already-married Sally Fairfax: "the World has no business to know the object of my Love, declared in this manner to—you when I want to conceal it."

Martha Washington. Only a few months after declaring his love for Sally Fairfax, the twenty-six-year-old Colonel Washington emerged from the wilderness, quit the military, and married one of Virginia's wealthiest widows, Martha Dandridge Custis.

From Wilderness to Mount Vernon. George and Martha Washington and her two small children settled at Mount Vernon, which Washington had acquired from his brother Lawrence's estate. For the next decade and a half, Washington lived the life of a tobacco planter and family man.

The Middle-Aged Washington in His Old French and Indian War Uniform. Fourteen years after emerging from the wilderness and quitting the military, Washington had his portrait painted wearing his old Virginia Regiment uniform, certainly not imagining that three years later the Continental Congress would name him commander in chief of a young army fighting to create a new nation.

the burial and services. He chose a spot along Braddock's Road where it dipped through a deep, wooded glen. A rivulet trickled in the dank woods—the draining life of a fallen army. He ordered a trench dug in the middle of the road where it crossed through this muddy low spot. A contingent of men lowered General Braddock, wrapped in two blankets, into the hole. Washington presided over the burial. Prayers were read. For fear of attracting Indians, he may have ordered no honorific gun salutes fired—a Royal Army officer of Braddock's rank would traditionally merit three shots from each of seven cannon. Ceremony completed, men shoveled dirt into the muddy hole. Once it was filled, the tattered remains of Braddock's army marched over the unmarked grave, the squishing hooves of horses, the grinding wheels of the remaining wagons, and the stumbling feet of soldiers disguising the burial location with their tracks. If pursuing Indians found it, they might dig up General Braddock's remains and subject him to a final indignity by scalping him.*

By any measure, it was a total, humiliating defeat. Out of 1,469 officers and soldiers in Braddock's flying column, 457 were killed and 519 wounded, for a staggering overall casualty rate of 66 percent, or two-thirds of General Braddock's force. The officers were nearly wiped out—sixty-three killed out of eighty-nine commissioned officers. A large, highly organized force of British Army regulars—the might of the British Empire—had been routed by a band of Indians and French soldiers. The shattered army had panicked, and now fled in retreat. Colonel Dunbar and the remains of the army would not stop retreating until they reached winter quarters in the safe haven of Philadelphia. Other men fled on their own, deserting, the regulars toward the Atlantic seaports that would bring them closer to home in Britain, the colonial soldiers heading for their families in Virginia, Maryland, and North Carolina. A

---

* In 1804, workers widening the old route of Braddock's Road into the National Highway struck human remains in a deep glen. Determined to be Braddock's, these were removed and reinterred on a nearby hilltop, marked today by a stone pillar along the National Pike (U.S. Highway 40) near Farmington, Pennsylvania.

wealth of weaponry, ammunition, wagons, and provisions, brought into the wilderness at enormous cost, had been destroyed.

News of this defeat shook the British Empire to its foundations. The immediate reaction in Britain was to look for some scapegoat to blame for this crushing loss. General Braddock soon came to fill the role. He was arrogant, unkind to his men, blunt and dismissive, conducted a poorly run campaign. He was not suitable to the task. High-flown British commentators across the Atlantic attributed Braddock's loss to moral weakness. American colonists blamed Braddock for not permitting the men—especially the American recruits—to fight from behind trees in the Indian style, and for cudgeling with his sword those who did so. Americans also blamed the British Army regulars for fleeing the battlefield, while it was the American troops—in the colonists' point of view—who stood fast and provided what protection had been offered to the panic-stricken British retreat. Even British authorities praised the American soldiers and noted the difference in fighting styles. "Now we must employ Americans to fight Americans," Newcastle remarked.

Plenty of blame flew on both sides of the Atlantic. But the hero of the battle—if there could be one—was one George Washington. Word spread among Virginians and other American colonists of his bravery in the face of battle, his recovery of the wounded Braddock, and his remarkable escape with bullet holes through hat and coat and horses shot out from under him. "Yor Name is more talked off in Pensylvenia then any Other person of the Army," Christopher Gist would soon write to Washington.

Here began the stories of Washington's bravery. Over the weeks, over the months, over the years, and finally over the decades, his bravery would become legendary. One preacher, giving a sermon five weeks after the battle, already interpreted this bravery as evidence of a divine mission for young Washington: ". . . I cannot but hope that Providence has hitherto preserved [Colonel Washington] in so signal a manner for some important service to his country."

Even the British were impressed. "Who is Mr. Washington?" wrote Lord Halifax. "I know nothing of him but that they say he behaved in Braddock's action as bravely as if he really loved the whistling of Bullets."

Washington himself helped promote the image of a young hero, writing letters in the aftermath of the battle that placed much of the blame for the loss on the regular British troops while praising the officers and the Virginia troops. Five days after General Braddock died, when his shredded army had retreated as far as Fort Cumberland, Washington took the time to write to his mother to let her know he had survived: "[T]he English soldiers . . . were struck with such a panick, that they behavd with more cowardice than it is possible to conceive; The Officers behavd Gallantly in order to encourage their Men, for which they sufferd greatly. . . ."

"The Virginians behavd like men, and died like Soldier's," he wrote to Governor Dinwiddie, who was at first incredulous at the reports he heard of the defeat. "In short the dastardly behaviour of the English Soldier's exposed all those who were inclin'd to do their duty, to almost certain Death; and at length . . . broke & run as Sheep before the Hounds . . . and when we endeavored to rally them . . . it was with as much success as if we had attempted to have stopd the wild Bears of the Mountains."

In all of these accounts, his and others, it was clear where young Washington had stood—bravely in the thick of the action. He had come out of the battle unscathed while so many of his comrades died, some horribly. Would he have felt some form of survivor's guilt? Or did his concern for honor—his own and especially the Virginians'—outweigh whatever guilt he felt? He had performed honorably on the battlefield—there was no doubt of that. But off the battlefield, as General Braddock's aide-de-camp familiar with the Ohio wilderness and its Indian tribes, had he given his commander good advice? He surely turned this over in his mind. He had urged the general to split his troops into a faster "flying column" and a slower baggage column. Perhaps the whole army would have fared better if the general had taken a different strategy—if the two columns had marched together and confronted the French and Indians with a considerably greater force.

Maybe the French would have avoided attacking a larger army. Or would the delay caused by bringing along the baggage train have allowed the French to gather more troops at Fort Duquesne, thus ensuring French victory?

These were all unanswerable questions. Honor in battle was a deeply sensitive issue for Washington. He wanted to know, and he wanted everyone else to know, that he had performed honorably. He cringed at the shame of what was now called Braddock's Defeat. "I doubt not but you have heard the particulars of our shameful defeat, which really was so scandalous that I hate to have it mentioned," he wrote to his brother Jack three weeks after the battle.

Washington no doubt asked himself whether he had warned General Braddock emphatically enough about Indian ambush and fighting methods. The general had brought him along for his knowledge of the wilderness and the Indian tribes. Had he been insistent enough that General Braddock send out enough scouts and beware of ambush? Why hadn't the general secured the high ground after crossing the river the second time? The officers and Braddock were overconfident that the French would no longer ambush, and Washington himself appeared as surprised as they, writing to Governor Dinwiddie that the flying column was attacked "very unexpectedly I must own." Where was Washington with his expert Indian advice then?

Years later, he wrote that he had used "every proper occasion" to warn General Braddock and his officers of the necessity of using a defense that could effectively counter an Indian ambush—in other words, letting the men break ranks and hide behind trees—and that early in the battle at the Monongahela he had offered to lead the provincials in this kind of unconventional counterattack. But he insisted that the general and his officers were so in favor of "regularity & discipline and in such absolute contempt were these people held, that the admonition was suggesed in vain," underlining his assertions for extra emphasis. Washington states this so emphatically that one can't help but think he felt sensitive about it.

After the nightmarish retreat, General Braddock's shattered army and Colonel Dunbar's retreating troops regrouped at Fort Cumberland at Wills Creek. Dunbar would soon continue the army's retreat toward Philadelphia, where he planned to spend the winter, out of action and harm's way. Washington, having recovered from the bloody flux and fever, left Fort Cumberland after a stay of a few days. As an aide-de-camp to Braddock serving on a volunteer basis, he had specifically arranged with the general at the outset of the expedition to leave whenever business called him home. And now business, of a sort, summoned him—the pull of the coast with its lush plantations and gentle tidal estuaries and his own Mount Vernon. Exhausted, battered, having barely survived this foray into wilderness and war, he had had enough.

He emerged over the mountains from the deep woods of the Ohio Valley. It must have come as a rare pleasure after so long in the wilds to see frontier farms with crops growing and animals grazing and to stay at taverns along the way with decent food and perhaps clean sheets. On July 26, just over two weeks after the disastrous and haunting battle, he rode up to his Mount Vernon plantation on the hill overlooking the broad Potomac.

Immediately upon his arrival, the ladies of nearby Belvoir Manor beckoned, clamoring to hear of his heroism. Messengers hurried back and forth between the plantation houses. Young Washington felt in a poor state and needed a night's rest before seeing anyone socially. The ladies threatened to arrive at Mount Vernon on foot if he neglected to call on them at Belvoir the next day, so eager were they to see him. Sally Fairfax wrote on behalf of her friends:

Belvoir, 26 July 1755
Dear Sir,

After thanking Heaven for your safe return I must accuse you of great unkindness in refusing us the pleasure of seeing you this night I do assure you nothing but our being satisfied that our company would be disagreeable should prevent us from trying if our Legs would not carry us to Mount Virnon this Night, but if you will not come to us tomorrow Morning very early we shall be at Mont Virnon.

S. Fairfax
Ann Spearing
Elizth Dent

How different this was from the world he had just left—the world of blood and death and violence, of cannon shots and musket balls, smoke and flame, heart-rending cries for help and blood-curdling screams of victory. The tents and the dysentery, the salt pork and hardtack, the endless smothering forest and the great humped mountains. Here instead lay a world of green lawns, tended flower gardens, crystal and silver, servants and slaves, a world of graciousness and order. How appealing it must have seemed to young Washington, a comfort to wrap around himself. And there were the graceful, beautiful ladies, eager to hear of his heroics.

That next day, July 27, a Sunday, he probably told them his stories, perhaps seated on Belvoir's front portico shaded from the July sun. Parts he no doubt left out—scenes of battle "more easily imagined than described," in the parlance of the day. The ladies would take away a sense of the awfulness of it all and his bravery in the midst of it, and he would go away from this genteel visit with the warmth of adulation and an infusion of female charm and regard. In a timeless ritual that extends across cultures and centuries, the women had sent their men off to war and welcomed them home with open arms, enthralled. It must have been a heady, intoxicating moment for Washington and one whose taste he never forgot.

Though still weak and recovering, Washington spent the next week socializing, checking in on the business of his plantation, and following up with military matters in the aftermath of the shocking defeat. He wrote a long letter to his brother Jack, now in Williamsburg, to whom he had left the management of his properties. Surprisingly for a newly minted hero, he complained at length to Jack about the sacrifices he had made in money and health for his military service, only to suffer defeat and humiliation, as well as the lingering bitterness of having lost his colonel's rank in the Virginia military the previous fall when Governor Dinwiddie put him on a par with other Virginia captains.

> *I was employ'd to go a journey in the Winter (when I believe few or none woud have undertaken it) and what did I get by it? my expences borne! I then was appointed with trifling Pay to conduct an handfull of Men to the Ohio. What did I get by this? Why, after putting myself to a considerable expence in equipping and providing Necessarys for the Campaigne—I went out, was soundly beaten, lost them all—came in, and had my Commission taken from me. . . . I then went out a Volunteer with Genl Braddock and lost all my Horses and many other things. . . . I have been upon the loosing order ever since I entered the Service. . . .*

Why would a hero complain so? With rumors circulating that he might be called to serve again, he had no desire to suffer more humiliating defeats compounded by personal financial loss. Already in Williamsburg an alarmed Governor Dinwiddie was calling the House of Burgesses back into session to raise more troops. Young Washington's name was bandied about as possible commander of these Virginia forces. One correspondent in Williamsburg wrote that "scarce any thing else is talk'd off here" but Washington's fine conduct at Braddock's Defeat. It floated in the air that Washington would be asked to accept the command. It looked to him, however, like a losing proposition. Exhausted, battered, ill, and short on funds, Washington did not wish to lose whatever measure of precious honor he had acquired from the terrible battle. "I believe . . . that no Man can gain any Honour by conductg our Forces at this time," he wrote to his planter friend Warner Lewis, who had encouraged him to take command. ". . . I shou'd loose what at present constitutes the chief part of my happiness, i.e. the esteem and notice the Country has been pleased to honour me with."

A cynic might say that young Washington put his personal reputation above service to his country—meaning the province of Virginia—and the British Crown. So unlike the selfless Washington of later decades, this young Washington's personal ambition outweighed his dedication to the common good. Washington refused to offer himself for service. While he had not officially been asked, he told acquaintances he would agree to serve again only if asked and only if certain conditions were met. He would insist on naming his own officers because he had seen how a commander's reputation could ride on the bad behavior of those below him, and he did not want to become the scapegoat for a crushing loss, like General Braddock. He would demand certain luxury items he considered necessary for an officer on a campaign, such as a small, portable chest of drawers, known as a military chest. He also pointed out that any future campaign would move very slowly across the mountains due to the difficulty of hiring local wagons and drivers, who still had not been paid for the loss of their wagons and horses during General Braddock's disaster. The commander of such an "indolent" campaign would likely meet with, in Washington's dire phrasing, "approbrious abuse."

Yet he also made clear to Lewis that, while he did not want to offer himself, he would consider the position if it were offered to him and certain conditions were met. Because he could so easily walk away from it, this allowed him to be coy and put him in a strong negotiating position. This ability—to take advantage of the situation that presented itself—marked young Washington from an early age. And from an early age he was offered positions of great responsibility. It is ironic that he placed himself at the epicenter of what would be pivotal moments in history yet at the same time begged off, striking a modest, humble note, whether real or feigned. "Let them not be deceiv'd, I am unequal to the Task, and assure you it requires more experience than I am master of to conduct an affair of the importance that this is now arisen to."

His mother opposed the idea. But in a letter to his mother, unlike his other missives, he struck no modest tone. He wrote to her that it would "dishonour" him if he refused to heed a general call for his service, as if honor were a finite thing, a treasure to hoard, and he did not want to squander the share he had earned. If the terms were right, and the call loud and clear, he would accept.

WHILE TWENTY-THREE-YEAR-OLD GEORGE WASHINGTON entertained Sally Fairfax and friends at Belvoir with his war stories, eighteen-year-old James Smith, deep in the Ohio wilderness, received his own warrior's reception from women. Smith left Fort Duquesne with his captors, the Susquehannock. After the battering of the gauntlet, he still could not walk well. His captors carried him by canoe to an Indian town on the Muskingum River inhabited by Susquehannock, Delaware, and Mohican. There, a Susquehannock man who had rubbed ashes on his fingers for extra grip pulled out Smith's hair, lock by lock, "as if he had been plucking a turkey." The small patch of hair left on top of his head was braided into a topknot and affixed with silver brooches and feathers. They bored holes in his ears and nose and fitted the apertures with earrings and a nose jewel. Ordering Smith to strip off his clothes, they gave him a breechclout to wear, painted his body, and wrapped silver bands around arms and wrists and a string of wampum around his neck. An old chief then led him out into the village and called out. The inhabitants gathered around, while the old chief held Smith by the arm and gave a long speech. Smith had seen how the Indians had shown no mercy at Braddock's Defeat; he was sure he was about to be executed. Three women took him to the riverbank. Smith resisted mightily as they tried to immerse him. The tribe, which had gathered on the shore, laughed, until one of the women spoke to him in broken English: "No hurt you."

"[O]n this I gave myself up to their ladyships," he remembered, "who were as good as their word." With this immersion into the Muskingum River, Smith realized he had been baptized into the Susquehannock tribe. He would replace a lost warrior.

"My son," said the old chief, as Smith later recounted the event in his memoirs, "you have now nothing to fear, we are now under the same obligations to love, support and defend you, that we are to love and defend one another, therefore you are to consider yourself as one of our people."

At a feast that evening of venison and boiled green corn, the warriors performed a dance—one that would have grave repercussions for young Washington. Smith described how they moved to the beat of a tambourine-like drum and sang. "On this the warriors began to advance . . . like well disciplined troops would march to the fife and drum. Each warrior had a tomahawk, spear or war-mallet in his hand, and they all moved regularly towards the east, or the way they intended to go to war. At length they all stretched their tomahawks towards the Potomack, and giving a hideous shout or yell, they wheeled quick about, and danced in the same manner back."

The next morning, reports Smith, the warriors assembled in the village wearing packs on their backs and set out, at first silently, then with sporadic reports from their guns, and finally with the traveling song. "*Hoo caughtainte heegana,*" it began.

It was now early August. They were headed toward the frontiers of Virginia, some two hundred miles away through the deep forest and mountain—to the same frontier that Washington would pledge to defend.

By the end of August, Washington had come to terms with doing another stint of military service after various noteworthy Virginians urged him to take on the task, even though he was convinced, he later said, that it would result in loss of honor. "But the Sollicitations of the Country overcame my objections."

He rode into Williamsburg on August 27, stopped for a hair touch-up at the town's "French Barber," and called on Governor Dinwiddie, accepting the commission with the rank of colonel to lead the Virginia Regiment. He then rode northward again from the big brick Governor's Palace through the Tidewater flats toward the mountains and forests of the frontier. His destination was Fort Cumberland at Wills Creek. He planned to stop at towns along the way to check on recruiting efforts. Funded by the House of Burgesses, the regiment was to add another one thousand men to bring its total force to twelve hundred, with sixteen companies, each led by a captain, based out of Fort Cumberland. His commission stated that as the regiment's new commander, Colonel Washington was responsible for the defense of His Majesty's colony and the Virginia frontier "& for repellg the unjust & hostile Invasions of the Fr. & their Indn Allies. . . ."

His difficulties began at the first stop, Fredericksburg, on the Rappahannock River about one hundred miles north of Williamsburg and near his mother at Ferry Farm. Detailed instructions from Washington specified that each of the Virginia Regiment's captains should recruit thirty men, each lieutenant eighteen, and so on. These should be fine physical specimens, none of them younger than sixteen or older than fifty, under five feet four inches tall, having old sores on their legs, or subject to fits. The recruits were to be gathered by October 1 and meet at a rendezvous place. But at Fredericksburg, Washington discovered that no volunteers had stepped forward. Instead, the authorities had drafted local vagrants by force. The draftees had resisted so strongly that they had been thrown in jail to keep them, but their friends had stormed the building and sprung them. So went the first recruiting efforts.

The next problem arose when he learned that the supplier of provisions to the troops at Fort Cumberland, one Charles Dick, had not been paid by the colony of Virginia. He now threatened to cut off the regiment's supplies. Washington advanced him some of his own money. When he reached his next stop, Alexandria, along the Potomac, he found that no new recruits had stepped forward from there, either. He also had to find a way to provide shoes, shirts, and stockings for the battered Virginia survivors of Braddock's Defeat who had remained at Fort Cumberland after the long retreat, or else the soldiers would not be able to train in winter. Within a week after he took command, writes historian Freeman, Colonel Washington had discovered that he had "No men, no discipline, no clothing, no organization, no money. . . ."

Riding on toward Fort Cumberland, he now left the coastal plain, crested the Blue Ridge, and dropped down to Winchester, the growing town in the Shenandoah Valley near the frontier's edge. During these long rides he mused about how to organize, train, and equip an army. At each stop, he wrote out orders to his officers and memoranda to himself—to teach the men the new platoon way of exercising, to have the men cook in their barracks, to practice target shooting in order to learn the skills of bush fighting.

Leaving Winchester, he arrived on September 17 at Wills Creek and Fort Cumberland. Only two months had elapsed since the horror of Braddock's Defeat and the long, nightmarish retreat. The retreat had finally ended right here, at Fort Cumberland. Since then, however, Colonel Dunbar had gone on to Philadelphia and into winter quarters, taking with him the shreds of Braddock's regular British Army Redcoats sent from Ireland—the late Sir Peter Halkett's 44th and his own 48th Regiment of Foot. This was a shameful withdrawal, to go into safe winter quarters during the summer, complained the Virginians. Of the colonial troops who had survived the debacle, many had deserted. But still about two hundred of these provincial soldiers had stayed at Fort Cumberland to fight another day instead of returning to their homes in Virginia, North Carolina, or Maryland.

As an aide-de-camp, young Washington had learned about the proper organization of an army at General Braddock's side. Promoted because of his dauntless performance at Braddock's Defeat, but at age twenty-three younger than many of his troops, he now attempted to bring order and discipline to the rough-and-tumble of Fort Cumberland. He asked for inventories of food, arms, and equipment. He made lists of officers and gave promotions. He organized recruiting efforts. He prohibited a soldier from selling liquor to the troops. He corresponded with Governor Dinwiddie about procuring supplies.

He wrote notes to himself:

> Have the Arms all clean'd.
> Not to enlist Felons.

He studied inventories:

> 1000 Barrls Flouer
> 50 Barrls Beef
> 27 Tun of Codd Fish

He banned "Swearing, getting Drunk, or using an Obscene Language" and, ever conscious of appearance to the point of vanity, designed a resplendent uniform for his officers and himself: "the [blue] Coat to be faced and cuffed with Scarlet, and trimmed with Silver: a Scarlet waistcoat, with silver Lace, blue Breeches, and a silver-laced Hat. . . ."

Such were the minutiae—of supplies and logistics, discipline and recruiting—that preoccupied young Washington now and in the months to come. He was attempting to recruit and train a regiment of over a thousand men. In the upcoming six months alone, he would write and receive some three hundred orders, memos, and letters, most of them about the Virginia Regiment: about recruits, muskets, powder, ammunition cartridges, powder horns, spare ramrods, tents, shovels, axes, leather buckets, wagons, horses, beef cattle, salt to preserve the beef, barrels to hold it, kettles to cook it, flour, the consequences of desertion, death by hanging, and many more about frontier defense itself. This exercise would prove invaluable for young Washington in learning to equip and train an army—an education that would come to fruition two decades in the future. But at the time, it was a task fraught with frustrations. It proved nearly impossible simply to find enough recruits, not to mention provisions, pay, or arms. Much more difficult still would be how to use the troops effectively to counter the dark, terrifying threat emerging from the Ohio wilderness.

Through September, Washington attempted to regulate life at Fort Cumberland, helped by his officers, such as Adam Stephen, the Scotsman and former Royal Navy doctor now made a lieutenant colonel in the Virginia Regiment. Washington also toured a frontier outpost one hundred and twenty miles to the south known as Fort Dinwiddie. In early October, he again left the mountains for the coast. He would head back to the Governor's Palace in Williamsburg to give a firsthand report to Governor Dinwiddie and the burgesses on the state of things on the frontier. The Virginia Regiment needed more of everything, starting with money and recruits.

After several days' ride, he was just short of Williamsburg and the Governor's Palace when an express rider caught up to him and delivered an urgent message. It was from Colonel Adam Stephen, his top officer back at Fort Cumberland.

> *Sir,*
>
> *Matters are in the most deplorable Situation at Fort Cumberland—Our Communication with the Inhabitants is Cut Off. By the best Judges of Indian Affairs, it's thought there are at least 150 Indians about us—They divided into Small parties, have Cut Off the Settlement of Patersons Creek, Potowmack, Above Cresops. . . . They go about and Commit their Outrages at all hours of the day and nothing is to be seen or heard of, but Desolation and murders heightened with all Barbarous Circumstances, and unheard of Instances of Cruelty. They Spare the Lives of the Young Women, and Carry them away to gratify the Brutal passions of Lawless Savages. The Smouk of the Burning Plantations darken the day, and hide the neighbouring mountains from our Sight. . . .*

The "Outrages" had begun. The raids launched by tribes deep within the Ohio wilderness had reached their targets on the upper Potomac and other rivers. The warriors from James Smith's village were among them. Unburied bodies of women, men, and boys lay scattered about burned homesteads. In one cornfield, soldiers discovered three bodies, struck in the skull with stakes, then half-burned in a fire. On another homestead, neighbors had buried the murdered master only to find that wolves had uncovered the body and eaten part of it. Farther south, it was just as bad. On Greenbrier River, Indians had raided a makeshift fort where settlers had taken refuge, killed about twelve settlers, captured two girls, burned eleven houses, and made off with hundreds of horses and cattle. At Penn's Creek, on the Pennsylvania frontier, they captured young Marie Le Roy and Barbara Leininger, leaving Marie's father with two tomahawks protruding from his head, lying feetfirst in the doorway of the burning Le Roy cabin.

Here would begin the truest test and most profound shaping of young Washington's character. He had proved his physical courage in battle but now would face a more subtle and psychological test. He bore personal responsibility not only for the welfare of his men but for the lives of hundreds and hundreds of settlers—men, women, and children—who had carved out homesteads along the Virginia frontier. Driven onward by his relentless ambition to succeed and his exalted need for personal honor, he would come to a place where frustration and defeat overwhelmed him. It was through this difficulty and these crises during the Outrages and events thereafter that young Washington began the transformation from vain, ambitious, self-absorbed youth to the mature and selfless leader he would become.

UPON RECEIVING ADAM STEPHEN'S URGENT MESSAGE, Washington turned his horse around. He sent messages to the Tidewater towns of Alexandria and Fredericksburg, summoning all possible recruits to fight on the frontier. He then rode quickly back over the Blue Ridge to Winchester—the last settlement before the frontier. Here, in this town of sixty or so cabins and houses strewn along a long main street, he discovered that panic had taken hold. Settlers had fled from outlying farms into what they hoped was the safety of the town. Many of those already in Winchester, however, were fleeing farther east, over the Blue Ridge, back to coastal Virginia, where they hoped they would remain beyond reach of the Indian attacks.

Washington planned to pursue the Indians with reinforcements from the local county militias but discovered that he did not have authority over them. The members of the militia "absolutely refus'd to stir, choosing as they say to die with their Wives and Family's," he wrote. Besides militiamen who would not leave their homes, he likewise found that he could not impress—or take for military use—the horses or wagons of the settlers except by force, and even then the settlers threatened to "blow out my brains." Writing of these difficulties to the governor, he threatened to resign unless the Assembly took action to give him more authority over the militia. "I must with great regret decline the honor that has been so generously intended me. . . ."

He had served as commander of the Virginia Regiment for barely a month, and already he threatened to quit. It was not the first time he threatened, nor would it be the last. A pattern had begun to emerge—one that perhaps did not escape the notice of Governor Dinwiddie. Washington did not want to go forward because he felt doomed to failure in the present circumstances, and failure would reflect poorly on him. His vanity outweighed the common good.

Besides needing more troops and more authority, Washington knew he needed the help of Indians in order to fight Indians. He wrote, for a second time, to Andrew Montour, the multilingual trader who was three-quarters Indian and who wore braided brass wire in his ears, a ring of bear grease around his face, and a fancy red damask waistcoat. "I desire you will bring some Indians along with you; . . . They shall be better used than they have been, and have all the kindness from us they can desire," he wrote in the first letter. In the second, growing more desperate for their help, he struck a note of obsequiousness and flattery, urging that Montour come at once with his followers: "I was greatly enraptur'd when I heard you were at the Head of 300 Indians on a March towards Venango. . . ."

The raids subsided. The Indians melted back into the Ohio wilderness, at least for the time being, allied with and encouraged by the French and sending a brutally clear message to the British: stay out of the Ohio.

It was now mid-October. Washington had served as commander for about six weeks. The Indian raids had killed about seventy settlers, burned homesteads, and caused hundreds to flee. Washington was learning that the Virginia frontier embraced a vast stretch of territory some three hundred miles long of forest and mountain, river and glade, punctuated here and there by the first cabins of settlers and their stump-filled fields. It represented the colonial equivalent of today's mountainous Afghani border, broken by countless hidden trails through which the enemy could advance unseen. Highly adept at moving quickly and silently, the enemy appeared suddenly and just as suddenly disappeared, far more nimble than conventional soldiers. Washington and the Virginia Regiment, however, had their own version of special operations in the men known as rangers. Initially trained by friendly Indians in the native techniques of warfare and woodcraft, these soldiers ranged through the forest as scouts and to counter attacks by hostile parties of Indians.

When the raids subsided in mid-October, young Washington again left Fort Cumberland and rode hard toward Williamsburg. Governor Dinwiddie had called the Virginia Assembly into session to strengthen the military laws; he wanted Washington present to explain the need. But Washington had another compelling reason for going to the capital. His power at Fort Cumberland, he had recently discovered, had been usurped by one Captain John Dagworthy, of Maryland. Again putting his personal reputation and pride above the common good, Washington threatened to resign unless the matter was resolved.

The problem centered on the indisputable fact that Captain Dagworthy, while a Marylander, had his officer's commission directly from the British Royal Army—known as a King's commission. Colonel Washington's commission came from Governor Dinwiddie and his Virginia provincial forces, as did Lieutenant Colonel Adam Stephen's, who served under Washington and was temporarily in charge of Fort Cumberland. While a colonel usually would outrank a captain, in this case the regular army commission of Captain Dagworthy outranked the provincial commission of Colonel Washington. Nevertheless, questions lingered about the legitimacy of Dagworthy's claim to power, whether Dagworthy's royal commission, now almost ten years old, had expired, and whether he should be considered a colonial officer instead of a Royal Army officer because he had been sent on this expedition by the governor of Maryland and not the king.

Washington arrived in Williamsburg on November 3 as the House of Burgesses and Governor's Council were in full swing, along with the capital's social season. Between meetings with the governor, burgesses, and councilors to strengthen the military laws, Washington attended balls and dances. One can picture him, tall, athletic, not given to brilliant conversation, perhaps even awkward or quiet, but a graceful mover and a favorite among the ladies in the swirl of the Apollo Room's dance floor for his widely touted heroism at Braddock's Defeat. Some reports circulating in London even claimed, erroneously, that eight hundred Virginians had held off sixteen hundred French and Indians for three hours after the regular British Army soldiers fled.

"The Accounts of Our Behaviour is much Exaggerated," wrote Adam Stephen to Washington. "[W]e must give them Credit—and pay the publick the Balance next Campaign."

"We have heard of General Braddock's Defeat," wrote Joseph Ball, Washington's uncle, from London. "Every Body Blames his Rash Conduct. Every body Commends the Courage of the Virginians and Carolina men. . . ."

"[A]ll the Publick Prints, private Letters, and Gentlemen from England say; that the behaviour of the Virginia Troops is greatly extold, and meets with public praises in all the Coffee Houses in London," wrote Washington to Adam Stephen.

A hero in so many eyes, young Washington found it doubly difficult to take a position beneath the insistent Captain Dagworthy. Subordination to Dagworthy did not square with Washington's image of himself as somehow protected by Providence and destined for greater things. Rather than dutifully knuckling under and following institutional protocol, he complained—loudly.

Washington's many complaints about Dagworthy finally prompted Governor Dinwiddie to write to Governor Shirley of Massachusetts, acting commander in chief of British forces in North America after the death of General Braddock, requesting that Washington be given a colonel's commission in the British Royal Army, and Adam Stephen one for a lieutenant colonel. This would provide Washington with a clear rank above the usurper. Captain Robert Orme, Washington's friend and fellow aide-de-camp to General Braddock, who did have a King's commission and was returning to England having survived his wounds from Braddock's Defeat, wrote to Washington that if he did not get the royal commission, "I think Mount Vernon would afford you more Happyness."

Washington impatiently waited for a reply from Shirley while attending meetings, dances, and dinners at the town's taverns along Duke of Gloucester Street. In the meantime, the Assembly met in the big brick building at the street's far end and responded to his wishes by giving him some disciplinary teeth to enforce orders, including the death penalty for deserters, as in the British Royal Army, and lifting the limit of twenty lashes for disobedience, profanity, and drunkenness. In mid-November, Washington left the Williamsburg swirl to head back to Fort Cumberland. But he could not make himself return to this base of operations at the foot of the mountains and edge of the wilderness. Captain Dagworthy held forth there. "... I can never submit to the command of Captain Dagworthy," he wrote to Governor Dinwiddie.

He also found it difficult to part with the social pleasures of the Tidewater region for the spartan life of a remote fort. He claimed that duty—recruiting more troops and overseeing supplies—kept him on the civilized side of the mountains. The recruiting, however, went very slowly. After much supposed effort, ten officers brought in only twenty new recruits. By early December, three weeks after leaving Williamsburg, Washington had proceeded no farther than the comfortable town of Alexandria on the Potomac. Over the mountains at Fort Cumberland, Captain Dagworthy stubbornly refused to yield, and no news had yet come from Governor Shirley about a possible King's commission for Washington.

Washington remained in Alexandria in early December, leading up to the December 11 elections for Virginia Burgesses. His friend and former traveling companion George William Fairfax—Sally's husband—was running for a seat from Fairfax County, the location of Alexandria, and young Colonel Washington campaigned for his friend. He apparently got into a fierce argument with a supporter of a rival candidate, who insulted him in some way. According to legend, the rival supporter hit Washington with a stick. Instead of challenging the man to a duel, the story goes, Washington later apologized for being in the wrong—a story held up decades later as proof of Washington's ability to control his temper and his pride. In reality, however, both were still a work in progress.

He showed no such flexibility in the Dagworthy matter, adamantly refusing to recognize the captain's authority but not daring to defy it directly by returning to Fort Cumberland. Should he disobey Dagworthy's orders, he ran the risk of defying one of the king's officers and could suffer the consequences of insubordination. He avoided Fort Cumberland instead. He wrote to Governor Dinwiddie that he saw no reason to return to Fort Cumberland, arguing that Indians would not attack during the winter, as snow blocked the mountain passes. Rather than returning to the outpost, he wrote to the governor, he could serve more usefully "riding from place to place," arranging things and forwarding provisions to the fort. This restless riding from place to place would mark much of Washington's early command—not always with favorable results.

CHRISTMAS OF 1755 found young Washington at Winchester, in the Shenandoah Valley over the Blue Ridge, where he had ridden after leaving Alexandria. He set up an office in a small log cabin near the town's center, or so tradition maintains about the still-standing structure. His communications during the holidays lacked all sentiment whatsoever and were preoccupied with two men who had deserted on Christmas Eve. He gave orders that soldiers who strayed more than a mile from town would be treated as deserters. "Any Soldier who shall desert, though he return again, shall be hanged without mercy," read the less-than-cheery orders for December 25.

A happier time was had by the officers of the Virginia Regiment at Fort Cumberland, under Adam Stephen, about eighty miles to the northwest by crude road. He hosted a dinner for the Virginia Regiment's officers—"24 fine Gentlemen."

> *We had an extreamly good dinner, and after drinking the loyal Healths, in a Ruff [drum roll] and Huzza at every Health we pass'd an hour in Singing and taking a Cheerful glass. We then amus'd ourselves with acting part of a Play, and spending the Night in mirth, Jollity and Dancing, we parted very affectionatly at 12 O'Clock, remembering all Absent Friends.*

It's hard to imagine the young, intensely self-conscious George Washington taking part with his fellow officers in something so lighthearted as a play, which often involved males playing female roles, although, at least in his later years, it was said he liked to relax with his subordinate officers by tossing a ball with them, known as he was for having a powerful throwing arm. Here was a leisure activity at which he could excel.

The holidays also brought news from farther north. Braddock's ill-fated campaign against Fort Duquesne represented only one of the four prongs of the British thrust against the French in North America during 1755, and word arrived that the other three prongs—against the French forts at Niagara and on Lake Champlain and against the French Acadians in Nova Scotia—were having decidedly mixed results. Meanwhile, by sailing ship, France had transported thousands more troops across the Atlantic to protect its North American holdings.

Still, by the holidays of 1755, neither Britain nor France had declared war on the other. A state of peace officially held between the two empires. But the clash of empires over the disputed lands of the Ohio, sparked by the opening shots fired by young George Washington at Jumonville Glen, had taken on a momentum of its own.

In London, readers of the poetry in *Gentleman's Magazine* learned that General Braddock had not died in vain, predicting that he would be known through the ages as an exemplar of British valor and that his voice from beyond his unmarked grave urged the loyal sons of Britain to take revenge.

> Then 'tis decreed—the vain exulting Gaul,
> In these ill-starred fields beholds my fall.
>
> To women leave the tribute of the tear:
> A brave revenge alone becomes the brave,
> A brave revenge these dying heroes crave.

CHAPTER SIXTEEN

As the year turned, young Washington remained preoccupied with Captain Dagworthy. He did not leave Winchester, as if he had set up a separate command there. He did not address or challenge Dagworthy directly but worked through Adam Stephen, his own man at Fort Cumberland, instructing him to sound out Dagworthy. Dagworthy refused to budge. He claimed that his decade-old King's commission gave him rank above Colonel Washington and other provincial officers and thus command over Fort Cumberland and all its troops, including the Virginia troops there.

Frustrated, Washington wrestled with disciplinary matters and desertions in this secondary post at Winchester, an early shaping of his leadership style when dealing with his officers. During a card game of Brag, an old British betting game that resembles poker, played among gentlemen officers at Winchester, Ensign Dekeyser was suspected of cheating. One player, Lieutenant Bacon, later testified that he spotted the nine of diamonds hidden under the thigh of Ensign Dekeyser. "[Lieutenant Bacon] then told the Gentlemen that there was a Scoundrel in Compa. who acted under the Character of a Gent. but did not deserve the Name or to rank with them."

The card game broke up. When Ensign Dekeyser stood up from the table, the nine of diamonds dropped to the floor. He reached down to pick it up and claimed he must have dropped it earlier while dealing, saying that he had been drinking heavily. But other witnesses later testified that he had not been drinking, unlike the other officers at the table.

It was an example of life at a military outpost and time hanging heavy. A court of inquiry comprised of fellow officers found Ensign Dekeyser guilty of behaving in a "scandalous Manner such as is unbecoming the Character of an Officer & a Gentleman."

The following day, January 8, Washington addressed his officers, first suspending Dekeyser and then warning the other young officers not to follow his example but instead to be inspired to do their duty. "Remember, that it is the actions, and not the commission, that make the Officer—and that there is more expected from him than the Title."

Urging his officers to reach for the highest standards, this showed the beginnings of inspirational leadership, even though it came from an officer so deeply concerned about his own commission and title. Yet Washington also injected a note of humility in his address when he pledged to his officers to be as impartial in his judgment as he was capable, as if to recognize his own human fallibility.

WASHINGTON REMAINED in a kind of restless exile in Winchester, so that thirty-four-year-old Captain Dagworthy and young Colonel Washington each ran their own fiefdoms a few days' ride apart, each presumably dedicated to protecting the frontier settlers. But as winter deepened, the stalemate ate away at Washington, confined to his little cabin just off the town's lone street, all slush and mud under dreary gray skies. Where was Captain Dagworthy when Washington trudged through the frozen wilderness to deliver the message to the French commandant? And when he and his men had suffered under bullets and rain at Fort Necessity? Dagworthy was comfortably back in Maryland, tending to his various businesses. And yet he, as a mere captain, claimed seniority. It rankled. Finally it became insufferable. On January 14, he wrote to Governor Dinwiddie. ". . . I have determined to resign a commission . . . rather than submit to the Command of a Person who I think has not such superlative Merit to balance the Inequality of Rank. . . ."

Instead of hastily quitting, however, Washington prudently suggested a solution that would at least quell his restlessness to take action. He would personally ride north to Boston to meet with General Shirley, the Royal Governor of Massachusetts, whom London had placed in charge of all British forces in the North American colonies. In Boston, Washington would petition Governor Shirley to rank him above Captain Dagworthy.

Governor Dinwiddie granted permission, and Washington set out in early February, accompanied by two servants and two of his subordinate officers, for a round-trip winter journey on horseback of over one thousand miles. With this sojourn he hoped to determine his fate. Yet he had to leave behind what command he did have, and with it his responsibility for guarding the frontier. His assumption that Indians would not cross the snowy mountains and attack the frontier settlers would prove false.

With his retinue, and traveling in a style he felt befitted the commander in chief of the Virginia Regiment, he rode first through Maryland and then the rolling, prosperous, orchard-dotted Pennsylvania countryside, now leafless in winter, to arrive in Philadelphia. With three thousand houses, paved streets, and streetlamps, Philadelphia was the largest city young Washington had yet visited, and with a population of twenty-five thousand, the largest in the colonies. It was the intellectual capital of British North America, with a college, two libraries, a statehouse, and the beginnings of the American Philosophical Society, founded by Benjamin Franklin and colleagues a decade earlier to advance scientific, geographic, and other knowledge. Young Washington spent four days in the stylish city, visiting tailors and hatters to refine his dress as an officer and a gentleman. Despite his advice to subordinates that actions make an officer, he loved luxurious dress and a stately appearance and showed a weakness for silver lace.

He and his servants and officers then rode on to New York, a city of about fifteen thousand, its low brick houses nestled at the lower tip of Manhattan Island and bisected by a tree-lined street named Broad Way. This ran from the fortress-like Battery on the island's southern tip about fifteen blocks northward to the end of the city's dwellings at the Negro Burial Ground and the Freshwater Pond, and continued into the countryside beyond. Docks and slips lined the East River, where the many slaves of the busy little seaport hefted and carted cargos. Tapping his network of Virginia gentlemen acquaintances, Washington stayed at the home of Beverley Robinson, brother of John Robinson, Speaker of the Virginia House of Burgesses. Beverley Robinson had wed a New York aristocrat of high standing, Susannah Philipse, daughter of the Lord of the Manor of Philipsburg, who had been granted royal charter for a huge piece of land along the Hudson just north of Manhattan. It happened that Susannah had a younger sister, twenty-six-year-old Polly, who was an eligible heiress to fifty-one thousand acres of the manorial lands—something that caught the attention of young Washington, who had both a passion for land and a fondness for ladies.

Having left his men at a string of small forts to guard Virginia's frontier, the gallant young officer with the new wardrobe escorted Polly and Susannah twice to a popular exhibit then showing in New York, "The Microcosm, or World in Miniature." This consisted of a simulated Roman temple animated with a set of mechanical devices that mimicked classical figures playing musical instruments, birds flying, chariots moving, and ships sailing. It was the kind of novelty that bespoke the dawning of a mechanical age. Despite the outings to this marvel, however, and her high and landed social status, no spark appears to have ignited between young Washington and Polly Philipse. Perhaps the vision of Sally Fairfax overwhelmed him still.

In other entertainment, he lost a little money gambling at cards, bought fresh horses, and set forth up Broad Way, splitting off on the High Road to Boston at the Common that now holds City Hall, passing the Block House, the Powder House, the Negro Cemetery, and the Poor House before exiting New York's gate near today's Brooklyn Bridge. After riding as far as New London, Connecticut, his group left their horses and boarded ship for the seaport of Boston, where they took rooms at Cromwell's Head Tavern on School Street, named after the severed head of Oliver Cromwell. Awaiting his audience with General Shirley, Washington went sightseeing in the city, which, according to a contemporary traveler, Andrew Burnaby, resembled one of the best country towns in England and had around twenty thousand inhabitants. He toured the Boston finery shops, buying a new hat, two pairs of gloves, and more silver lace. Despite admonishing his Virginia Regiment troops to curtail their gambling, Washington himself enjoyed it and he lost again gambling at cards, this time more heavily, including a game at the Governor's House. He had already achieved some measure of renown. The *Boston Gazette* noted that Colonel Washington was in town: "a gentleman who has a deservedly high reputation for military skill and valor, though success has not always attended his undertakings." Brave but unsuccessful—the judgment captured in a thumbnail Washington's young career.

When General Shirley received Washington, he said he needed time to think about the Dagworthy matter. A few days later, on March 5, he issued his verdict, handing to Washington a memo addressed to Governor Dinwiddie clearly stating that, if the case should arise and Captain Dagworthy and Colonel Washington were both at Fort Cumberland, "It is my orders that Colonel Washington shall take the Command."

Memo happily in hand, mission accomplished, young Washington and his retinue turned south toward Virginia. Along the way, however, he learned of a hitch. General Shirley, it turned out, had also placed Governor Sharpe of Maryland in charge of all troops from several southern provinces, including Virginia. If the British mounted a major attack against the French in the Ohio Valley wilderness, Governor Sharpe would take command and could well give advantage to his protégé and fellow Marylander, Captain Dagworthy, over Washington. Or so Washington suspected.

Despite the long journey to Boston to resolve the issue, Captain Dagworthy again had made Washington simmer with envy. Yet again, based on slender evidence but burgeoning pride, he decided to quit. But as soon as he rode into Williamsburg on March 30 with the intention of resigning, urgent messages arrived from the frontier. In Washington's long absence, Indians, accompanied by their French sponsors, had again attacked the Virginia frontier settlers he was supposed to protect.

He quickly dropped his plan to resign. He realized that British forces now would most likely mount another attack against Fort Duquesne, to stop the Indian raids. Here was his chance to play an instrumental role as a commander, unlike the aide-de-camp post he had under General Braddock. Changing strategy, instead of quitting he wrote directly to General Shirley and Governor Sharpe and asked that they make him second-in-command of any advance against the French in the Ohio. After his anxious imaginings about the Maryland governor favoring Captain Dagworthy, Washington was no doubt surprised to get a reply from Governor Sharpe endorsing him as "much esteemed in Virginia and really . . . a gentleman of merit." General Shirley confirmed that Washington would be second-in-command to Governor Sharpe in any advance against the French in the Ohio. After months of effort, he had won an honorable victory over the stubborn usurper, Captain Dagworthy. No one could now question his rank on a campaign into the Ohio.

But far more serious problems had just begun. Now that he returned his attention to the frontier, the responsibility that had been his all along—to protect the frontier settlers from massacre—landed suddenly and crushingly upon him. He rode fast over the Blue Ridge to Winchester in the Shenandoah Valley. He found that Indians threatened frontier settlers up and down the valley, formerly safe from Indian attacks. They had abandoned their homesteads and fled to small forts or to the town of Winchester itself. No one knew if the Indians would attack Winchester. Washington desperately tried to recruit more men. He had only forty armed men under his command at Winchester—the remainder were at Fort Cumberland at Wills Creek or spread among the small forts on the Virginia frontier. Washington sent out a plea through surrounding Frederick County for militia to rendezvous at Winchester. They would then force the French-allied Indian raiders out of the Shenandoah. Only fifteen men showed up.

"All my Ideal hopes, of raising a Number of Men, to scour the adjacent Mountains, have vanished into Nothing," complained a frustrated Washington to Governor Dinwiddie. "The timidity of the inhabitants of this County is to be equalled by nothing but their perverseness."

Washington realized that only a draft would provide enough men. The Virginia House of Burgesses had arrived at the same conclusion. But

Washington also knew that he needed Indian allies to defend the frontier. He hoped help would come from Cherokee traveling up the Great Warriors Path from their home in the Carolinas, but, despite past requests from Governor Dinwiddie, their help had amounted to little.

"Indians are the only match for Indians," he wrote to Speaker of the House Robinson, "and without these, we shall ever fight upon unequal Terms. . . . Their cunning is only to be equalled by that of the Fox; and, like them, they seize their prey by stealth."

While supremely frustrated by the raiding Indians, Washington would learn much from them—about patience, about waiting for opportunities, about avoiding major battles if possible—that would come to fruition in the distant future when he served as commander in chief of a far larger body of troops.

And still more problems bedeviled the young commander. Rumors flew in Williamsburg and "greatly inflamed" the House of Burgesses and Governor's Council that all sorts of drunkeness and immorality occurred among Washington's charges in the Virginia Regiment. His first instinct was to defend himself and his own reputation, horrified to think that these charges against the troops reflected poorly on him. He vehemently asserted to Governor Dinwiddie that he had given orders to forbid "gaming, drinking, swearing, and irregularities of every kind," then admitted that he could have actually enforced these orders if he had gone to Fort Cumberland, which he had avoided due to Captain Dagworthy's presence. Once again, his sense of pride had stood in the way of his effective command. Writing to House Speaker Robinson in his own defense, he grew increasingly indignant that these unseemly rumors about his troops maligned his public reputation. Remarkably prickly and sensitive, he yielded to the petulant part of himself—the part that felt he had sacrificed to take the post and wasn't being adequately appreciated—and threatened, for perhaps the fourth time, to quit. "It will give me the greatest pleasure to resign a command, which I solemnly declare I accepted against my will."

One imagines Governor Dinwiddie figuratively rolling his eyes at the adolescent behavior of this twenty-four-year-old, although one can understand as well the frustrations of Washington—enduring the privations of frontier living and the struggle to command recalcitrant men—at what must have felt like petty criticisms of his men's conduct.

Things only got worse along the frontier. On April 18, a mere twenty miles from Winchester, a contingent of troops from the Virginia Regiment left the small outpost called Edward's Fort and encountered an estimated one hundred Indians and French, some on horseback. The French and Indians captured or killed two Virginia officers and fifteen men. In the next two days, reports arrived of French and Indians surrounding more small frontier forts, and Indians massacring three frontier families. Panic spread. The residents of Winchester anticipated a full-scale attack at any moment. Townspeople, assisted by soldiers at Washington's orders, chopped down trees and bushes near the town's main street to prevent Indian ambushers from hiding. Washington urged nearby counties to call out the militia. He begged for volunteers to protect Winchester. Only twenty showed up. Both the frontier and town of Winchester had erupted into a state of full-blown crisis.

Like a brushfire gone wild, too many attacks and threats flared up at too many places to begin to stamp them out, a key element of Indian strategy. Outlying forts desperately needed troops and ammunition. Washington called a council of war of his top officers to decide whether to send the few troops remaining in Winchester to the outlying forts or keep them in place to guard the town. He may have learned from the example of General Braddock, who held several councils during the march to the Ohio, of the need to consult his subordinate officers on crucial decisions, and especially the importance—perhaps a lapse in Braddock—of listening closely and weighing carefully what they said. It was another practice Washington gained from the wilderness campaigns that would serve him well in the decades to come. The council of war advised that the troops stay in Winchester. Washington prudently agreed—a more cautious officer than the stubbornly bold young man who had attacked the breakfasting French party at Jumonville Glen and, rather than retreat, waited at ramshackle Fort Necessity for the fierce French and Indian counterattack. Along Winchester's dirt street dotted with cabins and crude board houses, he could find women cradling babies, boys and girls playing, dogs and pigs rooting, horses and cows. He now had civilians in his care. An attack on Winchester could result in a terrible, bloody loss, the slaughter and mutilation of women and children, the taking of captives, a great column of smoke rising over the burning town. Better to stay put and guard it. But this meant letting the outlying forts and settlers fend for themselves against the bands of Indian raiders sliding through the forest.

"[N]othing but a large and speedy reinforcement can save them from utter destruction!" Washington wrote to his mentor, Colonel William Fairfax, in charge of the militia for Fairfax County, urging him to send men immediately.

He scrambled, ordering up more powder, commanding some outlying forts to evacuate and join with others. Rumors flew of a body of four hundred Indians roaming the woods. He dispatched pleas for help, consolidated troops, banned the discharge of guns in Winchester without orders to avoid confusion with warning shots fired to signal an attack. Still the frontier deteriorated. After two and a half years of hard work, high danger, and political maneuvering, he had attained clear command of the Virginia Regiment. But within a mere two weeks of achieving this, he found himself in an impossible situation—hopelessly underequipped against an unseen foe, surrounded by mass panic, and awash in a bloody and gruesome reign of terror. Scalped bodies lay outside burning farmhouses, the corpses stuck full of arrows, the men's brains beaten out with their own rifle barrels, the barrels left sticking out of the hollow skulls, the farmstead's axes and scythes embedded in the mutilated torsos.

On April 22, as families perished, the full weight of Washington's predicament bore down on him and broke through his pride-hardened shell. The responsibility for protecting these frontier families lay on him, yet he could do nothing but stand by helplessly as they died gruesome deaths. In a state of near-hysteria, he scrawled an urgent message to

Governor Dinwiddie, displaying an odd blend of deep empathy for the families and an obsession with the way this horror would reflect on his personal reputation in genteel Tidewater circles—known as "below."

> *But what can I do? If bleeding, dying! would glut their insatiate revenge—I would be a willing offering to Savage Fury: and die by inches, to save a people! I see their situation, know their danger, and participate their Sufferings; without having it in my power to give them further relief, than uncertain promises. . . . In fine; the melancholy situation of the people; the little prospect of assistance[;] The gross and scandalous abuses cast upon the Officers in general—which is reflecting upon me in particular; for suffering misconducts of such extraordinary kinds—and the distant prospects, if any, that I can see, of gaining Honor and Reputation in the Service—are motives which cause me to lament the hour that gave me a Commission: and would induce me at any other time than this, of imminent danger; to resign without one hesitating moment, a command, which I never expect to reap either Honor or Benefit from: But, on the contrary, have almost an absolute certainty of incurring displeasure below: While the murder of poor innocent Babes, and helpless families, may be laid to my account here! The supplicating tears of the women; and moving petitions from the men, melt me into such deadly sorrow, that I solemnly declare, if I know my own mind—I could offer myself a willing Sacrifice to the butchering Enemy, provided that would contribute to the peoples ease.*

This passage remains one of the most remarkable and passionate Washington ever wrote. The crisis in April 1756 on the wilderness frontier engendered a parallel crisis in young Washington's own growth. His rank and personal reputation had obsessed him for many months, perhaps all the years that he had known brother Lawrence and the aristocratic Fairfaxes, men of reputation and rank. It obsessed him still, as he wrote so forceful a letter to Governor Dinwiddie. All the horror "*may be laid to my account here!*" So wrote the self-absorbed part of young Washington. But in his passion, something else came through, too. For the first time and in language stronger than any he had used before, Washington displayed a deep empathy for the sufferings of the common people. In these moments of crisis, surrounded by pleading frontier families, young Washington began to move beyond his self-absorption and obsession with rank and reputation. In his empathy for the people's suffering, he made a first step toward the selflessness and sacrifice for which he would one day become legendary.

CHAPTER SEVENTEEN

HIS MOST RECENT THREATS TO QUIT, SENT FOUR DAYS earlier to Speaker John Robinson and Governor Dinwiddie among others, had now reached Williamsburg, setting off alarm that he would actually go through with it. His friend Landon Carter got wind of Washington's threats and wrote him an impassioned letter to stay. If you quit now, he warned, those who wish you ill will say you quit because you're afraid. Your friends will have a difficult time defending you. Forget any insults to your honor. It's better to die a hero than quit. "No Sir," Carter wrote emphatically, "rather let Braddocks bed be your aim than anything that might discolour those Laurels that I promise my self are kept in store for you."

The warning came clearly—don't let your thin skin and self-absorption undermine your prospects of much longer-term fame. Carter, perhaps, sensed the beginnings of greatness in Washington even at this early age of twenty-four and urged him, instead of lapsing into petulant fits that begged for attention, to take a longer view of what might constitute his own self-interest and that of his native Virginia.

Washington also received a letter from Landon's brother, Charles Carter, who tried to soothe him and urged patience and forbearance. "When ever you are mentiond [in the House of Burgesses] tis with the greatest respect I hope you will therefore arm yrSelf with patience and dispise such reflections as May be cast by any Malevolent enimies to you. . . ."

Colonel William Fairfax, his mentor, also wrote to head off a resignation. The aristocrat tried to appeal both to Washington's sense of vanity and to his good manners, comparing him with great Roman generals and suggesting he take a lesson in gratitude from them. "Your good Health and Fortune is the Toast at every Table, Among the Romans such a general Acclamation and public Regard shown to any of their Chieftains were always esteemd a high Honor and gratefully accepted."

While these men wrote to him trying to convince him not to quit, Washington poured out pleas to Governor Dinwiddie and Speaker Robinson for more men from a draft, for ammunition, for provisions, for as many Cherokee and Catawba Indian allies as the governor could muster, and also to send for a supply of Irish beef stored at Alexandria. He outlined an elaborate plan to protect the frontier with a series of twenty or so small forts, each manned by about one hundred troops, and lamented that the Assembly had allowed for only fifteen hundred troops instead of the two thousand he desired. He expressed more concern about the allegations of misconduct and more about his helplessness to protect the frontier settlers. "Not an hour, nay, scarcely a minute passes, that does not produce fresh alarms and melancholy accounts. So that I am distracted what to do!"

There was a hint of grandiosity and narcissism in all this, and even paranoia. Washington imagined that he was the subject of slanderous remarks circulating through the capital, the target of some sort of conspiracy. With his repeated threats to quit, he tacitly invited the praise and reassurance of friends and acquaintances. And with his many protests of helplessness he implied that, given the right resources, he could be savior of an otherwise doomed people. He believed that so much revolved around him. This need for reassurance and praise caused him to attack the challenges that faced him with great energy, determination, and attention to detail. While, perhaps, the result of some streak of grandiosity, it, too, would serve him well in the years ahead.

As young Washington struggled to defend the Virginia frontier in the spring of 1756, James Smith's Indian captors took him deeper into the Ohio wilderness to a Wyandot (Huron) town, Sunyendeand, near the western end of Lake Erie. They had survived the deprivations of another winter. Here they enjoyed the fruits of the spring hunt, singing and dancing and visiting each other's bark-covered wigwams or longhouses to enjoy feasts of bear's oil mixed with dried venison and maple sugar. They passed the time playing a gambling game in which small stones that were black on one side and white on the other were shaken in a gourd cup and spilled out on the ground like dice, while the gambler called out "black" or "white." But one day the leisure ended, their food mostly gone except for a thin hominy soup, and the men and boys aged fifteen to sixty again prepared with songs and dances to go to war. Concerned he would flee if brought too near his own people, they left Smith behind with the women, the children, and two old men, who wanted to know Smith's predictions for the outcome of the fighting.

> The Indians were then in great hopes that they would drive all the Virginians over the lake, which is all the name they know for the sea. They had some cause for this hope, because at this time, the Americans were altogether unacquainted with war of any kind, and consequently very unfit to stand their hand with such subtil enemies as the Indians were. The two old Indians asked me if I did not think that the Indians and French would subdue all America, except New England, which they said they had tried in old times.

Smith told them that he did not think the Indians had it in their power to conquer the British in North America. The old men responded that the conquest was already under way, and there were reasons it would succeed: "[T]hey said they had already drove [the British colonists] all out of the mountains, and had chiefly laid waste the great valley betwixt the North and South mountain . . . and that the white people appeared to them like fools; they could neither guard against surprise, run, or fight."

This region that the Indians had laid waste was the same that young George Washington struggled so to defend—including the broad, beautiful Shenandoah Valley. The Indians had run the whites out of it. The two old men wanted to know Smith's reasoning for saying that the conquest of the British colonies would fail. Although Smith was young, only in his late teens, the old men told him to speak his mind freely. "I told them that the white people to the East were very numerous, like the trees, and though they appeared to them to be fools, as they were not acquainted with their way of war, yet they were not fools; therefore after some time they will learn your mode of war, and turn upon you."

Eventually, the traveling warriors returned to the Wyandot town where Smith lived bearing their spoils—many scalps, horses, meat, prisoners, and other plundered items, relieving the hunger that had wracked the village. They had separated into smaller parties and attacked the southern part of the Shenandoah Valley, in Augusta County. Washington, based in Winchester at the valley's northern end, was at his wit's end trying to defend it. He could not have imagined the Wyandot village deep in the wilderness that was the source of the attacks, or James Smith sitting before a fire in their longhouse discussing the strategic shortcomings of the British colonists. Nor could he have imagined, after so bloody an interlude, the taking of so many scalps and the mutilation of bodies, the leisure and ease the Indian warriors now enjoyed, as Smith described it, having plenty of meat and harvesting an abundant crop of corn, so that they lived in love, peace, and friendship together. They shared all food very carefully and equitably, even when they had little. There were no disputes among them. "In this respect," he wrote, "they shame those who profess Christianity."

As spring turned to summer in 1756, a year had passed since Braddock's Defeat. The few settlers who remained in the Shenandoah Valley had holed up in small forts, flocked to Winchester, or fled eastward over the Blue Ridge toward the Tidewater region and "below." Young Washington pleaded to Governor Dinwiddie and Speaker Robinson for help. "Desolation and murder still increase; and no prospects of Relief. The Blue-Ridge is now our Frontier."

They wrote back that help was on the way in the form of a draft to raise more men and permission from the Assembly to build a strong central fort at Winchester—as Washington had requested—along with a string of smaller forts along the frontier. By mid-May, nearby counties had answered pleas to send militia to Winchester, and officers and militia flowed into the little town. Washington now had close to nine hundred men in Winchester. He began to divvy them up among different frontier forts, warning the militia that if they attempted to desert they would be drafted into the Virginia Regiment. But when rumor spread of a large group of Indians stalking about on the South Branch of the Potomac, about forty miles west of Winchester, scores of the newly arrived militiamen panicked, deserted, and fled back over the Blue Ridge to safer ground "below." Washington realized that he could not rely on the militia for frontier defense, and he would have to focus on recruiting and training the Virginia Regiment instead. Still, he wanted to make an example of some of the deserters. He chose a particularly roguish deserter—"a most atrocious villain"—for a court-martial and execution by hanging or firing squad, along with a sergeant, Nathan Lewis, whom he deemed cowardly in a skirmish with the Indians.

From a house he rented in Winchester, he had his hands full ordering his troops to "scour the woods" for Indians, scraping up provisions, and dealing with discipline. He chastised one officer for allowing his men to take chickens from a Quaker family's homestead, tear down its fence for firewood, and set the soldiers' horses loose in the family's cornfield to feed. He gave orders to prohibit the selling of liquor to Indians, dealt with fistfights, held court-martials for cowardly officers. The men labored at building the fort at Winchester, on a low hill at the head of the main street, in the meantime staying at the Winchester courthouse, converted to a garrison and reinforced with breastworks and four cannon. In June, Washington made a quick trip to Williamsburg, where he found an ailing and worn-out Governor Dinwiddie. At sixty-three, he yearned to retire to England and take the waters at Bath. Increasingly, he gave Washington freedom to make his own decisions.

In mid-July, back at Winchester, Washington received news of the larger world in a letter from his mentor, Colonel William Fairfax. As if the terrain that he had trekked inscribed the limits of the world about which he cared, young Washington rarely if ever made reference in his letters to events outside the Virginia frontier and Ohio Valley. Yet nation-changing upheavals were occurring in Europe and elsewhere in North America. Empires clashed. Washington gave no hint that, in his rashness two years earlier against Jumonville's breakfasting party, he had fired the opening shot of the conflagration that now erupted into flames in North America and around the globe.

More British Royal Army regular troops—Redcoats—under General Abercromby had arrived from across the Atlantic, Colonel Fairfax informed him, and headed up the Hudson River valley toward Albany. Like the Ohio River valley, the Hudson Valley formed a key strategic link in the geography of empire in North America. One might imagine a giant T with its foot at New York City. The vertical line of the T was formed by the Hudson River, by long, skinny Lakes George and Champlain, and by the Richelieu River—all navigable water routes. The horizontal line of the T was formed by the Saint Lawrence River. Along it lay Montreal and Quebec and farther to the west the Great Lakes. This T of waterways thus provided a water link between major ports of British America and those of French America.

In the middle of that vertical line, along the shores of Lake George, the French and British had already clashed once, the previous autumn, in the indecisive Battle of Lake George. Now they vied to seize as much of that vertical water route as possible, attempting to establish forts along it. These were Fort Carillon, for the French (later renamed Ticonderoga), on southern Lake Champlain and, for the British, thirty miles farther south, Fort William Henry on the southern shore of Lake George and, just across a canoe portage and rise of ground, Fort Edward on the headwaters of the Hudson River. The troops stationed at the two British forts were intended to provide a bulwark against the French paddling down that T from the north, taking Albany from the British and—who knew—pushing onward down the Hudson Valley to overrun New York City, thereby severing New England from the more southern British colonies, such as Pennsylvania, New Jersey, Maryland, Virginia, and the Carolinas.

Colonel Fairfax wrote to Washington of other significant news from across the Atlantic. British authorities had named Lord Loudoun to replace the interim appointment, General Shirley of Massachusetts, as commander in chief of British forces in North America. Lord Loudoun was expected to arrive any day from England. A French fleet, meanwhile, had sailed out into the Mediterranean to the British-held island of Minorca. Some fifteen thousand French troops had landed on the island and besieged the British at their stone-walled Saint Philip's Castle. Admiral Byng and his fleet had been dispatched from the English Channel to rescue the besieged fort. Although Fairfax did not know it at the time, this French action against Minorca spurred Great Britain to issue a formal declaration of war against France on May 18, 1756, to which France would reply in kind three weeks later.

By early July, when Colonel Fairfax wrote his news-filled letter to Washington, British admiral Byng had already attempted to relieve Saint Philip's Castle. Not fully engaging the French line of battle and badly damaged, Admiral Byng sailed back to Gibraltar to repair his ships, leaving behind the besieged fort under its eighty-four-year-old commandant. For this action, in a case still infamous for the severity of its punishment, Admiral Byng some months later was court-martialed for not doing his utmost to fight or pursue the enemy. As he knelt blindfolded on a pillow on the quarterdeck of his Royal Navy ship, he was shot by a firing squad.*

Washington acknowledged none of the great events Fairfax mentioned in his letter or his role in kindling them. Rather, his attention focused on one thing—that Lord Loudoun, the new commander in chief, would soon set foot in British North America. Young Washington very much wished to be brought to Lord Loudoun's attention. He had already asked Governor Dinwiddie to write a letter of recommendation. Now Washington wrote to Lord Loudoun on his own behalf. He flattered his lordship with congratulations and praised the king's wisdom in appoint-

---

* This inspired Voltaire to write in his satiric novel *Candide,* which followed the travels of an eternal optimist, that "in this country, it is good to kill an admiral from time to time, in order to encourage the others."

ing him head of the military forces in the colonies. He then made sure that Lord Loudoun knew that Washington and his troops were the first to fight the French and Indians and understood their dark and devious art of wilderness warfare.

> *We humbly represent to your Lordship, that We were the first Troops in Action on the Continent on Occasion of the present Broils, And that by several Engagements and continual Skirmishes with the Enemy, We have to our Cost acquired a Knowledge of Them, and of their crafty and cruel Practices, which We are ready to testify with the greatest Chearfulness and Resolution whenever We are so happy as to be honoured with the Execution of your Lordship's Commands. In Behalf of the Corps*
>
> *Go: Washington, Colo.*

In another mark of his drive for acceptance as part of the Virginian elite, Washington took up fencing lessons from a sergeant who was in residence at Winchester. Learning that Lord Loudoun would visit Virginia, he urged his men at Winchester and those under Adam Stephen at Fort Cumberland to spend their leisure hours in training in order to "make the best appearance possible" when his lordship visited.

Six weeks after Lord Loudoun's arrival in North America, in late July, however, as Washington worked to sculpt and polish his public image in hopes of a King's commission, a stinging attack appeared in the September 3, 1756, edition of the *Virginia Gazette,* signed by an anonymous editorialist, The Centinel. Exquisitely sensitive, young Washington believed the attack took aim directly at him.

> The Profession of Soldier, especially at such a Time as this, is not only noble but benevolent; and worthy at once of universal Honor and Gratitude. . . . [But] When raw novices and rakes, spendthrifts and bankrupts, who have never been used to command . . . are honored with commissions in the army . . . when the officers give the men an example of all manner of debauch-

> ery, vice and idleness; when they lie skulking in forts, and there dissolving in pleasure . . . instead of searching out the enemy . . . when nothing brave is so much attempted . . . when men whose profession it is to endure hardships and dangers cautiously shun them, and suffer their country to be ravaged . . . then, certainly, censure cannot be silent; nor can the public receive much advantage from a Regiment of such dastardly debauchees.

By then, young Washington had commanded the Virginia Regiment for one year. One can imagine him reading it, blood flushing his pale cheeks with anger and humiliation, blue-gray eyes flashing, as he held the papery sheets in trembling hands. "What useless lumber," the writer had said of the Virginia Regiment. How could this cowardly writer dare impugn his honor like this? If instead of gratitude he was showered with abuse he would take action. He would quit. He would ride to Williamsburg and hand his commission back to Governor Dinwiddie. He would return to a planter's life at Mount Vernon. But then he paused. How would quitting look? Would this bring more dishonor still? He turned to his secretary, John Kirkpatrick, and asked him to gauge how his friends and his political acquaintances would react if he resigned.

The hesitation was, in a way, a sign of his growing maturity. In the past, he would have written petulant letters, thrown up his hands, and threatened to resign. But now he thought twice about those threats. How would it look if he up and quit his post as defender of the Virginia frontier while silent attacks emerged from the forest leaving behind slaughtered settlers and burning homesteads?

It would not look good, came the answer from Kirkpatrick, having conducted the eighteenth-century colonial equivalent of an opinion poll. "In the Short time I have had to dive into the Generall Opinion of Your resignation I find it disagreable, & unpleasant to all their inclinations, and woud certainly be at a Loss for Such another to fill your place. . . ."

Kirkpatrick sympathized with Washington for having to face this "Vain Babling of Worthless, Malicious, Envious Sycophants." He suggested that silence would serve as the best response. In the meantime,

Washington could continue to serve his country with zeal and await a meeting with Lord Loudoun. In the presence of his Lordship, presumably, Washington could highlight his record of service and petition for the coveted British Royal commission. But above all, understand that such scurrilous charges grow from envy, Kirkpatrick warned. That envy of him exists surely indicates that Washington has merit, suggested Kirkpatrick, quoting Alexander Pope's "Essay on Criticism":

> Envy will merit, as its shade, pursue,
> And like a shadow, proves the substance true;

Washington received much the same advice from a friend in Alexandria, Scottish merchant William Ramsay, to whom Kirkpatrick showed the letter. Ramsay believed that the Centinel referred not to Washington but to "gentlemen"—in other words, aristocrats—who had won their officers' commissions through connections. Washington, by contrast, had been appointed commander in chief of the Virginia Regiment on his merits and had shown "courage & bravery both in [Braddock's Defeat] & many other actions." Like Fitzpatrick, Ramsay, too, advised that silence was the best response to the Centinel's diatribe.

Still, Washington could not put down his pen. His reputation was at stake, he believed. Something had to be done. He wrote a letter, now missing, to his older half-brother Augustine Jr. (Austin) containing a response to the Centinel, with instructions to publish it in the *Virginia Gazette* if Austin thought appropriate. Twelve years older and a Virginia Burgess (as well as a member of the Ohio Company), Austin jumped into the task and rode from his plantation to Williamsburg to consult with various politicians and important persons there. He wrote back to Washington with the same conclusion. Knowing his brother and his sensitivity about his reputation, Austin sketched the dark consequences of Washington's quitting: his fellow officers would quit, soldiers would desert, and Virginia's frontier would be left defenseless, for which young Washington would be held to account. "I am sensible you will be blamed by your Country more for that than every other action of yr life," Austin wrote.

If you care about your reputation, do not quit: that was the advice from friends and family. Austin also knew how badly Washington coveted a British Royal officers' commission rather than a colonial one. He played on that, too, advising his younger half-brother to stay on the job until Lord Loudoun arrived, "then you will know what prospect you stand to be put on the British establishment. . . ."

Washington meanwhile supervised the building of the new fort at Winchester, now named Fort Loudoun after the new commander in chief, and stayed in his rented house, attended by his servant and cook, Thomas Bishop, and took fencing lessons. On August 16, he assembled the Virginia Regiment and important townspeople at the partially built fort and announced the official declaration of war against France—"for divers good causes," Washington editorialized, "but more particularly for their ambitious usurpations and encroachments on [the king's] American dominions." He marked the occasion with cannon shots, musket volleys, many toasts drunk to His Majesty's and others' health, and a parade of the three companies of Virginia troops around the little town.

At the same time, mid-August 1756, far north of the Virginia frontier, the British suffered a staggering loss on the shores of Lake Ontario at their Fort Oswego when the French besieged it under General Montcalm, a hard-headed, extremely vain, and fearless French nobleman who had been wounded five times in various European wars. In a fierce exchange of cannon fire, a French cannonball took off the head of the fort's British commander, Lieutenant Colonel James Mercer. Over one thousand British troops surrendered, also giving up one hundred and twenty-one cannon and control of Lake Ontario. A year after Braddock's Defeat, this came as another devastating and demoralizing defeat to the British in the newly declared war.

Washington kept in motion in his own arena and found equally discouraging news. At September's end, he heard reports of more Indian attacks in the southern Shenandoah Valley. Touring the southern Virginia frontier to see for himself, he found the small forts poorly manned, discipline lax, pay in dispute, the militia unreliable, and, generally, the string of small forts inadequate to the task of defense. And there were Indians about. Riding through a rainy forest, Washington and his colleagues narrowly missed an ambush by a few minutes, as an Indian party hiding in the woods let them pass by, waiting for a different party heading in the other direction.

"I passed & escaped almost certain destruction for the weather was raining and the few Carbines unfit for use if we had escaped the first fire," he later recalled, toting up yet another of his narrow escapes.

Heading back north toward Winchester, he again thanked Providence. Accompanied by a band of gentlemen militia officers who knew nothing about maintaining silence when traversing hostile territory, he believed that only divine intervention kept them from encountering any Indians on the seven-day journey; "otherwise we must have fallen a Sacrifice, thro' the indiscretion of these hooping, hallooing, Gentlemen-Soldiers!"

Even the friendly Indians proved a disappointment. Governor Dinwiddie had worked to recruit a large force of friendly Cherokee warriors—as many as four hundred—from North Carolina to join the British. But when the promised force turned up, it amounted to only seven men and three women. Reporting to Governor Dinwiddie, Washington lamented the sorry state of the frontier defense. "This Jaunt afforded me great opportunity of seeing the bad regulation of the militia; the disorderly proceedings of the Garrisons; and the unhappy circumstances of the Inhabitants," he wrote and tactlessly offended the governor by suggesting that much more could have been done to recruit Indian allies.

Arriving back at Winchester in late October after his three-week tour of the southern Shenandoah, Washington heard of another crisis brewing at Fort Cumberland. The officers there had finally received a copy of the Centinel's article, had flown into a rage at the charges leveled at them, and threatened to quit by November 20 if they had not received an apology by then. They believed that Governor Dinwiddie knew in advance of the Centinel's attack and that he agreed with it.

Washington now found himself soothing his officers about the same article that had offended him so. He realized that if they quit, the Virginia Regiment would fall apart, as would the frontier defense. He became the conduit and mediator between his men in the mountains and the higher authorities down in the Tidewater region. An advocate for those officers and men under his command and for the frontier settlers, he began to shift his focus of concern from himself to those around him. He was becoming a leader.

The shift did not occur all at once. The more time he spent isolated in the backcountry, such as during his three-week journey on the wild fringes of the southern frontier, the less self-centered he became. It was as if in his trips to the Tidewater region among the Virginia planter aristocracy he reverted to his role of aspiring gentleman, with his honor and reputation foremost in his mind, values shaped by the decades—the centuries—his family had spent aspiring toward the English and Virginian elite. But once back over the Blue Ridge on the frontier or in the deep wilderness, simple survival—of himself and those around him—became paramount. The layers of social pretensions sloughed away, no longer propped up by the fieflike plantations with their teams of slaves who took care of all daily personal needs. One never knew what the next bend of the forest path would bring, the next shift of the weather. Over the mountains, in the wilderness and on the frontier, he confronted the most elemental challenges—food, shelter, clothing, the threat of enemy attack. His ambitious quest for gentlemanly status paled beside these essential needs. The shared privations of the wilderness forged a deep bond between young Washington and those around him.

In early November, he rode from Winchester, where Fort Loudoun was under construction under his supervision, to Fort Cumberland at Wills Creek to listen to his officers' complaints, then rode on toward Williamsburg to present those grievances to the governor and his council. On reaching Alexandria and settling in for a few days with prosperous merchant friends, he received an angry letter from Governor Dinwiddie, who felt insulted that Washington seemed to blame him for poor recruiting of Indian allies, and complained of Washington's "unmannerly" tone. To Washington's distress, the governor also ordered one hundred men sent from Fort Loudoun at Winchester—favored by Washington as a central headquarters—to Fort Cumberland at Wills Creek, which he had hoped to downsize. Once again, Washington turned around from

Alexandria and headed back to the frontier, apologizing to Governor Dinwiddie that "If my open & disinterested way of writing and speaking, has the air of pertness & freedom, I shall redress my error by acting reservedly. . . ."

Washington's pledge that he would speak with greater reserve signaled another step on his path to leadership, and a greater understanding of the need to use cautious diplomacy when dealing with one's commanders. Silence could serve as an eloquent response. A new self-censorship and obliqueness would mark his future correspondence, a growing consciousness of the boundaries of propriety as a young officer addressing a superior. It was as if young Washington did not want to say too much for fear of offending, or of tarnishing his public image. This self-censorship would only grow in the upcoming weeks when a stolen copy of his journal, written during his attack on the French at Jumonville Glen and his retreat to Fort Necessity, was published for the general public.

CHAPTER EIGHTEEN

WASHINGTON'S DEVELOPING SENSE OF SELF-CONTROL—understanding how to work diplomatically, how to rein in his temper and indignation, how to look after his men and those around him—clashed at times with his ambition, which still knew few bounds. In February, he rode to Philadelphia to take all his complaints to Lord Loudoun, the new commander in chief of British forces in North America, and to give the new commander his advice for Virginia's frontier defense. But most of all, he hoped Lord Loudoun would grant him a British royal officer's commission.

His lordship had called a meeting in Philadelphia of provincial southern governors, including Governor Dinwiddie. While skeptical of Washington's usefulness at the meeting, Dinwiddie had reluctantly given in to his request that he be allowed to attend. "I cannot conceive what service you can be of in going there," wrote the governor to Washington; "however, as you seem so earnest to go, I now give you leave. . . ."

With Lord Loudoun's scheduled arrival in Philadelphia delayed, Washington had to entertain himself there for several weeks in late February and early March 1757. He dined and drank with Governor Sharpe of Maryland, gambled at cards, and went to dances. While the Philadelphia winter blanketed the brick and cobblestone city in cold and grayness, he found a far more festive and glamorous life indoors there than on the frontier in his rented house and half-built fort at Winchester or in the remote mountains of Fort Cumberland. To his surprise, he also found himself the author of an upcoming publication. An advertisement in the *Pennsylvania*

Gazette announced the publication of a book, translated from the French, of documents that purportedly blamed the hostilities between the two nations on Great Britain. Among these translated documents was a copy of young Washington's journal concerning the events of Jumonville Glen and Fort Necessity.

It turned out that the French had taken his journal from the belongings left behind at the surrender of Fort Necessity. Translated into French, the journal had been published in Paris as part of the collection of documents blaming the war on Britain. The British in turn had discovered a copy of this publication when they seized a French ship, and British and American publishers translated it back into English. Washington no doubt cringed to find that his journal had been used as propaganda against the British cause. He promptly denied the accuracy of the translation.

But still he did not hesitate to press aggressively for his own cause, although framing his self-promotion in the language of deference. In mid-January, he had written Lord Loudoun a voluminous letter that described in minute detail all the problems with the defense of the frontier—from the impossibility of defending it with a string of forts unless better manned and closer together, to the need to seize Fort Duquesne to stymie the attacks, to the lack of proper military regulations, the lack of pay, the lack of rewards for capturing deserters, the lack of tools, the lack of clothes, and the lack of shoes. "I have known a Soldier go upon Command with a new pair of Shoes, which perhaps, have cost him from 7/6 to 10/ and return back without any; so much do they wear in wadeing Creeks, Fording Rivers: climbing Mountains cover'd wt. Rocks &ca."

While acquainting Lord Loudoun with the myriad problems of frontier defense, so unlike those in European theaters of war, Washington, amid the requisite flattery, carefully closed his letter with requests for a royal commission for himself and greater recognition of the Virginia Regiment, saying that General Braddock had promised him a royal commission, as had Governor Shirley. But the former had died; the latter had gone off to England. "[D]on't think My Lord I am going to flatter," interjected Washington in a passage where he did just that. "I have exalted Sentiments of Your Lordships Character, and revere Your Rank; yet, mean not this, (coud I believe it acceptable). my nature is honest, and Free from Guile."

He sent this long missive with a cover letter to Lord Loudoun's aide-de-camp, James Cuninghame, in which, more boastful and bolder than he was in the letter to Lord Loudoun, he asserted that few people had better knowledge of the situation than he and briefly sketched a plan to take the Ohio from the French:

> *I am firmly perswaded that 3,000 Men under good regulation (and surely the 3 Middle Colonies coud easily raize, and support that Number) might Fortifie all the Passes between this and Ohio: Take possession of that River: cut of the Communication between Fort Duquisn and the Lakes, and with a middling Train of Artillery (with proper Officers & Enjineers) make themselves Masters of that Fortress, which is now become the Terror of these Colonies.*

But, if the British took no offensive action, Washington summarized the sorrowful situation on the Virginia frontier thus:

> *That an Offensive Scheme of Action is necessary if it can be Executed, is quite Obvious. Our All in a manner depends upon it. The French grow more and more Formidable by their Alliances, while Our Friendly Indians are deserting our Interest. Our Treasury is exhausting, and Our Country Depopulating—some of the Inhabitants fly intirely of, while others Assemble in small Forts destitute (almost) of the necessary's of Life; to see what Measures will be concerted to relieve their Distresses.*
>
> *This Sir, I assure you, is at present the Situation of Affairs in Virginia.*

When he was finally granted an audience with Lord Loudoun, a stocky, fifty-two-year-old Scottish bachelor known for his administrative skills, Washington received little satisfaction for his many requests. Some accounts say that Lord Loudoun brusquely dismissed the young officer who stood before him, others that he probably thanked Washington for his unsolicited advice. Whichever the case, he took none of Washington's many suggestions. Nor did he acknowledge Washington's wish for recognition of the Virginia Regiment as part of the regular army, nor, most painfully, did he grant the British royal officer's commission that Washington so coveted.

This proved a pivotal moment both for young Washington and, through him, for the tumultuous events that would unfold over the coming decades. Ever since his boyhood and his worship of his dashing half-brother Lawrence with his red-and-gold-braid uniform who had served as a colonial officer sailing aboard Royal Navy ships, Washington had dreamed of a royal officer's commission. Now he had requested it at least twice and possibly three times from the commanders of British forces in North America, and each time it had slipped through his grasp. Despite Washington's detailed letters and flattering requests, his long journey to Philadelphia, and his weeks of waiting for his lordship's arrival, on this last occasion Lord Loudoun had flatly denied or blithely ignored it, along with everything else Washington had suggested. "One can only imagine the burning fury and humiliation in Washington's breast as he rode back to Virginia," writes Washington historian John Ferling.

His resentment at unequal treatment had started long before his audience with Lord Loudoun, and would fester long after. While awaiting Lord Loudoun's arrival, Washington had written to Governor Dinwiddie, urging him to press Lord Loudoun for fair and equal treatment of American colonials to place them on a par with British subjects:

> *We cant conceive, that being Americans shoud deprive us of the benefits of British Subjects; nor lessen our claim to preferment. . . . As to those Idle Arguments which are often times us'd—namely, You are Defending your own properties; I look upon to be whimsical & absurd; We are Defending the Kings Dominions, and althô the Inhabitants of Gt Britain are removd from (this) Danger, they are*

> *yet, equally with Us, concernd and Interested in the Fate of the Country, and there can be no Sufficient reason given why we, who spend our blood and Treasure in Defence of the Country are not entitled to equal prefermt.*

He wrote this almost two decades before similar sentiments about equality exploded into the American Revolution. One cannot help wondering how differently events in the 1770s might have turned out if Lord Loudoun had granted young Washington his coveted royal officer's commission and made the Virginia Regiment a regular army establishment on that visit to Philadelphia in 1757. Washington would then have been firmly tied into the British military establishment. Perhaps he would have made it a lifelong career. As it was, he remained an outsider—close enough to feel envy yet never truly taken into the fold. Responding forcefully, he found the arguments against equal treatment "whimsical and absurd." His resentment sprang not only from his own rejection but on behalf of the men and, especially, the officers under him. He personally felt the discrimination that weighed against them, too—another sign of his maturing leadership and his move beyond a singular focus on his own ambitions. While Washington's beliefs and feelings about fair and equal treatment for American colonials would lie fallow for years, they never left him. As the American Revolution dawned, they would burst to life.

Frustration mounted for young Washington from the spring of 1757 onward. After his journey to Philadelphia and his long wait there for Lord Loudoun, the disappointed young officer returned to his frontier post in Virginia after an absence of nearly two months. Just before leaving Philadelphia, he had delivered to Lord Loudoun a carefully written "memorial," containing yet another plea for recognition and recounting the brave behavior of the Virginia Regiment at the Battle of Monongahela (Braddock's Defeat) and in many bloody skirmishes against French and Indian raiders on the frontier. He testified that the Virginians were "first in arms" in the war, having fought since the beginning, and had saved many frontier settlers from a cruel death. He apologized for bragging, but brag he must in order to bring their deeds to the attention of the king, who surely would recognize the Virginia Regiment if he knew of their heroism:

> *That the recounting those Services, is highly disagreeable to your [correspondent], as it is repugnant to the Modesty becoming the Brave; but that they are compelled thereto by the little Notice taken of Them; It being the General Opinion that they have not been properly represented to His Majesty: Otherwise the Best of Kings, would have graciously taken Notice of your Memorialists in Turn. . . .*

Washington mentioned nothing of the blunders committed under his leadership—the ambush at Jumonville Glen, the "assassination" of Jumonville, the humiliating surrender at Fort Necessity. Perhaps Lord Loudoun and the king knew of Washington and the Virginia Regiment in this less-flattering context more than for their unalloyed heroism, and had good reasons for denying a royal commission.

Letter delivered, Washington rode from Philadelphia back to Virginia, making a swing up over the Blue Ridge to Winchester and the Shenandoah Valley, where a force of Catawba and Cherokee Indian allies had begun to arrive, as well as a quick jaunt to Fort Cumberland at Wills Creek. In what amounted to a small victory for Washington, Lord Loudoun had decided that Fort Cumberland should belong to the Maryland troops and commanders, and the Virginia Regiment should complete Fort Loudoun at Winchester for its headquarters, as Washington had advocated. But despite the advantageous reorganization of forts, Washington worried that the Virginia Regiment would be weakened, because Governor Dinwiddie had offered to contribute some of its troops to help defend South Carolina, which feared an attack from French warships.

After cosmopolitan Philadelphia, he was stuck again in the crude frontier town of Winchester as spring thawed its one long street into a gumbo of mud and manure. His royal army commission shot down by Lord Loudoun, young Washington began to fantasize about the elegant life he would create on his own at Mount Vernon—a style that might match nearby Belvoir in Tidewater graciousness. In his spare time at Winchester, having moved from his rented house into quarters at the half-finished earth-and-log fort, he planned the remodeling and furnishing of his plantation house "below." In mid-April, he ordered from London an elegant mahogany dining set, wallpaper in various hues for five rooms, fine paper for a dining room with fifteen-foot-high ceilings, fancy door locks, two hundred and fifty panes of glass, and a marble fireplace, along with a specially sized landscape painting to mount above it. Enclosing his order with a covering letter to his London agent, Richard Washington, he specified that all items should be "neat"—meaning fashionable and well made. With his list of luxury goods, he also expounded to his agent on the hardships of his life on the frontier and the difficulties of its defense. "I have been posted then for twenty Months past upon our cold and Barren Frontiers, to perform I think I may say impossibilitys that is, to protect from the cruel Incursions of a Crafty Savage Enemy a line of Inhabitants of more then 350 Miles extent with a force inadequate to the taske, by this mean I am become in a manner an exile. . . ."

The wilderness had lost some of its charm for him, especially when serving there no longer seemed likely to earn him a royal commission. At the end of April 1757, having endured the frontier life at Winchester for only about three weeks since his Philadelphia sojourn, Washington rode out of the mountains and down to Williamsburg, where the Assembly was in session. Governor Dinwiddie had just announced to the lawmakers that he would return to England due to the strains of the office and his poor health, and the burgesses and councilors in turn thanked him for his service during a time more trying than anything his predecessors had faced.

Washington probably was delighted to see Sally Fairfax in the town, along with her husband, George William, and his father, Colonel William Fairfax. The Fairfaxes had traveled from Belvoir Manor south to Williamsburg for the Assembly session and the social life, as wealthy Virginia planters, many of them burgesses, congregated at the capital. Washington no doubt attended dinners and other social events at the

Fairfaxes' Williamsburg home, and perhaps danced with Sally there or at the public balls at the Apollo Room. He no doubt entertained the Fairfaxes with hair-raising stories from the frontier of Indian raids and mountain crossings and listened to the sound advice of Colonel Fairfax on how to conduct himself.

But Washington found little satisfaction from the Assembly itself. It could not decide at what strength—and expense—to maintain the Virginia Regiment. Another frustration befell him when the governor and Assembly, as a cost-saving measure, took away a bonus that he had enjoyed—a commission of 2 percent on all the funds for the troops that passed through his hands. This 2 percent commission paid for the expense of maintaining a well-laden dining table for himself and his officers and also defrayed the high cost of traveling about on the frontier.

"I . . . hope, that your Honor will not, after the repeated assurances given of your good inclination to better my Command, render it *worse,* by taking away the only perquisite I have," Washington wrote to Governor Dinwiddie, who instead would grant him a flat £200 yearly in table expenses, a substantial amount considering that he could buy about one hundred and fifty cows with that sum.

And travel he did. He rode hard in a great circuit between the mountains and the Tidewater region, between frontier homesteads and small forts and elegant plantations, between wilderness and civilization, as if the sheer energy of perpetual motion would solve all problems and bring him to the attention of those whose favor he sought. At home in both worlds, he crossed easily between them, bringing tales from one strange land to the next, the traveling hero on his quest to save a threatened people. He seemed to hate to sit still. When present at his post at Winchester's Fort Loudoun, he worked hard, churning out memos and orders and letters, monitoring the troops and trying to discipline them, dispatching them to the smaller forts strung along the frontier. But Washington found Winchester boring. ". . . I am tired of the place, the inhabitants, and the life I lead here," he had written a few months earlier.

In the four months between early February and the end of May, he ended up spending only a couple of weeks at his post in Winchester.

In mid-May, he left the social scene in Williamsburg and the business of the Assembly and rode back toward the frontier. Leaving one set of frustrations, he was about to ride into another. Parties of Indian allies had finally arrived at Winchester and Fort Loudoun from the nations of the Cherokee and Catawba in the Carolinas, as well as the Nottaway and Tuscarora. Recruited from regions to the south by various agents sent by Governor Dinwiddie, and including one hundred and forty-eight Cherokee warriors, they had walked hundreds of miles to Winchester, eager to receive the abundant "presents" that had been promised to lure them north to go raiding for French scalps. To their anger, they quickly discovered that the Virginia Regiment had no presents to offer them, nor money to pay for the French and Indian scalps they brought in, as promised by the Assembly. The Indians held an angry council. The Virginia officers remaining at Fort Loudoun in Washington's absence, including his aide Captain George Mercer, tried to make amends by offering goodwill belts of wampum. The Indians refused them. "They . . . told me the Govr knew not how to treat Indians," wrote Mercer to Washington, "that the French treated them always like Children, and gave them what Goods they wanted. . . . But they found from every Action, the Great Men of Virginia were Liars."

Mercer promised the Indians that the governor would send the presents and he would keep them safe and clean while the Indians sallied forth and made war against their mutual enemies. The Indians now relented. A Cherokee chief warrior, Youghtanno, said that he would accept the wampum and the promises of goodwill, accept the word for the deed. He had journeyed this long way to fight for King George, and fight he would unless the Virginians turned out to be liars. Another chief, Swallow Warrior, apologized for his earlier remarks about the Virginians but said he hoped that Governor Dinwiddie would take the message to heart and not make empty promises again. "In his own part," Mercer reported Swallow as saying, "he did not want Presents, but that it was his Promise of great Rewards from the Governour that engaged his young Men to come in, and that the Govr had now made him a Liar amongst his own Warriours: that made him angry."

Washington rode straight into this crisis. A Cherokee raiding party under first warrior Wawhatchee had recently brought four scalps and two prisoners to town. No presents had yet arrived from the governor. The Cherokee demanded reward immediately or threatened to walk back home. Everyone feared they might wreak havoc en route. Washington urged them to await the arrival of Edmond Atkin, the new Indian agent appointed by the London authorities to oversee Indian affairs in the southern colonies, which included Virginia. But they refused to listen to more promises, and Washington, despite his best arguments, fretted that the Cherokee and Catawba allies would abandon the Virginia forces any day. "[T]hey are the most insolent, most avaricious, and most dissatisfied wretches I have ever had to deal with," he wrote to Governor Dinwiddie, urging Atkin's—and the presents'—speedy arrival.

A few days later, the seething Catawba left for home. The Cherokee held on. A large party of seventy or eighty Cherokee warriors, along with Major Andrew Lewis and some of his soldiers of the Virginia Regiment, roamed the wilderness paths and forests, scanning the earth for the tracks of French and Indian parties to hunt down, and scalps and prisoners to take. Washington wrote with growing desperation to Governor Dinwiddie and Speaker John Robinson that Atkin had to arrive soon to handle the volatile situation. He also strongly recommended that one person, preferably his old companion Christopher Gist, be appointed under Atkin to handle all of Virginia's transactions with the Indians. There was a need to speak with one voice, Washington had come to understand, "as we are strangers to the . . . proper method of managing them."

"An Indian will never forget a promise made to him," Washington wrote to Governor Dinwiddie. "They are naturally suspicious, and, if they meet with delays, or disappointment, in their expectations; will scarcely ever be reconciled. For which reason, nothing ought ever to be promised but what is performed; and one only person be empowered to do either."

With prescience and perception, Washington understood early that this problem would intensify as the American frontier expanded westward, as more traders, settlers, land speculators, government agents, and others made promises and deals at cross-purposes with each other, or for which they had no authority, or which proved impossible to keep. A string of broken promises to the native inhabitants defined the western frontier as surely as the homesteader's ringing ax. And the consequences could be fatal. If something were not done soon, wrote Washington to the governor, "it will not be in my power to prevent their going off full of resentment!" Frontier families could pay the price.

More frustrations mounted for Washington at his Winchester post. For the second winter in a row, he had gone off in February on a mission of self-promotion—this past winter to call on Lord Loudoun at Philadelphia—and for the second spring in a row he had returned after a long absence to find the frontier in chaos and neglect. A tension existed between self-promotion and selflessness, between personal glory and the common good, and now he was seeing firsthand the trade-offs between them, the consequences of looking after his own welfare over that of his troops and the frontier families. His driving ambition was proving a dual-edged sword—one that helped propel him forward but could also inflict a great deal of harm on his own reputation and the troops and settler families around him.

As May turned to June, Washington struggled to keep the promises made to his soldiers, such as getting paid. The troops had received no pay recently, and none was expected for another six weeks. He considered himself lucky that his men had not mutinied and deserted, he confided to fellow commander Colonel John Stanwix, based in nearby Pennsylvania and in charge of five companies of British Army regulars in the Royal American Regiment. It helped, Washington wrote, that he kept his troops exhausted by the heavy earth-and-timber work of building Fort Loudoun at Winchester, rendering them too tired to mutiny. But that had its frustrations, too. Work on the fort proceeded so slowly that he did not expect to complete it anytime soon.

Washington's correspondent, Colonel Stanwix, as a British Army commander, ranked above Virginia colonel Washington, who was subject to any orders Stanwix gave. Then in his midsixties, Colonel Stanwix had deep roots in the elite echelons of the British military and political culture, having joined the Royal Army twenty-five years before Washington's birth; inherited the estate of his uncle, the wealthy former governor of Gibraltar; and served many years as a British member of Parliament. He had only recently arrived in America to command a thousand-man battalion in the four-thousand-man Royal American Regiment, a special force that Parliament had authorized in the wake of Braddock's Defeat, with funding of £81,000 and recruits drawn from British Army regulars, American colonials, and Swiss and German military experts who knew mountain warfare to counter the Indians' fighting techniques.

Always eager for an opportunity to impress his superiors, young Washington offered Colonel Stanwix his unsolicited advice on strategy, writing emphatically that now was the perfect time to take Fort Duquesne, as several escaped captives reported a weak force of only about three hundred French occupying it. He speculated that the French lacked the men to reinforce it, as they were too busy defending their territory closer to home against northern expeditions Lord Loudoun had sent toward Canada. "Surely, then, this is too precious an opportunity to be lost," Washington urged Colonel Stanwix, touching on another of his many frustrations. Stanwix apparently did not respond to this suggestion.

Another of Washington's frustrations arose from his quibbling with Governor Dinwiddie over extra expenses for himself and his Virginia Regiment, and the size of the servant staff allowed him. How many batmen could he have? These were men assigned to handle the packhorses that carried an officer's luggage. An exasperated Governor Dinwiddie wrote young Washington a forceful letter stating that he would receive the same allowance for these extras that the Royal Army officers received and no more—two batmen for a commander like himself plus two per company. In addition, he would receive provisions to feed his livery servant, if he had one. Also, provisions would be provided for no more than six women per company. A company in those days traveled with its complement of servants and women, whether officers' wives, nurses, washerwomen, or prostitutes, while a commanding officer such as Washington brought so much personal luggage it took two horses to carry it all.

Indian agent Atkin finally arrived in Winchester on June 2. Slothful, seemingly only semicompetent, the fifty-year-old Scot now took responsibility for negotiating with Virginia's Indian allies. Washington received a bit of other news that eased, for a moment, his frustrations. An express message from Fort Cumberland informed him that, after a long scouting expedition, Lieutenant James Baker and a group of Cherokee had just arrived. Five days earlier, deep in the Ohio wilderness, they had spotted two sets of human footprints on a narrow path leading toward Fort Cumberland. They had stealthily tracked the prints through the forest for eight or ten miles when their head scout suddenly motioned everyone to sit. He had glimpsed a party of ten French soldiers headed their way, returning from their own scouting trip.

"[W]hen the Frenchmen came within about fifty paces," reported Lieutenant Baker, "they saw our Men all Naked, and called to us and ask'd who we were, at which time we all rising together fired on them which they returned, we waited not to lode again, but run in with our Tomahawks the Frenchmen then making of as fast as they cou'd, but the Indians out runing them took two of them prisoners. . . ."

In all, the French lost six men—two dead, two wounded, and two taken prisoner. On the British side, a French musket ball through the head instantly killed the Swallow Warrior as he pounced with his tomahawk, while another shot wounded his son in both thighs. The Cherokee, enraged at Swallow Warrior's death, killed one of the French prisoners. The party then headed back to Fort Cumberland, enduring an excruciating four-day journey hauling their wounded man on their backs and eating nothing but wild onions. Such was Indian warfare in the Ohio wilderness.

Another frustration eased when, from Williamsburg, came the report that the burgesses had passed a bill providing for a draft to fill out the Virginia Regiment and £30,000 for back pay and future wages. In keeping with Virginia's strict class hierarchy, the draft targeted men at the lowest rungs of society and without a voice—"loiterers," "deserters," and those individuals without property who could not vote in the election of burgesses. The bad news was that the burgesses provided for only 1,272 troops—not nearly enough by Washington's thinking. Having witnessed the sorry state of some recruits, young Washington wrote himself a memo on just what kind of draftees he wanted: "Not to receive any but what is fit for the Service; reject all that are old—Subject to Fits—and otherwise infirm."

On the night of June 15, a state of high alarm erupted at Fort Loudoun when an express messenger galloped in from Washington's old rival, Captain Dagworthy, with the terse, shocking report that the French and Indians were on the march to attack the post Dagworthy commanded, Fort Cumberland. Six Cherokee scouts had just returned from the Forks. They had observed many French troops march out of Fort Duquesne accompanied by wagons and an artillery train and saluted by the boom of a large cannon. On the assumption that they were heading for Fort Cumberland, Captain Dagworthy had hastily summoned reinforcements from all quarters.

While Captain Dagworthy scrawled pleas for help, one of his officers, Major John Livingston, added a note to Washington with the Cherokee report that the French had bigger cannon than the British. He paused his pen for a moment—more intelligence was arriving as he wrote. With his scribbling quill betraying a breathless panic, Livingston then added a desperate, last-minute postscript to Colonel Washington: "[A] Safe Delivarence never was in Greater Jeopardy no men no provision &c. this is the Cry of This Garrison. . . . [T]he next Colol I belive may be from Montreal. . . . [I] am afraid the Great Guns wont be Loaded a second time by us."

The reality of command was proving far more complicated than a childhood dream of heroic action rewarded by unalloyed glory. Facing a difficult dilemma, Washington now had to ask whether he should heed the desperate pleas for help and send his Virginia Regiment troops from Fort Loudoun in a heroic attempt to defend Maryland's Fort Cumberland from imminent French attack or take a more cautious, conservative stance and keep his troops at Fort Loudoun and Winchester to hunker down and defend his own Virginia fort.

It had not gone well the last time he pushed ahead aggressively into an uncertain situation. As these urgent, late-night messages arrived, perhaps the memory surfaced of his men lying dead in trenches full of blood and mud and water. He decided to consult. He ordered his subordinate officers rousted from their beds at 2 A.M. to hold a council of war. With the question laid squarely before them in the candlelit log quarters, the impromptu council deliberated, Colonel Washington presiding.

The Cherokee scout report placed the French and Indian forces well en route to Fort Cumberland, having crossed the Monongahela River six days earlier. Washington and his officers discussed whether they had time enough to round up their own troops from outlying forts and march them to Fort Cumberland before the French and Indians struck. The answer was no. Furthermore, if they marched the Virginia Regiment to Fort Cumberland, they left Fort Loudoun defenseless. Should Fort Loudoun and the town of Winchester fall to the French and Indians, so would the entire Shenandoah Valley. From there, who could tell what might happen? The French could traverse the Blue Ridge into the rich Virginia coastal plain and its Tidewater plantations.

The Washington of three years earlier—the rash, impulsive one who stalked and ambushed the French party at Jumonville Glen, the heedless one who ignored Indian advice and heroically, disastrously, made a stand at Fort Necessity—might have chosen to push forward with all haste and all available troops to attempt the relief of Fort Cumberland. But this was a more mature Washington—now age twenty-five. He had experienced the consequences of moving too impulsively. He understood that his fellow officers might know more than he about the exact situation of the troops, and their judgment might prove valuable. He thought, perhaps, less about his own opportunities for heroism and more about the welfare of the whole. Rather than charge ahead, he paused to deliberate and consult. This would serve him famously in the decades ahead—the careful solicitation of advice, the close weighing of it, the thoughtful decision.

In the flickering candlelight, in the predawn hours, Washington and his council unanimously voted no.

But he still faced the dilemma of what to do about the troops and settler families taking refuge at the smaller, outlying Virginia forts. Should they send the women and children to safer settlements and concentrate all the men and soldiers to defend a single outlying fort? Or should they bring them all in to defend Fort Loudoun? Again, the council deliberated. It was decided, again unanimously, that holding one small, outlying fort would serve little use—far more important to bring in the men and the troops to defend Fort Loudoun and give the women and children refuge.

Sleepless, Washington scrambled, dispatching express messages to Governor Sharpe of Maryland, Colonel John Stanwix in Pennsylvania, and Governor Dinwiddie to notify them of the French invasion eastward from the Forks and the decisions of his council of war. He also summoned militia from three Virginia counties. In his message to Governor Sharpe, Washington made it abundantly clear that he thought Fort Cumberland was lost. "If the Enemy is coming down in Such Numbers, and with such a train of Artillery as we are bid to Expect Fort Cumberland Must inivitably fall in to their hands as no Efforts can be timely Made to save it. . . . It is Morally certain that the next Object Which the French have in view is Fort Loudoun. . . ."

While he wrote to Governor Sharpe that there was no hope for Fort Cumberland, he wrote to Captain Dagworthy at the fort itself to take hope—help was on the way. ". . . I have no manner of doubt, but a very considerable Force will be with you in a very little time."

One can speculate about whether Washington actually believed this or simply said it to boost morale in a doomed outpost. He clearly wanted to protect himself from charges of neglect. He found himself in a delicate, tension-filled position—saving his own fort while cutting Captain Dagworthy's fort loose to stand or fall on its own. His "moral certainty" that the French and Indians would aim for Virginia's Fort Loudoun after vanquishing Maryland's Fort Cumberland offered him a moral high ground on which to stand. The speedy arrival of reinforcements, however, could save both forts.

The moment express messengers bearing Washington's urgent calls for reinforcements rode up to Colonel John Stanwix at his headquarters about one hundred miles north in Carlisle, Pennsylvania, and to Colonel William Fairfax at Belvoir Manor, they, too, scrambled into motion. Colonel Stanwix arranged for wagons, ammunition, artillery, provisions, and a force of six hundred men to march down the Great Valley to Winchester to bolster Fort Loudoun. Colonel Fairfax rode hard from his Belvoir estate up the banks of the Potomac to Alexandria to meet with the militia captains of Fairfax County, hoping to raise one hundred men. They were to rendezvous at William West's ordinary and then to march to Winchester. Governor Dinwiddie, meanwhile, ordered out the militia from five other counties, which he hoped would total one thousand men.

For the next several days, under Washington's direction, the troops dug, hewed timbers, and raised ramparts feverishly day and night, apparently by torchlight, to complete the defenses of Fort Loudoun enough to withstand an imminent French attack of troops and cannon. Washington and Indian agent Atkin interrogated a French prisoner whom a scouting party had hauled in and debriefed Cherokee scouts who, employing a kind of hunter's measuring tape based on the dimensions of various animals, reported on the enormous size of the French guns that would level the British forts. "For, say they, your Guns are but muskets, compared with those the french have with them," wrote Washington to Colonel Stanwix. "Theirs will admit a *Fawn* in the muzzle, while yours will not take in a mans fist."

Washington's state of high alarm, however, suddenly dropped a notch the next day, June 21, when he received another message from Captain Dagworthy at Fort Cumberland confessing to confusion—actually, there was no French artillery train heading this way. Six Indian scouts had now arrived from Fort Duquesne to report that the false accounts had come from frightened young warriors who did not understand what they saw and heard. In fact, said the scouts, the French had sent out no wagons or artillery to attack the British forts but simply a large scouting party that had departed the fort to a send-off of cannon fire. "The storm which threatned us with such formidable appearances is, in a manner, blown over," Washington wrote to Colonel Fairfax.

With no love lost for Captain Dagworthy after their prolonged power struggle, Washington dismissively attributed the misunderstanding to Dagworthy's lack of a reliable interpreter. He wrote to his superior,

Speaker Robinson, that he found such a mistake "surprising" in an officer, implying that he would never make such a one.

Despite express messages sent by Washington to cancel the summoning of the militia, however, it was too late, and several hundred militiamen arrived at Winchester from the Virginia counties. Struggling to get them to work on the fort or to guard outlying settlers, Washington found them difficult—"obstinate and perverse," as he put it, and "egged on [encouraged] by the officers. . . ."

With the adrenalized alarm off and July wearing on in this summer of 1757, two years after Braddock's Defeat, with midsummer heat and buzzing flies smothering the drowsy frontier town, Washington's frustrations soared again. A number of Cherokee Indians wrongly jailed by Atkin nearly died in confinement due to bad food, which threatened to bring the retribution of their tribe down upon the Virginians. The incompetence of Peter Hog, commander of an outlying fort in the southern Shenandoah, caused Washington abruptly to remove him. The incredibly high rate of desertions was the worst of it. The burgesses had enacted a draft of vagrants and others who had no vote in order to boost the Virginia Regiment's strength to over twelve hundred from its current low of about four hundred. To enforce it they had renewed a law punishing desertion with death. En route from their home counties to Winchester's Fort Loudoun, however, the Virginia Regiment draftees disappeared at an alarming—and to Washington, infuriating—rate.

"I am greatly at a loss how to proceed . . . ," Washington wrote to Speaker Robinson. "[M]ore than one fourth of the draughts deserted before they reached this; and still continue to go off, notwithstanding I use every precaution I can possibly devise, to prevent this infamous practice."

He finally reached for the most draconian of measures—execution. What is it that can prompt a twenty-five-year-old to decide to hang some of his fellow soldiers? Where does frustration deepen into anger—anger at feeling helpless, ignored, and insignificant in the eyes of those upon whose favor he counted most? Even the militiamen refused to follow Washington's commands, and the draftees simply ran. Was it anger about the failure of an early promise and dream? The polished, dashing image embodied by his half-brother Lawrence, or a Royal Navy officer such as those he met in Barbados, standing astride the deck of a well-ordered ship with foam at its prow, lay so very far from this muddy, dusty, fly-bitten, ill-disciplined, isolated, ragged outpost on the edge of the frontier where he could make nothing work as he liked.

Washington would learn to contain and channel and use his anger, but not here, not yet. No doubt he saw the matter of executions as attention to duty, that he had to show the firmest hand in disciplining the troops. Without discipline, how could he possibly hold together the Virginia Regiment and protect the frontier settlers from the bloody massacres of prowling Indians, the hostage-taking, the scalping, and the dismemberment? He had to answer to those who pleaded to him for protection.

One imagines Washington's frustration and anger finally merging into resolve, stiffened by a sense of duty. He would solve this problem once and for all. He would bring order and discipline to his troops even if he had to take drastic measures. Once he had discipline, everything else would fall into place.

It's one measure of his anger and frustration that he ordered erected at Fort Loudoun a towering gallows forty feet in height. This in itself was something of a feat of engineering since it surely stood higher than any structure at Fort Loudoun or among the sixty houses of Winchester, perhaps including the flagpole. He requested that Governor Dinwiddie send blank orders of execution signed at the bottom. With these, Washington could simply fill in the name of the soldier to be executed rather than send all the way to Williamsburg for permission, as he had when he executed a deserter the previous year. He ordered court-martials to be held on July 25 and 26 for twenty-two prisoners held in the fort's prison. While not his first execution, this surely was his most dramatic.

"I have a Gallows near 40 feet high erected (which has terrified the *rest* exceedingly:)," he wrote to the distinguished older British officer Colonel Stanwix in Pennsylvania, "and I am determined, if I can be justified in the proceeding, to hang two or three on it, as an example to others."

Washington himself did not preside at the court-martials; Major

Andrew Lewis did. Washington oversaw the proceedings and testified against prisoners. There is no doubt that the members of the court knew his strong feelings. The first deserter brought forward, twenty-five-year-old Ignatius Edwards, the same age as Washington, was a carpenter and a good dancer and fiddler. They had both grown up in Stafford County. Perhaps they had known each other. Edwards had already deserted twice before. Colonel Washington testified before the court that he had specifically warned Edwards after his second desertion that if he deserted again and was caught, nothing could prevent his being hanged. This testimony was corroborated by a fellow officer. Edwards was then asked if he had anything to say.

> The Prisoner Ignatius Edwards says in his Defense that he deserted with an Intent to hire two Men to come in his Stead. He was then told that . . . no Pretence of that Kind wou'd be admitted as an Excuse for the Colo. himself had before he deserted told him that he wou'd not take Men in his [place]. The Prisoner then was asked whether he had any Thing else to offer in his Defence which might diminish the horrid Nature of his Offence or make it appear in a more favourable View to the Court: He answer'd NO.
>
> It is the Sentence of the Court that the above Prisoner Ignatius Edwards shall suffer Death by being hanged or shot.

And so it went, through all twenty-two prisoners held for desertion. Most received a sentence of death, either by shooting or hanging. A few testified that when they joined the Virginia Regiment they had not been read the Articles of War nor received a bonus on signing. Spared death, these prisoners received sentences nearly as harsh—one thousand or fifteen hundred lashes each, enough to kill a man.

In some ways this was not unexpected, as corporal punishment had a very long history in the British Army, with the aim not to administer justice but to immediately enforce discipline. Records exist back to the Crusades in the twelfth century, when Richard I laid down severe punishments for knife-fighting on board ship en route to the Holy Land (one hand chopped off), for killing a man aboard ship (perpetrator tied to the dead man's body and thrown into the sea), and for killing a man on land (murderer buried alive in the dead man's grave). Running the gauntlet, where the guilty party was beaten with rods by his fellow soldiers, was a common military discipline in the early days. Trends in punishment changed over the years and became more specific and refined in administering pain or humiliation, such as "riding the wooden horse"—the transgressor made to straddle a sharp-edged wooden board, sometimes with weighed legs. Even the sutlers and camp followers of the British Army, both male and female, had punishments prescribed for transgressions, such as a turn in the "whirligig." This was described in 1786 by Captain Francis Grose as "a circular wooden cage, which turned on a pivot, and when set in motion wheeled round with such amazing velocity that the delinquent became extremely sick, and commonly emptied his or her body through every aperture."

By the eighteenth century, floggings were routinely administered in the British Royal Navy and Army, often with the cat o' nine tails, a vicious lash formed of nine knotted strings. With a doctor present to ensure that the victim would not die, the flogging was administered before the assembled soldiers as an object lesson to all.

At Winchester, on July 29, 1757, two of the punishments went forward—the sentence for fiddler Ignatius Edwards, the first to appear before the court-martial, and William Smith, a twenty-year-old saddler and the last to appear. Washington summoned all the troops at Fort Loudoun to witness the event, including the many new draftees. Perhaps townspeople watched as well. No exact description exists of the moment. Did the gallows tower on the hilltop above fort and town while the drums beat mournfully? Was it a bright, fresh Shenandoah morning or a hazy, baking afternoon? Did they drop through a gallows trapdoor, the sudden jerk of the rope snapping their necks, killing them instantly? Or were they slowly hauled up, feet kicking, as they gradually choked to death? Did they dangle there, bodies limp, heads askew, at the top of the forty-foot-high gallows, as a warning for all to see the price of desertion?

And what did Washington feel as the execution under his command

and with his blessing unfolded? From the start of his military service four years earlier he had taken on responsibilities far beyond his age. Now, at twenty-five, he held in his hands the ultimate responsibility—the power to grant life or death. Did the gravity of the moment weigh down on him as the drums beat or as the last signs of life disappeared from the twitching bodies? Or did he brace himself with resolve not to let the executions affect him? Did he feel the deserters had received their just reward? Or did he feel a twinge of self-doubt over having in his hands the power to condemn his fellow men to death and whether he had exercised it well? He surely felt older and more weighted as the bodies ceased to swing and the crowds dispersed, as if the rough pull of the rope had strangled something innocent inside him as well.

Always conscious of presenting himself in the best possible light to his superiors, he kept up his efficient self-assuredness when he reported the executions to Colonel Stanwix and to Governor Dinwiddie: "Your Honor will, I hope excuse my hanging, instead of shooting them: It conveyed much more terror to others; and it was for example sake, we did it. They were proper objects to suffer: Edwards had deserted twice before, and Smith was accounted one of the greatest villains upon the continent. Those who were intended to be whipped, have received their punishment accordingly; and I should be glad to know what your Honor wou'd choose to have done with the rest?"

On the day of the hangings, July 29, Washington issued orders to his officers to ready their men to march to the various outlying forts. Although he had struggled with desertions, the new draft enacted by the burgesses had brought the Virginia Regiment up to about seven hundred men. This, however, fell well short of the twelve hundred they had agreed to fund, and far, far below the three thousand that Washington had estimated were needed to guard the frontier and take Fort Duquesne.

Making the spectacle a vivid forewarning to anyone else thinking of deserting, Washington had clearly timed the hangings to coincide with this issuing of orders and assigning of companies to respective forts. His detailed instructions to his captains covered everything from how to protect the frontier settlers by guarding the mountain passes through which the enemy might pass, to how much food to ration to each soldier (no more than one pound of flour and one pound of meat per day, and provide food for no more than six women per one hundred men). His instructions covered proper care of the uniform and the issuing of stockings and shirts (two pair of stockings and three good shirts per man). Emphatically, he wrote of the need to train the men in the use of firearms, in the proper discharge of orders, in the need to study their profession, and, most of all, in discipline.

"Discipline," he wrote prophetically, "is the soul of an army."

He first expressed it here at Fort Loudoun, in July 1757, in the wake of the hangings of the two deserters. But this would become George Washington's credo in the decades ahead.

CHAPTER NINETEEN

THREE MONTHS LATER, IN LATE AUTUMN OF 1757, WASHington's health had broken. He had battled the "bloody flux"—dysentery—in the past, notably during Braddock's march toward Fort Duquesne. At the same time as the hangings, the bloody flux reappeared.

At first, he ignored it. But it had grown steadily worse as summer turned to fall. In early August he had ridden over the Blue Ridge to Alexandria on personal matters. These included settling lingering issues with Lawrence's estate and checking in at Mount Vernon, where he awaited the arrival from England of his order of tools and fine goods to repair and furnish his plantation house—everything from broadaxes and socket chisels to wineglasses and yellow silk damask curtains. He planned to live like a gentleman at his Tidewater estate, regardless of what happened with his military service and royal commission.

September had brought him back over the Blue Ridge to Winchester and Fort Loudoun, where his condition worsened. He faced the same litany of problems—more Indian attacks, more desertions, and a rumor that his good name had been besmirched in Williamsburg to Governor Dinwiddie. It was said that the previous year Washington had exaggerated the Indian threat in order to extract extra money and men from the Assembly. Alarmed at this potential damage to his reputation, Washington had quickly written to Governor Dinwiddie to ask if he had heard the rumor and to deny it vigorously, dismissing it as a "stupid scandal" over a matter in which the facts were well known. He had perceived a change in attitude toward him on the governor's part, he wrote, wondering who the slanderer might be and anxiously trying to correct the impression: "[N]o man that ever was employed in a public capacity has endeavoured to discharge the trust reposed in him with greater honesty, and mor zeal for the country's interest, than I have done. . . ."

Governor Dinwiddie responded a few days later that he had not heard the rumor. However, he wrote, he had ample grounds to change his friendly attitude toward Washington, although he was trying to put them behind him. "My Conduct to Yo. from the Begining was always Friendly," Governor Dinwiddie wrote, "but Yo. know I had gt Reason to suspect Yo. of Ingratitude, which I'm convinc'd your own Conscience & reflection must allow I had Reason to be angry, but this I endeavour to forget. . . ."

Attempting to seize the moral high ground, Washington countered that he was simply being impartial and frank. "I do not know that I ever gave your Honor cause to suspect me of ingratitude, a crime I detest, and wou'd most carefully avoid. If an open, disinterested behaviour, carries offence, I may have offended: Because I have all along laid it down as a maxim, to represent facts, freely and impartially. . . ."

Worn out, in poor health, looking toward retirement in early November and a return home to England, Governor Dinwiddie did not reply to Washington's ongoing defense of his reputation. He had already cautioned Washington not to listen to every rumor or he would be driven to distraction. In late September, Washington made a hasty trip over the Blue Ridge to the Fairfax estate at Belvoir to attend the funeral of Colonel Fairfax, his longtime mentor, who had died suddenly on September 3. Returning to Fort Loudoun, he seemed unable to remain isolated at Winchester for more than a few weeks at a time. He made plans for his next trip to the Tidewater region—this in early November to settle matters with Governor Dinwiddie before he left for England and undoubtedly to polish any stains on his own reputation.

But it proved a difficult month on the frontier—a quartermaster stealing supplies, men trading equipment for liquor at Winchester's "tippling houses," scanty gifts and rewards to give the Cherokee who brought in French scalps, an officer vanishing during a wilderness scout, preserved beef going bad in its casks. And French-allied Indians relentlessly attacking the frontier settlers.

"I exert every means in my power to protect a much distressed country: But it is a task too arduous!" wrote Washington to the distinguished Colonel Stanwix in Pennsylvania. "To think of defending a frontier as ours is, of more than 350 miles extent, with only 700 men, is vain and idle." And to Speaker Robinson in Williamsburg, "No troops in the universe can guard against the cunning and wiles of Indians."

The war had gone badly in other theaters, too, since its declaration the previous year. Washington received letters from friends out in the wider world reporting setbacks and defeats. From Beverley Robinson, his correspondent in New York, Washington heard that Lord Loudoun had sailed with a heavily armed fleet and thousands of troops from New York to Halifax, Nova Scotia, aiming to seize the major French fort at Louisburg, on Cape Breton Island. He planned to use the fort as a British base to blockade French ships from entering the Saint Lawrence and replenishing France's Canadian forces. When a powerful French fleet appeared on the horizon from across the Atlantic, however, Lord Loudoun backed off. Meanwhile, from his friend Joseph Chew, also in the port of New York, Washington heard the alarming news of enemy assaults on British strongholds. The French had massed troops and Indian allies at Montreal, traveled south along the traditional river and lake route—the vertical line of that T—past their Fort Carillon on Lake Champlain until they reached neighboring Lake George and its Fort William Henry. With five thousand French soldiers and two thousand Indians, they had besieged the British fort—a prize that could give access to the Upper Hudson Valley. France would then be poised to push down the Hudson toward Albany and New York City, less than two hundred miles away, and sever New England from the southern provinces.

After the French, led by the vain and fearless Marquis de Montcalm, dug trenches to approach within a few hundred yards of the fort, knocked out its cannon, and began to batter down its stout walls with blasts of eighteen-pounder cannon fire, Montcalm offered the British commander, Lieutenant Colonel George Monro, the chance to surrender. Colonel

Monro readily accepted, taking up Montcalm on the traditional offer of departing with the full honors of war—carrying their personal possessions, one symbolic cannon, and saluted by the conquering French troops and officers. The Indian allies of the French, however, had no such European "rules of war." As the defeated British column marched out of the fort with flags flying and drums playing, the Indians pounced on the rearmost members, killing, scalping, plundering for trophies, and taking up to five hundred captives, in what became known as the "massacre of Fort William Henry."

Bad news arrived from Europe, too. Washington's friend George Mercer wrote a chatty letter from South Carolina, where among other topics he voiced his disappointment with the "bad shape" of Charleston's ladies—"many of Them are crooked & have a very bad Air & not those enticing heaving throbbing alluring Letch exciting plump Breasts common with our Northern Belles." Mercer also added the latest war news, much of it discouraging, arriving on ships from across the Atlantic. British-allied Prussia, under Frederick the Great, had routed the French-allied Austrian army at Prague, killing nine thousand, taking seven thousand prisoners, and capturing two hundred cannon. It turned out, however, that the Prussians had suffered heavy losses in the battle, too, and that the Austrians had then cut off their supply lines. In a massive battle a few weeks later at Kolín, thirty-two thousand of the British-allied Prussians faced off against forty-four thousand of the French-allied Austrians. Despite Frederick's pithy exhortations to his hesitating Prussian guards—"Rascals, would you live forever?"—he suffered a crushing defeat, with fourteen thousand casualties, his first loss. He pleaded desperately to his ally King George II of Great Britain for help.

Washington did not respond to this news from abroad, despite having had a role in setting in motion these massive armies. He focused on his own small world—the Virginia frontier and Tidewater society, making pendulum swings between wilderness and civilization. With Governor Dinwiddie he simply could not let matters rest—making known the never-ending needs of his regiment, presenting a constant defense of his own reputation, and issuing a stream of requests for leaves of absence. As fall wore on, his letters to Dinwiddie took on a more complaining, even whiny, tone. It was as if, with his deepening illness, he fell back into a more adolescent, self-centered version of his twenty-five-year-old self. In the isolation of his frontier post, he had no higher-ups to look after him, none of the warmth and support he had received when ill as a member of General Braddock's family or among the ladies at Tidewater and Belvoir Manor. The governor's letters to him grew brusque and dismissive. Ill and exhausted himself, soon to cross the Atlantic to "home," he had finally lost patience with Washington's endless needs and sensitivities. Asked by Washington for yet another leave of absence to come down to Williamsburg to settle accounts, Governor Dinwiddie answered with a flat-out no:

> *[Y]ou have been frequently indulg'd with Leave of Absence, You know the Fort is to be finish'd & I fear in Your Absence Little will be done & Surely the Commanding Officer Should not be Absent when daily Alarm'd with the Enemys Intents. to invade our frontiers, I think you are in the wrong to ask it, You have no Accots as I know of to Settle with me. . . .*

Washington responded that it was not for a "party of pleasure" that he asked for a leave of absence, adding sulkily, "I have been indulged with few of those, winter or Summer!"

He then complained to Speaker Robinson, skirting Dinwiddie and appealing to another superior, needing to justify himself to what he hoped was a sympathetic ear. It was necessary to come to Williamsburg, he wrote, not only to settle accounts, but also to explain in person the depopulated state of the frontier and the plight of the settlers. "But His Honor was pleased to deny his leave, thinking my request unreasonable; and that I had some party of pleasure in view."

Washington again wrote Governor Dinwiddie on November 5, begging to trouble His Honor on the subject of Indian affairs, as it was indispensably necessary. He had no presents to give to their Cherokee allies for the scalps they brought in, he wrote, and the British might pay a terrible price if the Cherokee should turn on them. Working further on the governor's sense of guilt, Washington said he had wanted to come to Williamsburg to help the poor frontier settlers get money owed to them for supplying provisions to the Virginia Regiment. But this request for leave, he noted gratuitously, the governor had denied him.

After this letter, his third complaint in ten days about the governor's denial of his leave, Washington suddenly lapsed into silence. The next news from Washington to Governor Dinwiddie came as a total surprise. It was not from Washington himself, but a subofficer, Robert Stewart, writing on his behalf. Washington's body had suddenly given out; his health had broken down:

> *For upwards of three Months past Colo. Washington has labour'd under a Bloudy Flux, about a week ago his Disorder greatly increas'd attended with bad Fevers, the day before yesterday he was seiz'd with Stitches & violent Pleuretick Pains upon which the Docr Bled him and yesterday he twice repeated the same operation. This Complication of Disorders greatly perplexes the Doctr as what is good for him in one respect hurts him in another, the Docr has strongly Recommended his immediatly changing his air and going to some place where he can be kept quiet (a thing impossible here) being the best chance that now remains for his Recovery, the Colo. objected to following this advice before he could procure Yr Honrs Liberty but the Docr gave him such reasons as convinc'd him it might then be too late and he has at length with reluctance agreed to it, therefore has Directed me to acquaint Yr Honr (as he's not in condition to write himself) of his resolution of leaving this immediatly. . . .*

One can only guess how much Washington's incapacitating illness was due to physiological causes—camp fever, amoebic dysentery, or whatever undiagnosed malady—and how much to psychological and emotional ones. Modern medicine would see a deeper link between the sudden breakdown in his health and the emotional stress, exhaustion, frustration, and despair that had depleted both his psychological resiliency and his body's immune-system defenses.

For months he had struggled with the futile task of defending the frontier while at the same time frantically scrambling to supply his regiment, pay his men, discipline his troops, punish deserters, gain a royal commission from his higher-ups, and guard his honorable reputation against what he saw as deliberate attempts to undermine and slander him. For months his personal ambition and relentless drive to succeed had outstripped his ability to perform these difficult and sometimes impossible tasks. He had driven himself hard. He had kept up a steady stream of letters, begging, pleading, complaining. He had undertaken long journeys through the wilderness and along the frontier to examine conditions and fortifications, and to Boston and Philadelphia seeking a royal commission. He had tried to discipline and drill the Virginia Regiment, bring order and professionalism to the troops, and curtail the infuriating desertions. He had thrown himself into these tasks—for the good of Virginia, for the glory of His Majesty, both of which boosted his own honor and advancement—but even his powerful will could not overcome the obstacles that sat like a great unmovable boulder blocking his path.

The next months remain hazy ones in the life of George Washington. Incapacitated for much of the time, he took refuge at his home at Mount Vernon. He wrote few letters, or few that survive—less than a dozen over the next three months. Those that do survive are terse. They deal with his illness, with furnishings for Mount Vernon, with doctors' orders, with requests to Sally Fairfax at nearby Belvoir Manor for recipes for comforting dishes and ingredients for cures—Hyson tea and Malaga wine—as if in his misery he turned to her for tender care.* With a strained and distant relationship to his mother, who lived forty miles south at his boyhood home

* He worried about the shipment of luxury goods for Mount Vernon that he had ordered the previous April, and added to it. "[S]end me . . . 2 dozn Dishes (properly sorted)," he wrote to his London agent, "2 dozn deep Plates, and 4 dozn Shallow Ditto that allowance may be made for breakage pray let them be neat and fashionable or send none." [George Washington to Richard Washington, January 8, 1758.]

at Ferry Farm, he had few other close women in his life. He consulted doctor after doctor. They gave him strict orders to avoid stress, avoid the business of public life. Unable to travel, he was so weak at times he could barely walk. "I have been reduced," as he put it, "to great extremity."

At Christmas, he was incapacitated. As winter passed, he did not improve. By early March, four months after returning home, he had fallen again into a state of despair. His condition was now complicated by symptoms of what he called "Decay"—or consumption, meaning tuberculosis, typically characterized by coughing, spitting up blood, and the body's wasting away. He wrote to his confidant on military matters, the distinguished British colonel John Stanwix in Pennsylvania, that he believed his life in the military had come to an end. He thought, in some way, that he had failed. His health spiraling downward, no chance of gaining a royal commission, having failed at his task, he was about to give up. He had no further interest in the military life or public business.

> *My Constitution I believe has receivd great Injury, and as nothing can retrieve it but the greatest care, & most circumspect Conduct—As I now see no prospect of preferment in a Military Life—and as I despair of rendering that immediate Service which this Colony may require of the Person Commanding their Troops, I have some thoughts of quitting my Command & retiring from all Publick Business, leaving my Post to be filld by others more Capable of the Task; and who may perhaps, have their Endeavours crownd with better success than mine has been.*

After four years in the wilderness and on the frontier, after helping to spark the French and Indian War, and surviving Braddock's Defeat, and countless hours trying to build, train, and discipline the Virginia Regiment and protect frontier settlers, it sounded like he was finished and ready to retire to Mount Vernon. It sounded like the farewell address of George Washington, age twenty-six.

Already he had lived a lot.

CHAPTER TWENTY

On March 5, the day after writing that despairing letter, Washington left Mount Vernon to ride the one hundred and twenty or so miles to Williamsburg and consult another doctor about his debilitating illness. Turned back by his condition in his previous attempts to travel, this time he rode slowly, making stops along the way, so that what was normally a trip of a few days took him nearly two weeks. After a brief stay at the plantation of Speaker Robinson, Washington rode the remaining distance south to Williamsburg. Here he consulted with Dr. John Amson, who apparently had expertise with the bloody flux. Dr. Amson listened to Washington describe his symptoms and the course of his illness. He gave his diagnosis: Washington's health was improving and he would eventually recover. Evidently, as the next few weeks would attest, the doctor did not caution Washington to avoid vigorous activity.

Having wallowed in despair for months, a palpable sense of relief at Dr. Amson's prognosis seems to have flooded over Washington. He soon rode from Williamsburg about thirty miles north to the Virginia Tidewater plantation known as the White House, on the Pamunkey River. Part of the vast holdings of the Custis family, here resided a newly widowed young woman with two small children, twenty-six-year-old Martha Dandridge Custis, whose claims through her late husband to the Custis holdings made her the wealthiest widow in Virginia. With his endless hunger for land and his ambitions to rise in the Virginia aristocracy, these details were not lost on the young officer who came calling, ill as he was.

It is not clear if they had met before or just what transpired between them on that brief visit in mid-March of 1758. He almost surely knew of her in the small world of Eastern Virginia society, and probably had made her acquaintance, as well as that of her late husband, Daniel Parke Custis. Descended from one of Virginia's most notable and landed families, Daniel Custis had died suddenly only the previous summer at the age of forty-five.

Young Washington did not shy from pursuing an opportunity when he saw it, and Martha Custis offered an opportune match. He had just spent the previous four months as a semi-invalid at his home at Mount Vernon. While ill, he had asked for help and medicinal remedies from his close neighbor Sally Fairfax. At the same time, her husband, George William, had sailed off to England to attend to business matters, not intending to return to Virginia until spring. How much time Sally Fairfax and George Washington spent together that winter of his illness and in what capacity—if any at all—does not appear from his few existing letters during those months or any other historical record. Perhaps Washington finally reached the conclusion—or Sally encouraged him to reach the conclusion—that his infatuation for her was doomed. She may or may not have loved him in return but at some point told him that it could never work. Perhaps he sought out Martha Custis on his own. Or perhaps Sally, looking after his best interests as well as her own, pointed Washington toward the twenty-six-year-old widow with the huge Virginia landholdings. Perhaps someone else did. All of it is speculation.*

He may well have heard the notorious stories that accompanied those holdings. Daniel Parke Custis had inherited through the Parke and Custis family lines seventeen thousand acres, which on his death came into Martha's hands. The holdings were beset by an old scandal and ridden with lawsuits, however. These dated back half a century to Colonel Daniel Parke, who, though born in Virginia, spent much of his time in England. An aide-de-camp to the Duke of Marlborough, he had delivered the news

* Martha apparently burned all her own correspondence with Washington after he died, leaving scant direct evidence of their relationship.

to Queen Anne of Marlborough's great victory at Blenheim in 1704. Winning the queen's favor, Parke was rewarded with a diamond-encrusted portrait of her and appointment to the governorship of the Caribbean's Leeward Islands, based at Antigua.

Five years later, enraged by his arrogant and "debauched" governorship, an angry mob dragged Governor Parke from Antigua's governor's mansion, stripped him naked, and murdered him in the street, apparently leaving his mutilated corpse lying there for a week. He left behind a wife and two young daughters back in Virginia and a trail of scandal in Antigua. This included love letters from young women on the island, both married and unmarried, discovered by the rioters among his personal effects. One of the married women, Catherine Chester, had given birth to a girl named Lucy, said to be Governor Parke's illegitimate daughter. When his will was read, it shocked both islanders and family members to learn that he had willed his entire Antigua estate, worth thirty thousand pounds, far more than his Virginia estate, to Little Lucy Chester.

That had been nearly fifty years earlier. But the lawsuits continued—a protracted battle over far-flung colonial wealth between the husband of Little Lucy Chester, now grown up and married, and the husband of the strong-willed Frances Parke Custis, one of Governor Parke's legitimate Virginia daughters, who had married an equally temperamental planter named John Custis. At issue was which part of the estate, Antigua or Virginia, would inherit Governor Parke's many debts. Although the case had been kicked back and forth for years between courts on both sides of the Atlantic by famous and expensive lawyers, and the original litigants had died, by March of 1758, when Washington showed up at the White House on the Pamunkey River, it had still not been settled by the heirs. This meant that the entire Custis estate, valued at over £23,000 and seventeen thousand acres as well as three hundred slaves, inherited by Martha and her two young children, could be stripped away to satisfy the accumulated debts of their rakish great-grandfather, Governor Parke of Virginia and Antigua.

The dark cloud hanging over Martha's sprawling landholdings did not deter young Washington. Nor, apparently, did his illness. Perhaps he had concluded, as he wrote to Colonel Stanwix, that his military and public life had come to an end. He may have come to believe—or perhaps the doctors suggested it, or Sally herself—that his best hope of recovering his health was a tranquil domestic life, and for this, he would need a good wife. Martha was an eligible young widow, belonged to Virginia's elite, and, by all accounts, including young Washington's, had an "agreeable" personality, although perhaps not the teasing, flirtatious magnetism of Sally Fairfax. Petite—only five feet tall, compared with his six feet—and with delicate hands and feet and dark hair, she projected a natural warmth and modesty noted by many visitors to her home. Possessing good skills for managing the domestic life of a large plantation and its in-house staff of slaves, she liked to read the Bible and practice intricate needlework, holding both her own work and the work of those in her sewing circle to a high standard. Like young Washington himself, she had an eye for elegant outfits and velvets and silks imported from London. It was common among Virginia's elite to remarry quickly after the death of a spouse, so for Martha, already pursued by other suitors, to remarry quickly would not cause scandal.

So much turned around in the course of a few weeks. At the same time that Dr. Amson gave Washington the good prognosis about his health and Washington called on the Custis White House, a startling piece of news came across the Atlantic to New York, raced down the east coast, and arrived at Williamsburg. Great Britain's new secretary of state, William Pitt, had relieved Lord Loudoun of the command of North American forces after his miserable failures the previous summer and fall at Louisbourg and Fort William Henry. Pitt himself would now take charge of the war. For the coming fighting season of 1758, he had devised a British strategy against the French on several continents. In North America, he would focus on three major objectives—take the fort at Louisbourg on Cape Breton Island, take Fort Carillon on Lake Champlain, and take Fort Duquesne at the Forks of the Ohio.

With this stunning news, a sudden sense of urgency overtook Washington. For two years or more he had urged the capture of Fort Duquesne as the only solution to the relentless French and Indian raids. Ignored by Lord Loudoun and other higher-ups, he had despaired of ever seeing it happen. But now the highest British authorities, embodied by William Pitt, had launched a great plan to take back the Ohio. How could he miss it? He felt a surge of excitement, and perhaps a sense of personal ownership and even envy. That was *his* plan Pitt had proposed. He could produce copies of letters pushing for this very thing, to Lord Loudoun, Governor Dinwiddie, and others. He had been off the field of action for months, wasting away at Mount Vernon, ordering china plates from London. Now the action would take a dramatic turn. Scores of other young officers would fight in this glorious campaign, many of them Virginians. Heroes aplenty would emerge. He possessed a rare expertise—he knew the Ohio wilderness terrain and the French and Indian tactics. He was determined not to be left behind.

Within two weeks, Washington made a remarkable turnaround. In early March, he had appeared utterly forlorn, depressed, and gravely ill. He was contemplating quitting the military and public life forever, his health having suffered a "great Injury" and needing the "greatest care" to restore it. Darkly embracing a sense of failure, he was ready to surrender his post to abler hands than his own. Less than a month later, however, in early April, he had banished dark thoughts about quitting, his health had made a startling recovery, and he was riding eagerly away from the comforts of Mount Vernon back to his post on the frontier at Winchester.

What accounted for such a quick rebound from a deep personal crisis? These three simultaneous events—Dr. Amson's prognosis for his recovery, his acquaintance with Martha Custis, and William Pitt's plans for a campaign into the Ohio—sparked new life in a moribund Washington. He simply could not bear to miss the great Ohio expedition. A potent mix of duty to his country and hunger for personal glory motivated Washington to abandon Mount Vernon and early retirement from public life to resume his frontier post.

While still in Williamsburg at the end of March he rode again to the White House to pay a second call to Martha. Apparently it was then that Washington proposed marriage to Martha or asked that she consider it. Receipts show that a few days after this second visit, Washington ordered from London formal clothing suitable for a wedding ceremony, and Martha Custis appears to have done the same.*

What else could have taken place then? What did he tell Martha about his future role in the military, if, in fact, he was to have one, and what did she tell him? Did he say he was quitting, or was he then ready to hurry back to the frontier? Did she encourage him to return to Mount Vernon and the quiet life of a planter, or did she urge him onward, to do his duty for king and country, to protect their blessed Virginia and its mountainous western frontier from being overrun by the bloodthirsty French and Indians? Standing before Martha, dressed perhaps in his military uniform, did Washington see future glory reflected in her shining eyes?

Whatever the case, they apparently agreed to postpone matrimony until the campaign in the Ohio wilderness had ended. Martha seems to have inspired him—or at least not discouraged him—to take the path of bravery, glory, and possible death, for within a week of seeing her, and after an absence of five months, he rode back over the Blue Ridge to the Shenandoah and his post at Winchester and enthusiastically prepared to march over the next mountain ridges into the Ohio and take Fort Duquesne.

* * *

Those few weeks in March 1758 would set the course for much of what followed in Washington's life. He believed in Providence as a guiding force in human events. Few events, however, unfolded as planned over the next nine months—neither in love, nor in battle, nor in constantly shifting allegiances.

It was assumed that he had quit. Word that Washington had resigned had reached British general John Forbes, newly appointed by William

---

* Washington ordered from his London agent "the best superfine blue cotton velvet" to make a coat, waistcoat, and breeches along with fine silk buttons, while Martha ordered from her agent a "genteel" suit of clothes "to be grave but not to be extravagant and not to be mournful." [George Washington to Richard Washington, April 5, 1758, and Martha Custis to Cary and Co., 1758 (undated), both quoted in Freeman, 2:302.]

Pitt to command the massive force marching to the Ohio. This was hardly surprising, as Washington had not appeared at his post for five months. Or perhaps General Forbes heard it from Colonel Stanwix, who had received Washington's dark correspondence.

Under Pitt's orders, General Forbes was to lead into the Ohio a force of two thousand British Army regulars, five thousand provincials from the Middle Colonies, and as many Indian allies as they could recruit. While this massive body of troops exceeded by several times Braddock's army, it counted as only a fraction of the fifty thousand or so troops that Pitt had called up for all his North American campaigns, which formed only a single facet of his grand, global strategy against the French. He would tie up France where her armies were strongest—on the European continent—by financing his Prussian allies like Frederick the Great to fight the French on Great Britain's behalf. At the same time, British forces would harass the French by conducting raids across the English Channel on the coasts and ports of Brittany and Normandy. In addition, the powerful British Royal Navy would both dominate the Atlantic and target French possessions around the globe—in the Caribbean, West Africa, and India. Pitt would save his most powerful blow against the French for the place where they were weakest—North America. Here France had an enormous land mass to defend with only half the troops, many of them untrained militia, that Pitt was sending into action in his three great North American campaigns.

George Washington represented only one small cog in this vast global effort, despite the young officer's part in igniting the sprawling war. However, General Forbes knew of Washington and thought he could prove useful with his knowledge of the wild mountain and forest terrain. ". . . I should be extreamly sorry [if he has resigned], as he has the Character of a good and knowing Officer in the back Countries," wrote General Forbes to John Blair, serving as acting governor of Virginia after Governor Dinwiddie's recent return to England. "If he therefore would serve this Campaign I should be glad that you orderd his regt to repair to Winchester directly. . . ."

Spurred on by news of the impending attack against the Ohio, Washington soon arrived back at his old post at Winchester. Never slow to promote himself, he wrote to his regular correspondent and distinguished Royal Army officer, John Stanwix, whom General Forbes had recently promoted to brigadier-general, and asked that Stanwix recommend him to Forbes. Not because he sought a royal commission, Washington insisted, perhaps disingenuously—he had already given up hope of that, he said—but because he wanted to be distinguished "from the common run of provincial Officers; as I understand there will be a motley herd of us."

Washington felt a particular need to distinguish himself from the ordinary run of officers because William Pitt, showing an understanding of the colonial mind-set, had ensured there would be so many of them. Unlike Lord Loudoun's heavy-handed demands for troops and resources from the colonial legislatures, Pitt took a cooperative approach, suggesting that the Crown would reimburse the colonies for expenses incurred in fighting on the British Empire's behalf. Instead of resisting, as they had under Lord Loudoun, the colonial governments responded enthusiastically, with Connecticut authorizing five thousand men, New York twenty-seven hundred, while in Virginia the burgesses voted to up the number of provincial troops to two thousand. These, the burgesses decided, would be split into two Virginia regiments, the 1st Regiment under Washington and the 2nd Regiment under twenty-nine-year-old William Byrd III, an aristocratic member of the Governor's Council and a compulsive gambler and horse bettor.

No longer standing alone at the head of the Virginia Regiment, Washington had to work all the harder to distinguish himself from the "common run" of officer. Back at Winchester and Fort Loudoun, however, he found himself beset by many of the same problems that had stymied him earlier. Some four hundred Cherokee had shown up in Winchester to help in the fight against the French, finally giving the British a sizable Indian ally, but Washington feared they would drift away to their southern home if the campaign did not begin soon. Nor, as ever, were there enough presents with which to reward them. He urged General Stanwix as well as his own Virginia superiors in Williamsburg to launch the big expedition into the Ohio without delay.

But delay followed delay. His frustration built anew, despite his lengthy rest at Mount Vernon. By June, two months after he had arrived, eager to join General Forbes's campaign against the French, he and his troops still remained at Fort Loudoun. He believed that Fort Loudoun would serve as the ideal staging ground for the great thrust into the Ohio, located as it was at a major Indian crossroads and with ready access to the start of Braddock's Road. To his great distress, however, he learned that General Forbes eyed an altogether different route. Rather than follow the existing road, the general planned to assemble all troops at Raystown, about sixty miles north in Pennsylvania, and then chop a *new* road through the wilderness to Fort Duquesne. This news sent Washington into a frenzy of alarm. If General Forbes attempted this different route, wrote Washington to his friend Francis Halkett, who served as secretary to the general, "all is lost!—All is lost by Heavens!—our Enterprize Ruind. . . ."

Washington claimed to have no self-interest in the choice of route—"I am uninfluencd by Prejudice"—but, in fact, he had a great deal. If the general marched along Braddock's Road, Virginia merchants would stand to gain all the commerce for supplying many thousands of troops—trade that would originate in Virginia ports but that it would lose to Pennsylvania with the alternative route. Furthermore, wealthy Virginians had already speculated in Ohio Valley land through the Ohio Company, whose shareholders included Washington's late brother, Lawrence, and his other half-brother, Augustine Jr. The use of Braddock's Road, rather than a route originating in Pennsylvania, would help solidify Virginia's and the Ohio Company's claim to these prime lands at the Forks of the Ohio. The province, and the investors, could stand to gain enormously.

Washington argued forcefully in favor of Braddock's Road to Colonel Henry Bouquet, General Forbes's field commander, not on the basis of these benefits for Washington's province of Virginia but on the pressing need to arrive at Fort Duquesne before winter's snow and ice slammed the mountain ridges. To chop a new route from Pennsylvania to Fort Duquesne would require many months, he asserted. Having experienced it firsthand, he named the exact point where deep mountain snows would stop the expedition dead in its plunging, struggling tracks—the seventy-mile-long ridge known as Laurel Hill. Rising to elevations of three thousand feet, the great ridge reaching up into the winter clouds captured heavy snowfalls from storms sweeping in from either west or east, burying the unwary traveler and, Washington was sure, any hope for victory. His observations about Laurel Ridge, however, marked only the beginning of Washington's personal battle over General Forbes's route.

As he idled at Winchester awaiting orders from London, or Williamsburg, or the quill pen of General Forbes on route and troop movement, and after the previous year's frustrations, Washington may have realized more clearly than ever that the power to fund and set armies in motion resides in heads of state and governments. He wanted a role in those decisions. Several weeks earlier, having attended the Assembly session in Williamsburg, he had decided to take another run at a seat in the Virginia House of Burgesses. He now pursued it.

As in his failed attempt nearly three years earlier, he would seek a seat from Frederick County. This encompassed Winchester and part of the Shenandoah Valley, where Washington as a teenage surveyor had purchased land, and the current site of Fort Loudoun. Despite his frequent absences from his post at Fort Loudoun, the local people of Frederick County knew him much better than they had three years before. Unlike on his first attempt, he made certain to announce his candidacy early and to line up support. Some of the county's leading citizens and landowners enthusiastically backed Washington, including members of the Fairfax family, while his Virginia Regiment officers worked the voters on his behalf against incumbent Thomas Swearingen.

But an obstacle arose when orders from Colonel Bouquet finally arrived, sending him and his troops to Fort Cumberland and out of his voting district. Washington left Winchester on June 24 at the head of some six hundred men of the 1st Virginia Regiment—his first move in Forbes's Ohio expedition—arriving at Fort Cumberland after a slow, nine-day march bogged down by bad roads and twenty-eight wagonloads of fodder. As the July 24 election date neared, however, his supporters in Frederick County urged him to return to Winchester to "show your face" and ensure a win. Washington hesitated. What if the French and Indians launched a surprise attack on Fort Cumberland while he was away campaigning for himself in Winchester? Whether he won the election or not, his reputation might never recover. Showing a new sense of diplomacy and deftness in handling this thorny problem, he applied to Colonel Bouquet, the Swiss mountain warfare expert under General Forbes, for permission to return to Winchester for the election. At the same time he sought the private advice of those close to Colonel Bouquet at Raystown about how the colonel received the request. "Col. Bouquet was at first in a great dillemma," replied Adam Stephen, Washington's old acquaintance, now with Bouquet, "betwixt his great inclination to serve you, & the Attachment he has to regularity, duty & discipline."

Despite Colonel Bouquet's reluctant willingness to grant permission, Washington prudently decided to stay put at Fort Cumberland, but he did everything else within his power to ensure that he would win the election. In this era, candidates for office routinely bought drinks for voters on election day—and were expected to. The normally expenditure-conscious Washington instructed Lieutenant Charles Smith, whom he had left in charge of Fort Loudoun, to spare no expense when it came to buying rounds in the "tippling houses" of Winchester, whether the voter had pledged himself to Washington or not.

"I hope no exception were taken to any that voted against me," wrote Washington a few days later, "but that all were alike treated and all had enough it is what I much desird—my only fear is that you spent with too sparing a hand."

He won handily with 309 votes, while rival Swearingen received 45.* The liquor bill to win those voters was more impressive still: 28 gallons of rum, 50 gallons of rum punch, 34 gallons of wine, plus abundant beer and cider, consumed by 391 drinkers. (Apparently, Washington lost 82

* Four candidates ran for two seats. The totals were Washington, 309; Thomas Bryan Martin, 239; Hugh West, 199; Thomas Swearingen, 45. Martin, a young nephew of Lord Fairfax who had received large landholdings in Frederick County, was an ally of Washington's. [Freeman, 2:317–20.]

of those well-lubricated voters to rivals.) The bill totaled close to £40, almost as much as a schoolteacher's annual salary. Swearingen's expenditures for free drinks, if anything, went unrecorded.

Washington's first election victory demonstrated the advantages of preparedness and cultivating leading citizens, strategies he would remember in the years ahead. The headiness of his win may have emboldened him to advocate even more fiercely for the Braddock's Road route to Fort Duquesne. He had shown a modicum of tact when he approached Colonel Bouquet about returning to Winchester for the election, but now he reverted to his old stubbornness. Unhappily for Washington, the letter he wrote to Francis Halkett about the folly of General Forbes's new route—that all would be lost if the general followed it—accidentally came into the general's own hands. When he read it, his wrath fell on Washington and fellow Virginia Regiment commander Colonel William Byrd III. "I am now at the bottom of their scheme against this new road," General Forbes wrote scathingly to Bouquet about the pair, "a scheme that I think was a shame for any officer to be connected with. . . ."

Still Washington pushed.

AS SUMMER'S THICK HEAT ROSE in Virginia's wooded backcountry, and Washington focused narrowly on county elections and wilderness roads, events in the wider world accelerated apace. In the quest to strike at France's weakest links, William Pitt's other campaigns against New France got under way.

A crucial divide between the British and French in North America lay in what is now upstate New York. There sat two long, skinny lakes, running north to south and separated by a narrow neck of land. Nestled in one of the long, trenchlike valleys of the Appalachian Mountains, these lakes formed a vertical part of that T-shaped river-and-lake route that ran between the port town of New York and New France's Montreal. The French claimed the northern of these two skinny, connected bodies of water, Lake Champlain, named for the great French explorer, and the British claimed the southern one, or Lake George, named for the British king.

The British forces had suffered a stinging loss the previous year on

Lake George when the French besieged their Fort William Henry, reducing it to ashes while the Indians scalped the stragglers in the surrendering column. Now Pitt was determined to avenge that loss by taking the French stronghold on Lake Champlain. Known to the French as Fort Carillon, the star-shaped bastion guarded the lake's waters from a peninsula at its southern end—a peninsula known to the British as Ticonderoga.

Led by General James Abercromby, whom Pitt had appointed in charge of all British forces in North America, some sixteen thousand regular and provincial troops massed in early July at the charred ruins of Fort William Henry on Lake George—the largest force ever assembled in North America. Known as "Granny" to his men, overweight and fifty-two years old, General Abercromby cut less than a prepossessing figure—"an aged gentleman, infirm in body and mind," recorded one seventeen-year-old Massachusetts soldier in his diary. The actual assault on Ticonderoga was to be led by a much younger and more energetic British officer, thirty-four-year-old Lord George Howe. He had spent much of the past year studying bush warfare techniques and had gone out on scouting expeditions with a New Hampshire bush fighter and former counterfeiter, Robert Rogers, whose strategies derived from Indian warriors. Inspired by Rogers, Lord Howe ordered his British troops to cut their long hair, trim off the flapping tails of their coats and wide brims of their tricornered hats down to two and a half inches, and wear Indian leggings to slip quickly through the forest.

"No women follow the camp to wash our linen," wrote one soldier. "Lord Howe has already shown an example by going to the brook and washing his own."

At least initially, the assault followed conventional British strategy. With flags flying, drums beating, and bugles playing, an enormous British flotilla of one thousand bateaux, whaleboats, and barges rowed north up mountain-rimmed Lake George early on the sunny morning of July 5, 1758. When viewed at a distance of three miles, one spectator reported, the fleet entirely covered the lake's surface, the morning sun flashing on oars and muskets while music echoed from the wooded mountains. After rowing all that day and through the night, the flotilla landed at the north end of the lake the next morning. Lake Champlain and its star-shaped French fort at Ticonderoga lay only four miles away through the gorge of a steeply falling stream and brushy forest. A British advance party led by Lord Howe became disoriented in these dense woods and bumped into a retreating French party, also lost. Musket fire thudded through the undergrowth. A ball tore through Lord Howe's chest, and he fell dead. With his death, the great expedition against Ticonderoga lost its guiding force. "In Lord Howe," wrote one officer, "the soul of General Abercromby's army seemed to expire. . . . [A] strange kind of infatuation usurped the place of resolution."

Though equipped with the largest fighting force North America had ever seen, General Abercromby hesitated. It took another two days for his troops to march to the nearby forested peninsula of Ticonderoga. By then, the French at the fort, under the Marquis de Montcalm, had time to build a nine-foot-tall breastwork of logs across the peninsula, blocking access to the fort. Using the materials at hand, they also improvised a defense called an *abatis* by chopping down hundreds of trees in front of the breastworks to the distance of a musket shot, felling them so their tops faced outward in a dense thicket designed to slow a British attack. In the words of one British officer, it looked like a forest flattened by a hurricane. They hurriedly wove a dense mat of boughs just beneath the breastwork itself, hewing the green wood of the limbs to sharp points with their axes, to snag any British infantry who bravely attempted to storm the wall. Then Montcalm rested his men, gave them food and drink, and waited.

Headquartered at a small, abandoned French sawmill a mile and a half to the rear, in the river gorge that linked the two lakes, General Abercromby possessed a range of options, from besieging the fort and starving out the defenders, to rolling cannon up a nearby hill and shelling it into submission. But with a vastly superior British force of sixteen thousand men compared with about thirty-five hundred French defenders, General Abercromby, who still had not set eyes on the fort, elected to take it head-on. At twelve thirty on the sunny, warm afternoon of July 8, the first rangers sent by the general emerged from the forest, along with about one thousand light infantry and provincial forces in their blue, all of them firing over the downed and leafy treetops at the log breastwork where the French soldiers in white remained hidden. Then a mass of Redcoats—well-disciplined British regulars—emerged from the forest shadows, a full seven thousand of them, and marched forward resolutely in orderly rows, bayonets fixed on their musket barrels, to storm the log wall.

The felled trees disrupted the order of their march, and as the careful columns broke to twist and turn their way forward through stumps, logs, and leafy masses, a sudden explosion of musket fire and smoke erupted from hundreds of loopholes in the log breastwork. Musket balls whizzed about the marching Redcoats. Men dropped—"Cut Down Like Grass," in the words of one Massachusetts provincial looking on from the forest's edge. Returning fire, the most intrepid Redcoats pushed forward, until stopped by the mat of sharpened sticks beneath the barricade wall and the thundering barrage of musket fire from above. Then they retreated.

". . . I could hear the men screaming, and see them dying all around me," wrote one soldier, who hid behind a pine stump while musket balls thudded into the earth inches away. "I lay there some time. A man could not stand erect without being hit, any more than he could stand out in a shower, without having drops of rain fall upon him. . . ."

Sending orders from his sawmill at the rear, General Abercromby again and again ordered the Redcoats to storm the barricade with bayonets—six times, by an admiring Montcalm's counting, between one in the afternoon and seven in the evening. Finally, in the last British advance, as dusk fell, the soldiers' objective was to recover the hundreds of wounded. The dead remained strewn about the downed treetops and stumps and dangling from the sharpened limbs.

In all, the British lost 1,944 officers and men, killed, wounded, or missing, while the French lost 377. Almost as humiliating to the British was their retreat. That night, General Abercromby, concerned about a counterattack from Montcalm and unsure of the French numbers, ordered his troops to retreat through the river gorge and forest to the thousand boats beached on the shore of nearby Lake George. But rumors soon raced among the men that a murderous French counterattack had already begun, and they stampeded in panic, some leaving behind even their shoes, snagged in the mud of the swamps they scurried across. They jumped into their boats as the sun rose, and what had once been a great, proud flotilla with drums playing and flags flying now rowed madly down the lake, not pausing to rest until they reached the lake's far end and the ruins of Fort William Henry.

British regulars and provincials alike expressed disgust at General Abercromby's leadership. That the general could simply have planted two cannon on a nearby hill and decimated the French troops holding the barricade, wrote one British officer, would have been perfectly obvious "to any blockhead who was not absolutely so far sunk in Idiotism as to be oblig'd to wear a bib and bells."

Other events occurred in the wider world that George Washington knew little about but that would soon affect him profoundly. Singularly fixated on the route of attack against Fort Duquesne, he relentlessly advocated for the Virginia approach and Braddock's Road. Never one to embrace the subtleties of Indian culture and political alliances—rather seeing Native Americans as crafty and lethal woods warriors—Washington did not appear to fully grasp the significance of an undercover diplomatic mission that General Forbes and Pennsylvania's Governor Denny had launched to contact the enemy Ohio Indians.

Smart, methodical, conscientious, although chronically ill, General Forbes, unlike his predecessors, made a special point of understanding the complex political and cultural dynamics of the Indians of the Ohio and the frontier. Forbes had grown up on his Scottish family's estate in Dunfermline, home of Scottish kings dating back to 1000 AD, and had studied medicine before joining the British Army. He helped defeat the Jacobite Rebellion of 1745, perhaps acquiring a unique insight into tribal dynamics among Highland clans that joined to fight against the British Crown. In the Ohio campaign, Forbes received special permission from his superior, General Abercromby, to negotiate directly with the Western Delaware and other enemy Indians in the Ohio wilderness.

This came as a departure from normal protocol. Normally, British negotiations with Ohio Indians would have been conducted by William Johnson, the Indian agent based near Albany in New York province. Born in Ireland, Johnson had learned Mohawk as a young trader in the Upper Hudson Valley and then served as Indian agent and primary British contact with the Iroquois Confederacy, the powerful confederation of six nations that claimed it ruled over the tribes in the Ohio—the Delaware, Shawnee, and Mingoes. At the outbreak of fighting between the British and French, the Mingoes had remained friendly to the British, while the Shawnee had joined the French and launched raids on frontier settlers, as did the two tribes known as the Eastern and Western Delaware. The Eastern Delaware, originally inhabitants of the Delaware River valley in today's eastern Pennsylvania and New Jersey, had special reasons to resent the British, having already been cheated, by the sons of William Penn in the Walking Purchase scam, out of huge chunks of their native homelands.

It was the pacifist Quakers of Pennsylvania, unwilling to join the fight against the French and their Indian allies, who first reached out to the hostile Eastern Delaware and their leader, Teedyuscung. Through him, the Quaker negotiators also hoped to reach the Western Delaware and other enemy tribes deep in the Ohio wilderness. The nonviolent Quakers found Teedyuscung willing to listen to peace overtures. After meeting with Pennsylvania's Governor Denny in the summers of 1756 and 1757, Teedyuscung agreed to switch allegiances from the French side to the British. In return, he received promises from Pennsylvania to build settlements and a trading post for the Eastern Delaware in the rich Wyoming Valley, about one hundred miles north of Philadelphia, and to review the Walking Purchase.

Now, in July 1758, a year after these meetings, and while General Abercromby met humiliation at Ticonderoga and Washington harangued General Forbes about Braddock's Road, Teedyuscung made a sojourn far into the Ohio. He returned to Philadelphia escorting two key leaders of the Western Delaware, one of them a famous chief named Pisquetomen. Resentful of the meager rewards from the French for their help in fighting the British, the Western Delaware leaders, too, were open to proposals to switch their allegiance. But Pisquetomen requested an emissary from the British to come to the Indian towns far back in the Ohio forest and persuade their fellow Western Delaware leaders to side with the British. For a British emissary to journey into hostile territory controlled by the French and their Indian allies was clearly a highly dangerous and possibly fatal mission.

Pennsylvania's Governor Denny chose Christian Frederick Post for the job. Born in Prussia and trained as a cabinetmaker, Post had become a passionate convert to the Moravian Church, attracted by its de-emphasis on doctrine and its message to follow Christ's living example of "gentleness, humility, patience, and love for our enemies." Post came to Pennsylvania as a missionary for the wide-ranging Moravians, married a Delaware Indian woman, and developed a fluency in several Indian languages and an ease in dealing with various Indian tribes. A plain German of simple character, direct and honest, with a sense of duty and trust in God, as Francis Parkman describes him, Post came to the attention of Governor Denny and General Forbes as a possible emissary to the enemy tribes.

"[Post] now accepted his terrible mission," writes Parkman, "and calmly prepared to place himself in the clutches of the tiger."

POST AND HIS WESTERN DELAWARE ESCORT, Pisquetomen, left Philadelphia and struck out into the Ohio wilderness toward the end of July. It is unclear how much Washington, training his troops at Fort Cumberland for the impending attack on Fort Duquesne, knew of Post's diplomatic mission into the wilds. Summer was passing and, with it, seasonable weather for crossing the mountains. Washington railed about the choice of the wrong road, although he detected a glimmer of hope when he received orders to clear a stretch of Braddock's Road. As August waned, General Forbes's troops crawled slowly toward Fort Duquesne, clearing the Pennsylvania route. Unlike Braddock, who had tried to move quickly and carry everything with him, Forbes ordered a methodical advance by building fortified outposts at regular intervals along the way. His troops, under Colonel Bouquet, chopped their way over the Allegheny crest, and then over the long ridge known as Laurel Hill—where Washington had predicted winter would stop them dead—to establish an advance base at Loyal Hannon, some forty miles from the Forks. Even then, as the Pennsylvania road proceeded, Washington remained fixated on what he believed was a colossal mistake.

Stuck at Fort Cumberland while the road proceeded through Pennsylvania, Washington's frustration climbed to a dramatic pitch, a mixture of lamentation laced with anger. In a letter to his ally, Virginia Speaker Robinson, he blamed Pennsylvania for deceiving General Forbes without quite spelling out the guilty colony's name:

> *We are still Incampd here—very sickly—and quite dispirited at the prospect before Us—That appearance of Glory once in view—that hope—that laudable Ambition of Serving Our Country, and meriting its applause, is now no more! Tis dwindled into ease—Sloth—and fatal inactivity—and in a Word, All is lost. . . . We seem then—to act under an evil Geni—the conduct of our Leaders . . . is temperd with something—I dont care to give a name to—indeed I will go further, and say they are d——s, or something worse to P—s—v—n Artifice—to whose selfish views I attribute the miscarriage of this Expedition, for nothing now but a miracle can bring this Campaigne to a happy Issue.*

So outraged was Washington at this suspected behavior on the part of Virginia's "Crafty Neighbours" (in his words) that he proposed to Speaker Robinson they take the matter to the king himself, so that His Majesty personally would know how grossly the royal honor and public funds had been "prostituted." Washington's own ambition and self-interest—intimately intertwined with that of Virginia—had again blinded him to the greater good. Thwarted, unable to bend reality into a shape it would not readily take, he exploded into anger. His fondest vision placed him at the center of the battle, in the hero's role. But stuck in a backwater at Fort Cumberland with General Forbes's troops moving along the wrong road, Washington's cherished illusions—that "appearance of Glory once in view"—had been shattered. He found a scapegoat for ruining his vision—crafty Pennsylvania.

The matter of the ruinous Pennsylvania road still had not been resolved, at least in Washington's mind, by mid-September. He rode the thirty miles north to Raystown, the staging area for the Pennsylvania route (today's Bedford, Pennsylvania). Meeting with young Colonel Washington, General Forbes ordered him to proceed with his men from Fort Cumberland to Raystown. This made it clear that Forbes had decided the entire expedition would use the Pennsylvania route. Nevertheless, ongoing complaints from Virginia officers reached General Forbes's ears. He called before him Colonel Washington and his Virginia co-commander, Colonel William Byrd III.

General Forbes angrily lacerated the twosome with a sharp verbal warning. He suspected they would be glad if the Pennsylvania route failed, he told them—as he recalled the conversation later in a letter to Bouquet—because the two so favored the Braddock route. His decision, General Forbes said, had proceeded from the best intelligence available, without favor to one province or another. The two of them before him, Colonel Washington and Colonel Byrd, were the only men who had shown their "weakness" by so favoring their own province without knowing one thing about the other route. "As for myself," General Forbes had told the two, "I could safely say—and believed I might answer for you—that the good of the service was the only view we had at heart, not valuing the provincial interest, jealousys, or suspicions, one single twopence. . . . I fancy what I said more on this subject will cure them from coming upon this topic again."

And so, after months of complaining and cajoling, it finally came to rest, at least in what Washington uttered publicly, although he kept up his complaints and pessimism in private correspondence. His unyielding resistance had earned him the description "obstinate" from his fellow Pennsylvania officers, as well as the wrath and contempt of General Forbes, who at one point referred to the Virginia officers' advocacy of the Braddock route as a shameful scheme. Whatever else Forbes had said to

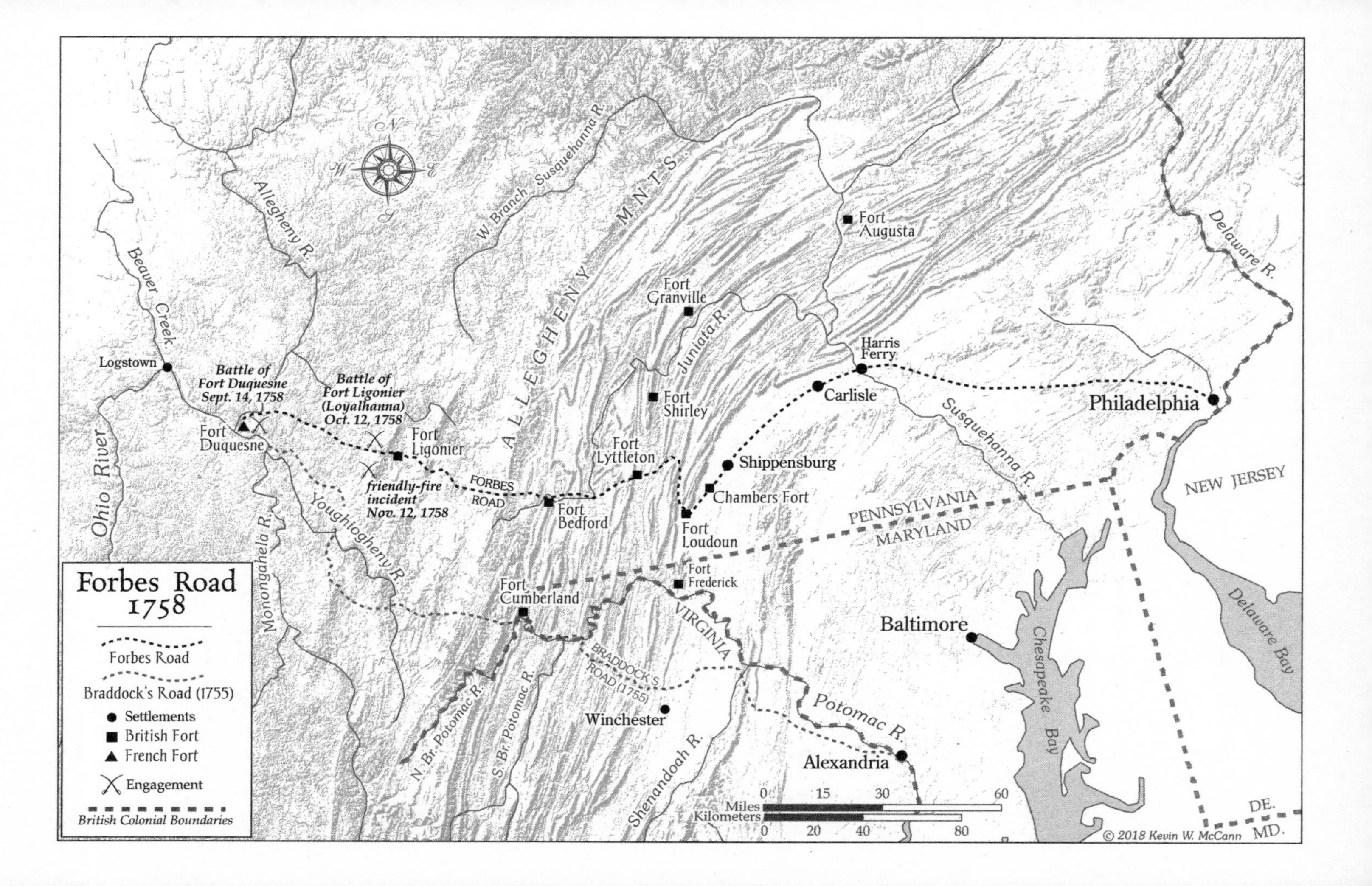

Forbes Road
1758
Forbes Road
Braddock's Road (1755)
Settlements
British Fort
French Fort
Engagement
British Colonial Boundaries
Logstown
Beaver Creek.
Ohio River
Allegheny R.
W. Branch Susquehanna R.
ALLEGHENY MNTS.
Battle of Fort Duquesne Sept. 14, 1758
Battle of Fort Ligonier (Loyalhanna) Oct. 12, 1758
Fort Duquesne
Fort Ligonier
friendly-fire incident Nov. 12, 1758
FORBES ROAD
Monongahela R.
Youghiogheny R.
Fort Bedford
Fort Lyttleton
Fort Granville
Fort Shirley
Juniata R.
Fort Augusta
Harris Ferry
Carlisle
Shippensburg
Chambers Fort
Fort Loudoun
Fort Frederick
Fort Cumberland
BRADDOCK'S ROAD (1755)
VIRGINIA
N. Br. Potomac R.
S. Br. Potomac R.
Shenandoah R.
Winchester
Potomac R.
Alexandria
Susquehanna R.
PENNSYLVANIA
MARYLAND
Baltimore
Chesapeake Bay
Philadelphia
NEW JERSEY
Delaware R.
Delaware Bay
DE.
MD.
Miles
Kilometers
0 15 30 60
0 20 40 80
© 2018 Kevin W. McCann

"cure" Washington, he had delivered the stinging rebuke that a true and honorable officer was not in this for himself, for his interests, or for his province. His duty was to dedicate himself to a greater cause. It was the kind of incident one could look back on years later and cringe.

WASHINGTON WAS NOT THE ONLY ONE of General Forbes's officers whose vision of glory propelled him into unhappy circumstances that September of 1758. Far worse consequences befell Major James Grant, when Grant, as General Forbes later put it, ran "headlong to grasp at a name and public applause. . . ." Yet Washington still did not fully absorb the concept.

A British Royal Army officer from the Scottish gentry, Grant, the embodiment of British arrogance toward colonials, admitted to a "repugnance" toward provincial officers. In early September, he and the fifteen hundred kilted clansmen of the 77th Regiment of Scottish Highlanders, recruited from the remotest mountains of Scotland and brought across the Atlantic, traversed the Allegheny Mountains on the partially cut Pennsylvania road to join Colonel Bouquet at Loyal Hannon. Here, about forty miles from the walls of Fort Duquesne, Bouquet's troops were establishing a forward base from which to launch General Forbes's attack on the fort. But enemy Indian parties roamed the woods nearby scalping or capturing soldiers who strayed from Bouquet's camp. While Colonel Bouquet wanted to send out small parties to track down the raiders, Major Grant had a much grander idea. Convinced by reports that the French had only a small force at Fort Duquesne, he proposed taking five hundred men, including a contingent of his fierce Scottish Highlanders, and making a stealth attack on the Indians who were said to be camping around the fort's walls.

Major Grant eventually talked Colonel Bouquet into his plan, despite warnings against it from experienced provincial bush fighters. Bouquet insisted that Grant take seven hundred and fifty instead of five hundred men, however. Like General Braddock, Major Grant put great faith in the superiority of the British Royal Army and his Scottish Highlanders. With three hundred of his Highlanders at its core, augmented by Virginians, Pennsylvanians, Marylanders, and a scattering of Indians, the force of seven hundred and fifty marched for three days. On the night of September 13, they arrived under cover of darkness atop a wooded river bluff a few hundred yards above the fort's walls. The plan formulated by Major Grant and Colonel Bouquet, both without significant experience fighting Indians, called for Grant's hidden force on the bluff to first identify Indian campfires blazing in the night around the walls. He would then send smaller detachments of troops to sneak down the hill and surprise the Indians who presumably huddled around each campfire. Major Grant and his Highlanders of the 77th would thus vanquish the enemy Indians.

The reality of the situation did not match Major Grant's glorious vision of it. The many Indian campfires did not blaze in the night as planned. Instead, the Highlanders spotted only two or three of them flickering around the fort's walls. Grant abandoned the plan to send small parties to attack each campfire. Instead, he assembled a force of two hundred Highlanders carrying shortened muskets, bayonets, and the Scottish broadsword, or claymore, and two hundred American provincials and ordered them to move down the hill, get as close to the enemy as possible without firing a shot, discharge their muskets just once, and then, without reloading, to charge with claymores and bayonets and dispatch the Indians. To prevent confusion, Grant's troops wore special white shirts to avoid skewering each other in the darkness.

Convinced his plan would work, Major Grant set it in motion. The four hundred Highlanders and provincials struck off downhill into the darkness . . . and nothing happened. Sometime later, their leader, Andrew Lewis, a provincial bushfighter who had been skeptical of the whole plan, climbed back uphill and reported that logs and other obstacles blocked the way down, separating the men, who were in danger of shooting each other in the darkness. Major Grant strode down to investigate, found the troops scattered in confusion in the forest, and, determined to engage the enemy somehow, gathered together fifty of them and sent them off to attack whatever they could find where the campfires had flickered. The party found no camps, but they did discover a storehouse standing outside the fort's walls. Wishing to leave their mark, they set the storehouse on fire.

This had the effect of kicking a hornet's nest—which might have been Major Grant's intention. With dawn's mist now wreathing the wooded bluff that loomed above the junction of the two rivers and the fort, Grant arrayed most of his seven hundred and fifty troops along it, while sending the Virginians and Royal Americans toward the rear under Andrew Lewis to secure a possible route of retreat. Enemy Indian scouts racing from the fort tracked Grant's soldiers and quickly discovered the detachment of Scottish Highlanders on the bluff. Major Grant ordered his drummers to play reveille to alert the other soldiers to the Indians and infuse the men with martial spirit.

"I must own I thought we had nothing to fear," he later remarked.*

It was a debacle. French soldiers and their Indian allies swarmed from the fort's gates, broke into small parties that stalked tree to tree in the misty forest, and sighted their muskets at the kilted and bereted Highlander officers wearing their dirks—long daggers—who strode about in the open directing their troops. Having dropped the Scottish officers, the Indians and French went to work on the Highlander troops, who, without their leaders, fell into confusion, then panic, and then headlong flight, despite Major Grant's efforts to rally them on the high ground. Major Grant kept fighting until he and a small body of troops were backed downhill to the river's edge.

"My heart is broke," he said, according to an officer who escaped. "I shall never outlive this day."

Then, calling his name, which they somehow knew, the French and Indians closed in on Major Grant, too.

---

* This was not the last time James Grant would overestimate his own prowess and underestimate his opponent's. Returning to England, he became a fiercely anti-American member of Parliament in the 1770s and, when the American colonists rebelled in 1775, disparaged their fighting ability, declaring in a speech that he could "go from one end of America to other and geld all the males."

CHAPTER TWENTY-ONE

As the Forbes expedition launched into the wilderness, Washington and Sally Fairfax renewed their correspondence. At the end of August, Washington had written to her husband and his friend, George William Fairfax, about getting new plank floors and other furnishings for Mount Vernon. (George William now went by the name William, since the death the previous fall of his father, Colonel William Fairfax.) In early September, both William and Sally had responded to Washington's letter with separate letters. (Neither of the two Fairfax letters survive.) Having received her brief message of good greetings, Washington wrote to Sally separately on September 12, probably taking up his quill at a writing desk in his quarters at Fort Cumberland, and expressed how happy he was that she had written to him after a long hiatus, fearing that she had "disrelishd" their correspondence. "In silence I now express my Joy.—Silence which in some cases—I wish the present—speaks more Intelligably than the sweetest Eloquence."

Swept up in the human drama of going off to war, Washington carried within him a precious, silent joy more powerful than the "sweetest Eloquence." This is a side of Washington—emotional, vulnerable, yearning—that remained hidden in the vast majority of his correspondence and is so at odds with the public image he learned to cultivate so carefully. It surfaced here, in September 1758, at a moment of reckoning that gave voice to his emotional vulnerability. Events were moving toward a climax that he knew would bring seismic changes to his life. General

Forbes had launched his great expedition to strike at the French. Washington now knew that he and his men would participate. Ahead lay an epic battle in which Washington saw himself reaping glory and honor or dying doing it.

And if he survived? That, too, would mean a dramatic altering of his course. The prospect of marriage arose. During his visits to the wealthy widow Martha Custis, twice in the spring and again in early summer, he had most likely already proposed marriage, on the condition, of course, that he survived the campaign into the Ohio. Either death in battle or marriage to Martha—two possibilities loomed. His life would either end or change unalterably. This the twenty-six-year-old Washington grasped in September of 1758. A note of finality entered his correspondence with Sally, as if he prepared himself for what lay ahead, and there were important things he wanted to say and things he wanted to know.

Seizing on the opening she had given him, he poured forth his heart. In her brief letter, Sally had apparently teased him about his courtship of Martha and perhaps his nervousness about the prospect of an upcoming marriage. He now responded that whatever "anxiety" he felt had not to do with the "annimating prospect of possessing Mrs Custis." He confessed that the prospect of going to war and gaining honor for himself and doing good for his province should stir his emotions. The emotions that so stirred him, however, were not due to the excitement of winning honor in the upcoming campaign. He admitted that they were, in fact, due to a lady. That lady was not Martha, but Sally herself:

> *Tis true, I profess myself a Votary to Love—I acknowledge that a Lady is in the Case—and further I confess, that this Lady is known to you.—Yes Madam, as well as she is to one, who is too sensible of her Charms to deny the Power, whose Influence he feels and must ever Submit to. I feel the force of her amiable beauties in the recollection of a thousand tender passages that I coud wish to obliterate, till I am bid to revive them.—but experience alas! sadly reminds me how Impossible this is.*

Couched obliquely, as if he could not so directly address his friend's wife, it is a remarkable confession.* Washington lamented the impossibility of his love for Sally ever coming to fruition. That it could not blossom confirmed for him that fate, or a higher power, determined the outcome of human affairs. "[This] evinces an Opinion which I have long entertaind, that there is a Destiny, which has the Sovereign controul of our Actions—not to be resisted by the strongest efforts of Human Nature."

He paused for a moment, owning up to the confession he had just made, letting it sink in. He had confided to her a "Simple Fact"—as if his love were as unalterable and natural as the sun or the moon. He wanted to ensure that Sally clearly understood his words. But she should also keep his declaration a closely guarded secret. "[M]isconstrue not my meaning—'tis obvious—doubt it not, nor expose it,—the World has no business to know the object of my Love, declard in this manner to—you when I want to conceal it. . . ."

He had given himself over to destiny—he sometimes called it Providence—and realized that this love for her would remain unfulfilled. But, he wrote her, he had to know one thing. Only she could answer it and only she could guess the question he wished to ask. "One thing, above all things in this World I wish to know, and only one person of your Acquaintance can solve me that, or guess my meaning."

On the brink of battle, or marriage, or death, Washington had to know whether Sally loved him. He could not ask it directly, of course, should the letter fall into some other hands. In the context of everything he wrote before, it is difficult to come to any other conclusion than that this is what he so desperately wished to know. Only Sally could tell him. He made that clear. Then he left it there, and abruptly moved on, with

---

* One could interpret these impassioned passages as perhaps referring to Martha, his prospective wife, instead of Sally, but this seems highly improbable. It is generally accepted by historians that this letter refers to his love for Sally and his wish to know if she loved him, and that he felt he had to write it obliquely.

a deep note of resignation resonating over his longing, mixed with the faintest echo of hope: "—but adieu to this, till happier times, if I ever shall see them."

SALLY RESPONDED TO HIM a short time later—unsatisfactorily. Whatever she told him in this second letter in less than a month (it also does not survive), she did not directly address his confession of love for her, nor did she acknowledge a love for him in return, in answer to the one thing he wished to know "above all things in this World." Rather, she apparently wrote a newsy letter of the goings-on at Belvoir Manor and an amateur theatrical production that Sally and her friends were staging, a popular drama of the time based on events in war, *Cato*.

She did not write what he wanted to hear. Ever persistent, he again wrote to her two weeks later, wondering if she had understood. "Do we still misunderstand the true meaning of each others Letters?" he started this letter of September 25. "I think it must appear so, thô I woud feign hope the contrary as I cannot speak plainer without—but I'll say no more, and leave you to guess the rest."

He then switched, at least momentarily, to newsier topics and especially the disastrous night attack that Major Grant had just attempted against the Indians camped outside the walls of Fort Duquesne. Without mentioning that Major Grant had come under deep criticism, including from General Forbes, for his ambition and drive for glory in rashly going forward with a raid that resulted in losses of one-third of his seven hundred and fifty troops to death, wounds, or capture, Washington made it abundantly clear to Sally that he, too, would have fought valiantly to the very end. "[W]ho is there," he wrote, "that does not rather Envy, than regret a Death that gives birth to Honour & Glorious memory. . . ."

Washington held a deeply romantic vision of love in time of war—at least as he portrayed it to Sally—that surely helped him embrace the prospect of a heroic death on the battlefield. In his September 25 letter to Sally, he reverted to his gloomy predictions about winter stopping General Forbes's expedition and the many inevitable losses due to "the Sword, Cold, and Perhaps Famine." The army would then have to make a humiliating retreat until spring, to universal condemnation. Much happier than this fate, wrote Washington, would be to spend it with Sally and her friends in their amateur theatrical productions. "I shoud th[ink] my time more agreable spent believe me, in playing a part in Cato with the Company you mention, & myself doubly happy in being the Juba to such a Marcia as you must make."

Using literary allusion to blunt his directness, Washington boldly stated to Sally that he would very much like to play the part of a star-crossed lover opposite Sally. The play *Cato, A Tragedy,* written in verse by Joseph Addison in 1712 and based on events in the Roman Republic, enjoyed great popularity in Britain and its colonies during the eighteenth century. Cato, upright Roman senator and military leader, is about to go to war against the invading Caesar, dictatorial enemy of the Republic. Juba, prince of the Numidians, who are allies of Cato, is in love with Cato's daughter, Marcia. Although his love for her is forbidden and he must hide it, Marcia nevertheless inspires Juba into battle against Caesar.

> O Marcia, let me hope thy kind Concerns
> And gentle Wishes follow me to Battle!
> The Thought will give new Vigour to my Arms,
> Add Strength and Weight to my descending Sword,
> And drive it in a Tempest on the Foe.

Marcia eventually hears that Juba has been killed in battle and finally declares her love for him aloud. Juba, it turns out, is alive and overhears Marcia's declaration of love. He then proclaims that he would rather have Marcia than victory.

> Juba will never at his Fate repine;
> Let Cæsar have the World, if Marcia's mine.

Biographers have noted that Washington loved the theater, and especially the play *Cato,* and seemed to think himself an actor on the stage of great events. In reading his letters and following his unfolding life, one senses that much of what he did, he did with one eye on posterity. He leaves the impression that he both participated in events and at the same time watched himself participating, monitoring the effectiveness of his performance. Striding across the stage, he thrived on the adulation of an audience. Later in life, he kept a count of the ladies who attended events held in his honor.

One can well imagine Washington as a young man craving especially the admiration of Sally Fairfax. As Marcia's "gentle wishes" had given new vigor to Juba's arms and strength and weight to his descending sword, so could Sally's admiration inspire Washington in battle. He had told her that he envied a glorious death on the battlefield. He had confided to her his deepest emotions. In the drama unfolding in 1758, it seems that Sally would willingly play the part of Washington's female admirer—but not, as he so desperately wished, his true love.

IN MID-OCTOBER, three weeks after writing to Sally, Washington and his men began the march from the Raystown staging area toward Fort Duquesne. Following General Forbes's orders, this took them on the Pennsylvania and not the Virginia route. They carried the general himself, who had fallen so ill with the bloody flux that he could not sit in a saddle and rode in a sling between two horses. Their immediate goal was Loyalhanna, the forward base forty miles short of Fort Duquesne, where General Forbes had massed the Scottish Highlanders and his other British Army troops as well as provincials for the final assault.

Rain and mud bedeviled the route, bogging down horses pulling heavy wagons in the mire of sucking mud and leaving Washington unconvinced that General Forbes had made the right choice. The general himself prayed for dry weather. "If the weather does not favor," he wrote to an acquaintance, "I shall be absolutely locked in the mountains."

The towering ridge known as Laurel Hill nearly stopped the expedition dead, but at the last moment the discovery of a shortcut saved them from the worst of its steep rocks, muddy gullies, broken stumps, and fallen logs. Colonel Washington and his men made the crossing and reached the advance base camp at Loyalhanna on October 23. Washington remained convinced that winter would stop the campaign before Fort Duquesne. Nevertheless, drawing on his past experiences fighting the French and Indians in dense forests, he sketched for the benefit of the general's officers a diagram of how a large body of four thousand troops, such as General Forbes's army, could be arranged to travel most safely in wooded terrain.

Although still badly ill when he reached the Loyalhanna advance camp, General Forbes regrouped and prepared his troops for the final assault on Fort Duquesne. It was now early November. Deadlines loomed: Winter would soon descend on the mountains. Thousands of the general's provincial troops might walk away toward home in Pennsylvania, Maryland, North Carolina, and Virginia on December 1, when their pay ran out or terms of enlistment expired. Every delay of General Forbes's army also meant that the French had more time to reinforce their troops at Fort Duquesne. The moment of commitment had arrived.

"[Time is] a thing at present so precious to me I have none to spare," wrote General Forbes to Secretary William Pitt in London. "[I] must in a day or two choose either to risk everything and march to the enemy's fort, retreat across the Allegheny if the provincials leave, or maintain myself where I am to the spring."

With preparations under way at the advance base, one of the most traumatic events in Washington's life occurred, a strange and tragic episode in the woods on November 12. That afternoon scouts outside the camp reported that a body of two hundred French and Indians were approaching. Similarly, a few weeks earlier, an enemy party had approached the advance camp, exchanged fire with British troops, and disappeared with many of the camp's beef cattle and horses. Now, with these reports of a new French and Indian advance on the camp, General Forbes ordered a party of five hundred Virginian troops under Washington to move into the forest to confront the enemy head-on and another party of five hundred Virginians led by George Mercer to surround them.

Washington's party advanced two miles into the forest and came on a campfire surrounded by a number of French soldiers and Indians. The parties fired on each other, the French and Indians fled, but Washington's men captured three prisoners. By now dusk had fallen. An evening mist crept through the forest and wrapped around the dark tree trunks. Someone spotted movement. Shadowy figures slid through the dimming, misty forest. It was an ambush. A musket thudded. With cascading effect, firing erupted throughout the forest, muzzle flashes lighting the dusk, billows of gunpowder smoke infusing the mist and rolling over the leafy floor.

Officers shouted orders. Troops grouped in rows. The cries of "Fire!" and the ripping crash of volleys echoed through the forest. Shadows dropped.

Then a figure ran into the open, waving his hat. "*Stop! Stop! Stop!*" he seemed to be shouting. Washington, too, realized that something had gone terribly wrong. Through the dim mist he saw that the shadowy figures wore the blue uniforms of fellow Virginians. Still the firing erupted—from ranks of soldiers firing volleys and from individual men.

Washington shouted to cease firing, but through the crash of musketry, the din and adrenaline of battle, the smoke and mist, the men did not hear. He drew his sword. He strode forward, advancing down the row of men with muskets leveled toward the shadowy forest figures. With the blade of his sword he knocked them upward, one after the other, so that the weapons aimed harmlessly up in the air.

Finally the firing stopped. In the silence, groans and cries for help rose out of mist and settling smoke. The two sides moved toward each other, arms down. They realized their fatal mistake. They saw the faces of the dead and wounded who lay sprawled on the forest floor. They had shot their own friends.

What did Washington think that night as he approached the flickering fires of General Forbes's camp? How would he tell the general what had happened? No matter which party had fired first, it reflected terribly on the discipline of Washington's troops. Did he feel the sharp clench of anguish at the loss of his own men? Or was it the cost of war? Were his hands full simply accounting for the troops and bringing back the wounded? Or did thoughts of Sally Fairfax surface as he moved through the darkness? He had hoped to play a heroic Juba to her admiring Marcia. He had confided that he sought glory on the battlefield and envied a glorious death. And now this—under his own eyes, the death of his fellow Virginia troops at the hands of his own Virginia troops and vice versa. Did he feel humiliation, or shame, or a wish to distance himself from it insofar as possible?

He may have felt all these things roiling inside him, and also the powerful urge to suppress them, either out of horror at his own men killing their fellow soldiers or fear that the incident would destroy his reputation or both. It appears in none of his surviving personal or official correspondence, as if it lingered inside him unresolved and too painful to commit to words. He mentioned it only obliquely at the time, late that night, writing an addendum to the next day's orders, sending a detail of four hundred men "to the Ground where the Skirmish was this Evening and to Carry . . . Spades in Order to Enter the Dead Bodies." It was as if he wished to bury the incident as quickly and thoroughly as possible.

Washington touched on the incident in one other surviving document. Many years later, his friend David Humphreys wrote a biographical manuscript about Washington, who responded with a brief set of "Remarks," correcting and amplifying various points. In these "Remarks," Washington detailed some of his experiences in the French and Indian War, including this friendly fire incident. He cast himself in a favorable, if not heroic light, suggesting that Mercer's troops mistook Washington's men for the enemy and fired first. In other words, it wasn't his fault. His life had never been in so much danger, before or since, he wrote, caught as he was between two lines of fire as he tried to stop the melee, finally knocking up the men's muskets with his sword.

Another participant, Captain Thomas Bullitt, remembered Washington in quite a different light. According to stories handed down through Bullitt's family and published in a memoir, Bullitt had run between the lines of fire waving his hat and shouting for the men to stop. He claimed that Washington did not react promptly to the eruption of friendly fire.

Bullitt openly criticized Washington, according to these family stories, and Washington forever held that against him. "It was thought and said by several of the officers, and among others by Captain Bullitt, that Colonel Washington did not discover his usual activity and presence of mind upon this occasion," according to the published family stories. "This censure thrown by Captain Bullitt upon his superior officer gave rise to a resentment in the mind of [Washington] which never subsided."

The tally of the dead was not known until the burial detail the next day. One officer and thirteen soldiers died in the friendly fire, and twenty-six were wounded, according to one count. This, like the sudden massacre at Jumonville Glen that shook the empires of Europe, came as the price of haste and impulsiveness. It is impossible to know whether Washington reflected on this at the time, or in the aftermath, or only years later. One imagines General Forbes—as conscientious as they came—talking Washington through a detailed review of the mishap, explaining how to prevent it from happening again. Perhaps he comforted his young officer, noting that incidents like this were the price of war. It seems likely that Washington internalized something important and lasting from this tragic incident. Years later, as commander of the Continental Army, he was anything but hasty and impulsive but rather known for his deliberate and methodical calculations and careful consultations with his fellow officers before taking action.

WASHINGTON DELIVERED THE BAD NEWS that night to General Forbes in the firelit camp, but he also brought back a prize from the bloody mistake in the forest—three prisoners, two of them Indian. The third turned out to be a British soldier named Johnson who had deserted and fled to the French at Fort Duquesne. Under threat of torture and death and the promise of reward if he told the truth, Johnson was interrogated by officers at General Forbes's camp. The French possessed only a weak force at Fort Duquesne, he told his interrogators, and they had made the recent advance against General Forbes's camp to disguise that fact. He also claimed that many Indian allies had abandoned the French and returned to their forest villages. This seemed an unlikely story—that the French had lost their Indian allies—but the two Indian prisoners corroborated it.

Time was running out. With winter coming, the French at the Forks reportedly weak, and part of his army about to disband on December 1, General Forbes now struck fast—or as fast as his deliberate method allowed. He had his officers handpick the twenty-five hundred strongest men, ordered them to carry only a pack, a blanket, their arms, and ammunition, and divided them into three brigades. Two were commanded by two regular British Army officers—Lieutenant Colonel Archibald Montgomery of the 77th Highlanders and the Swiss-born officer Colonel Bouquet. The honor of commanding the third brigade, about a thousand strong and the only one led by a provincial officer, went to Colonel Washington. His brigade included the provincial troops of his 1st Virginia Regiment and provincials from Maryland, North Carolina, and Delaware. Although not officially promoted, he received the honorific "Brigadier," which must have pleased him greatly.

General Forbes methodically sent out these brigades in leapfrog fashion to forge the way to Fort Duquesne, despite being so incapacitated by bloody flux that he was "quite as feeble as a child," as he put it, and often "in bed wearied like a dog." First, an advance force of Pennsylvanians would mark the route by slashing blazes on trees and build defensive structures at regular intervals to provide safety in case of a retreat. Tracing the advance party's blazes through the forest, Washington's ax-wielding force would follow, chopping trees, rolling away logs, hacking underbrush, prying out rocks, and laying log bridges over streams and marshes, to forge a road wide enough for light artillery and provision wagons, each laden with a ton of flour, salt beef, ammunition, and other necessities. Colonel Montgomery of the 77th Highlanders would then lead the main body of troops and artillery along the newly cut road. One, two, three, the brigades would leapfrog each other, soldiers' feet squishing into wet, low ground, wagon wheels battering over stumps and grinding over rocks.

They moved quickly. To simplify the procession, General Forbes banned women from the march.

Despite a late start on the morning of November 15, a shortage of axes, and the desertion of five soldiers, Washington's men cleared six miles of road from General Forbes's advance camp at Loyalhanna and camped on Chestnut Ridge, nearly two thousand feet in elevation. The next day, working dawn to dusk, with one man injured by a falling tree, they cut their way over the top of the long mountain ridge, making another six miles before nightfall, and the following day covered another seven or eight miles, although the supply of fresh meat ran low, depleted by the huge caloric needs of heavily laboring men swinging axes and prying rocks. Pausing to slaughter fifteen bullocks to feed his thousand-plus men, Washington ordered a 3 A.M. start to make up for lost mileage.

And so they proceeded, with Washington's brigade leapfrogging with Montgomery's brigade to cut the way. On November 23 they had arrived within a mere twelve miles or so of Fort Duquesne. Without strict precautions and good intelligence, they could now fall victim, as Braddock had, to a French and Indian surprise attack. They paused to reconnoiter. General Forbes ordered officers to meticulously inspect every weapon, ordered complete silence at night with firearms kept within reach, and two hundred lashes to any soldier who fired his musket without orders. The barking of camp dogs broke the silence of the forested night. The dogs had to be tied up, General Forbes ordered, sent back to Loyalhanna, or hanged.

Scouts slipped into the forest to see what lay ahead. Washington wrote no letters. The tension hung thick through the night. If the silent, stalking enemy suddenly broke forth, the men were to fall back fifty yards behind the rows of blazing campfires, lie on their bellies, arms at the ready, and await a command.

Nothing stirred in the woods. The thousands of men waited in the silenced camp.

As the soldiers cooked their daily pound of meat over campfires in the chill November night, the missionary Christian Frederick Post was another fifty miles farther back in the forest trying to evade the French. He had left Philadelphia four months earlier, in mid-July, dispatched by

Governor Denny and General Forbes on a mission to contact the hostile Indian tribes at their villages in the Ohio wilderness—the Western Delaware, the Shawnee, and the Mingoes. His goal was to convince them to switch from the French to the British, despite the three years of depredations—scalpings, dismemberings, hostage-taking—they had committed against British frontier settlers.

Passing through the frontier settlements on the way into the wilderness, Post spotted abundant evidence of the Indian raids. With his Indian guide, Pisquetomen of the Delaware, brother of Chief Shingas, whose people had committed many of the depredations, he saw deserted and burned homesteads and tall red-painted poles where Indian captors bound white captives to secure them during the night. From bushes beside the trail dangled scalp hoops entwined with strands of long white hair. Traveling westward on paths that bypassed Fort Duquesne, they crossed mountains and rivers and slept without tents in swarms of mosquitos and under heavy rains.

"The Heavens alone were our Covering," Post wrote in his diary, "and we accepted of all that was poured down from thence."

Despite the extreme risk and difficulty of his mission, Post remained remarkably calm. As much as he was trying to help the British, he also believed himself on a mission from God to help the Indians and save them from slaughter. "Therefore I was resolved to go forward," he told Chief Teedyuscung, who feared Post would die in the attempt, "taking my Life in my Hand, as one ready to part with it for [the Indians'] good."

After ten days of travel on circuitous forest paths to avoid Fort Duquesne, they had arrived in the vicinity of Venango, the French fort and Indian post situated where Rivière Le Boeuf joined the Allegheny. Post knew that if his presence and his mission to convert the Ohio Indians to the British side became known to the French within, he could be captured, tortured, executed—or likely all three. "I prayed the Lord to blind them, as he did the Enemies of Lot and Elisha, that I might pass unknown."

They camped that night within "half a Gun Shot" of the fort. In the morning, Post hunted for his grazing horse within ten yards of the fort's outer walls. "The Lord heard my Prayer, and I passed unknown till we had mounted our Horses to go off, when two Frenchmen came to take leave of the Indians, and were surprised at seeing me, but said nothing."

Soon they reached Kuskuski, a large Indian village in the center of the Ohio wilderness, on the Ohio River about twenty-five miles downstream from Fort Duquesne. Knowledgeable in Indian customs after his two marriages to Indian women and years in their villages, Post, along with Pisquetomen, dispatched a messenger to announce their presence and say that they carried great news from the governor of Pennsylvania. (It was considered rude behavior simply to show up unannounced.) The chief of the village, King Beaver, welcomed them personally, showing them to a special longhouse reserved for visitors. Sixty young warriors entered the longhouse, shook hands, and sat with Post and Pisquetomen around a council fire. Then King Beaver entered.

"Boys, hearken," he said, addressing the young warriors. "We sat here without ever expecting again to see our Brethren the English; but now one of them is brought before you." Turning to Post, he said, "Brother, I am very glad to see you; I never thought we should have had the Opportunity to see one another more; but now I am very glad, and thank God who has brought you to us."

For five days Post stayed as King Beaver's guest in the longhouse at Kuskuski, waiting for other Indian kings and chiefs to gather to hear his message. Receiving warm hospitality from these people who had committed the most heinous acts against British settlers for the last three years, Post feasted at King Beaver's lodge, talked late to warriors around the fire in the guesthouse, and watched them dance far into the night. Some white captives also lived with Indian families in the town. Here the two girls Barbara Leininger and Marie Le Roy, captured three years earlier at Penn's Creek, labored in the cornfields outside the town and helped with other domestic tasks, like servants. They saw Post but could not speak to him.

"The Indians gave us plainly to understand that any attempt to do this would be taken amiss," they recalled. "He himself, by the reserve with which he treated us, let us see that this was not the time to talk over our afflictions."

When the kings and chiefs finally assembled on the fifth day, they deliberated among themselves and informed Post that they alone could not render a decision whether to renew old ties to the English. They must consult other Ohio Indian leaders. Messengers dispatched to Fort Duquesne returned with the news that eight Indian nations were present at the fort. Their leaders wanted Post to come to them to deliver his message.

Post balked at going, concerned that the French would capture him and end his mission. But King Beaver and his fellow leaders insisted, pledging to protect him. Two days later, King Beaver's contingent of seventy Indian leaders and warriors, with Post and Pisquetomen among them, arrived on the riverbank across from Fort Duquesne. Post now found himself in the center of a kind of tug-of-war: the French commandant demanded him delivered blindfolded across the river to the fort, while Indian leaders urged him to remain in camp on the opposite riverbank and deliver his message from Governor Denny and the British.

"Some of my Party desired me not to stir from the Fire," he recorded, "for that the French had offered a great Reward for my Scalp, and that there were several Parties out on that Purpose. Accordingly, I stuck constantly as close to the Fire as if I had been chained there."

When the Indian leaders at Fort Duquesne paddled across to King Beaver's camp, three hundred French officers, soldiers, and Indian warriors accompanied them. Setting up a writing table at the council fire, French scribes sat with pens, ink, and paper to record exactly what Post said. Post spoke freely to the great assembly, feeling protected by King Beaver and by God, first wanting to clear away the bad blood between the British and the Ohio Indian tribes. He placed on the earth before the Delaware chiefs a wampum belt and, with symbolic gestures of picking up objects, metaphorically gathered up the bones from both sides and piled them in one heap for God to dispose of, "that they may never be remembered by us while we live, nor our Children, nor Grand-Children hereafter."

And so he continued at the council fire on the riverbank, laying down on the ground another wampum belt with each major point to emphasize his sincerity, while observing to himself that the French looked unhappy at his words. They will bury the hatchet, Post told the seated circle of Indian leaders, and—now holding up a great eight-beaded belt—whoever holds this belt will have the promise from the "English Nation" of a return to the peace enjoyed by their grandfathers. We love you and care for you, Post assured the Indian leaders, while cautioning them that the king of England has sent a great number of warriors into the Ohio—meaning General Forbes and his army. They do not intend to go to war with you, the Indians, he said, but with the French. "And by this belt I take you by the Hand, and lead you at a Distance from the French, for your own safety, that your Legs may not be stained with Blood."

Post laid the peace belt before the Delaware. When he had finished speaking, the Delaware rose, picked up all the beaded wampum belts, and laid them before the Mingoes. The Mingoes in turn rose and laid the belts before the Shawnee.

"You brought the Hatchet to us from the French," the Mingoes told the Shawnee, "and persuaded us to strike our Brothers the English; you may consider wherefore you have done this."

Acknowledging that they had taken up the hatchet from the French, the Shawnee leaders said they would send the wampum belts to other Indian tribes and in twelve days return with a response. The great meeting broke up. Informed by his long residence among the Delaware, fearless in his approach to them, and deep in his sincerity, Post had grasped the subtleties of Indian diplomacy and the relations among them in a way that few other whites had. Certainly George Washington had not mastered such subtle diplomacy or attempted to understand the complexity of their cultures or embrace their humanity, seeing them as subtle, cunning, ferocious savages—the ultimate forest warriors, like a particularly sophisticated piece of weaponry.

After the meeting broke up, canoes ferried back and forth across the river, returning the French to the fort and quickly bringing back word from the French commanders demanding that Post be turned over to them. At dawn the next day, Indian guides secretly led Post over little-known trails back to Kuskuski, where King Beaver, Shingas, Pisquetomen, and other Indian leaders feasted, considered the peace offer, and probed Post further. They told Post that the Delaware had begun to draw away from the French but also harbored deep doubts about the English. "[W]e have great Reason to believe, you [English] intend to drive us away and settle the Country, or else why do you come to fight in the Land that God has given us."

"I said," Post recorded, "we did not intend to take the Land from them, but only to drive the French away. . . ."

The Delaware leaders remained unconvinced. "'Tis plain that you white People are the Cause of this War; why don't you and the French fight in the old Country, and on the Sea? Why do you come to fight on our Land? This makes every Body believe you want to take the Land from us, by force, and settle it."

Back and forth went the discussion, with the Delaware leaders convinced of Post's sincerity but suspicious of English intentions about Indian lands in the Ohio, especially with a large body of British troops marching toward it. Summoning up ancient European rivalries that paralleled American tribal ones, the Protestant missionary Post claimed that wilderness traders of Irish Catholic origins had poisoned Indian minds against the good intentions of the Protestant English. But the Indians allowed they were glad that Post and the English had dug up a lost friendship that had lain buried between them. They urged Post to tell all the English that this friendship was being revived. Once all the English agreed to it, they said, send the great peace belt to let the Indian chiefs know, and they would send it to all Indian nations and hold to it fast. "[T]hen the day will be still," the chiefs said, "and no Wind or Storm will come over us to disturb us."

Sneaking along remote forest paths to avoid the French and the Indian scalp hunters seeking the bounty on his head, Post, Pisquetomen, and their companions headed back toward the English settlements, struggling through the mire, up thorny mountainsides, fighting lack of food, and spending nights in hiding without a warming fire. Post finally arrived at Fort Augusta.

"Thirty-two Nights I did lay in the Woods; the Heavens were my Covering," he wrote, emerging safely from the Ohio wilds. "Praise and Glory be to the Lamb that has been slain, and brought me through the Country of dreadful Jealousy and Mistrust."

AFTER THE LONG, TENSE, WATCHFUL NIGHT of November 23, 1758, morning dawned peacefully if anxiously in General Forbes's forest camp a mere twelve miles from Fort Duquesne. No surprise attack of screaming Indians and firing Frenchmen had launched out of the darkened woods during the night, no scouts had returned reporting enemy movement. The thousands of troops arrayed in their symmetrical rows roused and breakfasted. Tense hours of waiting followed. The camp in the thick forest breathed heavily with idleness and anxiety. Indian scouts were expected to return at any time. General Forbes issued orders. These were passed on to their respective troops by the three brigade commanders, Colonels Bouquet, Washington, and Montgomery of the 77th Highlanders. The army should prepare to march on Fort Duquesne at 7 A.M. the next morning in three divisions, left, center, right, each led and flanked by light horsemen with heavy artillery wheeling behind. For several months now, they had cut their way through the forest toward Fort Duquesne. It had been five years since Washington had first raised the alarm about the French in the Ohio, three years since bullets whizzed through his hat and coat at the slaughter of British troops at Braddock's Defeat. Everyone knew that the decisive moment had now arrived.

A voice shouted from the woods. "Johnston!"

Men's heads turned in alarm. The sentries posted at the camp's peripheries listened closely. They heard the proper password, *Johnston*. They shouted back into the forest, giving permission to pass.

An Indian scout walked softly out of the woods through the camp to General Forbes's tent. The scout reported what he had seen to the general and his assembled officers—from a distance, he had spotted a towering column of smoke spiraling into the heavens from the Forks.

The dull, distant thump of an explosion tapped eardrums and chest cavities.

Another Indian scout slipped out of the forest, offering a similar report. The French had set fire and blown up Fort Duquesne, he said, loaded their cannon aboard bateaux, and shoved off, riding the current down the Ohio River toward the distant Mississippi. Was this true? General Forbes wondered. Or was this somehow a ruse? He immediately dispatched a force of light horse to investigate.

Much had occurred behind the scenes since Christian Frederick Post had first met the enemy tribes in the Ohio wilderness. In mid-October, a great conference got under way at Easton, sixty miles north of Philadelphia. Indian leaders from the Iroquois Confederacy, the Delaware, and other tribes gathered with representatives from the family of William Penn, which still held a great deal of land, as well as other British to seek some resolution of the hostilities. In a complicated equation of power, all parties wanted peace, but each party had a price.

The powerful Iroquois Confederacy served as the fulcrum. The Iroquois, or Six Nations, the confederacy based near the Great Lakes in what is now upstate New York and longtime allies of the British, had for years past claimed suzerainty over many other of the region's tribes, having subjugated them in past battles. With all parties seeking a peaceful solution, the Iroquois, with help from their British allies, now reasserted their dominance over the Ohio Valley and nearby tribes. Most important of these, in terms of finding a peace, were the Eastern and Western Delaware, who had relentlessly raided frontier settlements. Teedyuscung, leader of the Eastern Delaware, had already allied himself with the British and had helped the British contact the hostile Western Delaware. But he now found himself in a position of little leverage. Perhaps with British and Iroquois encouragement, he spent much of the conference off to the side—and drunk. He was offered a deal. In return for promises from the British (later proved empty) to review the fraudulent Walking Purchase that took much of his land, in addition to promises from the Iroquois (also empty) to create a homeland in the rich Wyoming Valley for his homeless Eastern Delaware, Teedyuscung in the end acknowledged his tribe's subordination to the Iroquois.

"I sit there as a Bird on a Bough," he said. "I look about, and do not know where to go; let me therefore come down upon the Ground, and make that my own by a good Deed, and I shall then have a Home for ever. . . ."

The Western Delaware, represented by Pisquetomen, would not accept a peace agreement with the British unless it came with a guarantee that the Ohio lands would remain in Indian—not British—hands. To achieve this guarantee, or at least the appearance of it, the Penn family, represented at the conference by Conrad Weiser, thus gave back to the Iroquois all lands that the Penns had claimed in the Ohio Valley. This move took the Ohio Valley lands out of British ownership, at least in theory. In order to guarantee a voice to the Western Delaware and other Ohio tribes in future Ohio Valley land transactions, the Penn family agreed that the Ohio Indians such as the Delaware and Shawnee could henceforth negotiate not through the Iroquois, who were infamous for selling out other tribes' land to the British, but with the Penn family directly. This great realignment of Indian powers in the Ohio Valley triggered by the British quest for peace had all the delicacy and unspoken nuances of European diplomacy.

"Thus the form of Iroquois dominance over the Ohio Country was revived," writes historian Fred Anderson, "but the substance of Iroquois control over the Ohio Indians was not, for the Ohioans would be free to act for themselves in future dealings with the Penns. With these concessions secured, Pisquetomen agreed to peace on behalf of the western Delawares and the other Ohio bands for whom he spoke."

With the agreement reached, Christian Frederick Post had rushed from the council fires of the Easton conference, traveled once again southwestward into the forest along the freshly hacked Forbes Road, and found General Forbes grouping for the assault. Receiving the good news, General Forbes in turn had given Post a letter addressed to the Ohio chiefs in their forest villages who still might be tempted to side with the French. "[R]eturn forthwith to your Towns; where you may sit by y[ou]r

Fires, w[i]th y[ou]r Wives and Children, quiet and undisturbed, and smoke your Pipes in safety. Let the French fight their own Battles . . . restrain y[ou]r young Men . . . , as it will be impossible for me to distinguish them from our Enemies. . . ."

At 6 P.M. on the evening of November 24, the detachment of light horse that General Forbes had sent to investigate the Indian scouts' reports arrived at the Forks. On the strategic tip of land where the Allegheny and Monongahela rivers joined, they discovered that only heavily reinforced earthworks still stood. Smoke wreathed around the smoldering ruins of barracks, officers' quarters, and other buildings. Thirty chimneys, immune to flame, still stood among the ashen rubble, reaching up eerily into the November dusk like the obelisks of a lost civilization.

The Ohio tribes had kept to the terms of the peace. As Forbes's army had approached, chopping, clearing, hauling, marching, mile by mile and hour by hour ever closer, they had walked away from the French and Fort Duquesne and returned to their villages. Lacking Indian allies, the French had made their fateful decision. They put torches to the fort they had so painstakingly built, planted kegs of gunpowder under the structures and walls, and touched it off with a tremendous explosion, sending splintered logs and clods of earth arcing into the air. By the light of the burning fort, reported Indians camped on a midriver island, one party of French had loaded the fort's cannon on bateaux and shoved off into the current, bound for the Illinois country hundreds of miles downstream, while another paddled canoes up the Allegheny toward the Great Lakes. The French had abandoned the heart of the Ohio Valley.

MESSENGERS FROM THE PARTY OF LIGHT HORSE investigating the wreckage at the Forks hurried back to General Forbes's camp. They may have arrived the same evening. One imagines the cries of jubilation ringing through the dark, bare-branched forest as word spread among the rows of fires that defined the camp. They had forced the French to run, had seized victory without even a fight. They were now certain of returning home alive to their families. General Forbes, though sick and badly weakened, and his fellow officers must have celebrated joyfully. But for some young men, eager to make a name or test themselves in battle, a strain of disappointment may have tempered the joy.

One pictures twenty-six-year-old Colonel Washington, erect in his bearing, his uniform impeccably kept by a servant, slumping slightly with the news. He had prepared for this moment since he was a boy. He had thrilled to the whiz of bullets while under fire at Jumonville Glen, reveled in lionization for his bravery at Braddock's Defeat, boasted to Sally Fairfax that he favored a hero's death on the battlefield. He had marked himself for glory or death—or both. This was to have been the climactic moment. The might of the British Empire would clash against the might of the French deep in the American wilderness for the prize of the enormous Ohio Valley. Here lay the chance for a thousand heroic gestures, for noble deeds that would be sung for a hundred years. And now this—the French simply walked away. What glory was there in that?

If he slumped at the news, physically or emotionally, he no doubt pulled himself upright again. He would not show his disappointment to his men. A thousand details remained to be attended to for his 1st Virginia Regiment—pay, discharge, clothing, food—the never-ending and relentless logistical needs of an army. These matters had consumed him for much of the last three years. Would he leave a legacy as an administrator rather than a hero? What did this ending mean? Was this ending enough to bring him the honor and distinction he craved?

Yet another part of him may simply have had enough. Along with disappointment, he may have felt relief. The life in the wilderness and on the frontier may have finally left him exhausted, worn to the bone. The way sailors dream of farms when tossed in storms at sea, perhaps a quiet domestic life at Mount Vernon, with his schemes of cultivation, fine china, damask wallpaper, walnut furniture, and droves of slaves, may have held everything that Washington desired, in this moment in the cold and dark and bare November forest.

The future that he had imagined and expressed to Sally Fairfax—glory or death—was not to be. Instead, he saw unfolding before him another future, an alternative future, a quieter future. The evaporation of the French and lack of a climactic battle had buried, in some metaphorical sense, his years of yearning for the thousand amiable beauties and tendernesses of Sally Fairfax. He would no longer play the role of young hero to her admiring glances. He would return to the sprawling estate overlooking the Potomac at Mount Vernon and marry the warm and wealthy widow Martha Custis. There would be tranquility. There would be children—hers, definitely, and perhaps theirs. There would be prosperity with her huge tracts of land. There would be his seat in the House of Burgesses.

In the bare November forest deep in the Ohio wilderness, twenty-six-year-old Washington's ambition may have finally burned itself out—if only temporarily. He could go home now with distinction, if not a hero's glory.

EPILOGUE

WITH AMAZING SPEED, IN A MERE SIX WEEKS' TIME Washington transformed from gung-ho wilderness soldier to solid Tidewater family man. Even at his young age, he had shown a chameleon-like ability to assume new roles, to absorb new cultural rules, social bearing, outward dress, as if taking on a new part in a stage play: young surveyor toting his precision instruments and jotting mathematical notations, eager student of gentlemanly behavior under the tutelage of Colonel Fairfax, wilderness messenger striding through the snowy forest in his moccasins and Indian leggings, uniformed young Virginia officer with sword, riding boots, and perfect posture in the saddle, commanding troops in the field or entertaining the admiring ladies at Belvoir Manor. He had played all these parts in the space of ten years or less, and now, in the days immediately after the fall of Fort Duquesne, he switched into a new role that, at least personally, suited him well—Tidewater planter and family head.

After his army discovered the smoldering ruins of Fort Duquesne abandoned by the French, General Forbes assembled a procession into the forest to the site of Braddock's Defeat to bury nearly five hundred skulls of the victims and whatever animal-gnawed bones they could find strewn about. Assigning troops to winter at the ruined fort, General Forbes chose some of the Virginia Regiment, although they lacked warm clothes and heavy blankets. He and Colonel Washington agreed that the colonel should ride immediately to Williamsburg to ask for help from the governor and burgesses in equipping the Virginians for the winter. Eager to make the long journey over the mountains, and trading for fresh horses as he wore them out, young Washington arrived at Belvoir Manor in time to spend Christmas with Sally and William Fairfax, and by December 27 was at Williamsburg, where he kissed the hand of Governor Fauquier, Dinwiddie's replacement, and petitioned for help equipping the Virginia Regiment for winter. His duty completed, at least insofar as he saw it, Colonel Washington resigned his Virginia officer's commission.

Whatever his shortcomings, Washington had earned the deep respect of many of his officers. In a letter they signed on New Year's Eve, 1758, twenty-seven officers asked him to stay on another year. Praising his "impartial justice" and "regard to merit," they wrote that Colonel Washington had set an inspiring example of "true honor and passion for glory. . . . In you we place the most implicit confidence. Your presence only will cause a steady firmness and vigor to actuate every breast, despising the greatest dangers and thinking light of toils and hardships, while led on by the man we know and love."

Washington responded graciously. "I must esteem [your approval] an honor that will constitute the greatest happiness of my life, and afford in my latest hours the most pleasing reflections."

But he wasn't coming back. He had left that life behind. On January 6, 1759—a week after his officers sent their praise of his command and four months after writing the letter that simmered with his hidden, and doomed, passion for Sally Fairfax—George Washington married Martha Dandridge Custis at the White House plantation on Pamunkey River. He and Martha then moved into Mount Vernon. Just turning twenty-seven years old, Washington had already lived, for many, the equivalent of a lifetime.

For the next seventeen years Washington lived the life of a prosperous planter. Thanks to Martha's wealth, the young patriarch of Mount Vernon immediately achieved membership in the Virginia aristocracy. He oversaw thousands of acres of lands and hundreds of slaves. Probably sterile, never having children of his own, he lived in close companionship with Martha and her two children from her earlier marriage to Daniel Parke Custis—her son, Jackie, four years old at the time of George and Martha's marriage, and two-year-old Patsy. He slowly developed Mount Vernon into an exquisite estate. It was appointed with British luxury goods—fine china and mahogany chairs and a marble fireplace—and surrounded by well-tilled gardens. Slaves worked plantation fields scattered for miles around. The couple entertained hundreds of guests and attended theatrical events and balls in Williamsburg and at neighboring estates, where Washington loved to dance. He played cards for money, passionately rode on foxhunts, and served dutifully in the House of Burgesses, as a vestryman in the local parish, and as a justice in the Fairfax County Court. A meticulous businessman and an innovator in farming, he studied methods of agriculture, examined soils, and read manuals. Unable to grow a first-rate tobacco crop—a mark against him among the uppermost Virginia gentry—he instead experimented with crops and products other than tobacco. In short, among Virginians he attained the prosperity and respectability for which he had strived so hard among the British in his early twenties.

The war against the French had moved far beyond the Ohio Valley wilderness to other arenas—Europe, Asia, Africa, South America. In North America, the Ohio Valley remained firmly in British possession, with General Forbes naming its new stronghold at the Forks of the Ohio after the British secretary of state—Fort Pitt (later Pittsburgh). General Forbes himself had grown weaker and more fragile after the taking of the Forks. Carried to Philadelphia, knowing it was the end, he put his affairs in order and died only a few months later, in March 1759, with much praise for his handling of the campaign against Fort Duquesne. That next fighting season of 1759 came to be known to the British as the *Annus Mirabilis*—the "Year of Wonders." The fight against the French in

North America shifted northward, with the British pushing ever closer to Canada and taking the French forts at Niagara, Ticonderoga, and Crown Point, securing for the British a key travel route. The severing of Niagara and its crucial portage between Great Lakes Ontario and Erie left the French without easy water access to the vast interior and their outposts and Indian allies there.

The decisive North American battle in that "Year of Wonders" occurred at the fortified French city of Quebec, just eight months after Washington's resignation. The city's guns and thick walls crowned a bluff overlooking the critical, funnel-like waterway where the broad Gulf of Saint Lawrence narrows to the Saint Lawrence River. Here, in what has become a legendary battle, two enormous egos fueled by massive vanity faced off—the forty-seven-year-old and much experienced Marquis de Montcalm for France and a younger but equally determined British commander, General James Wolfe. Scores of British ships carrying about five thousand troops massed in the Saint Lawrence. After several unsuccessful attempts to take the fortified city, the thirty-two-year-old General Wolfe carefully glassed the steep face of the bluffs until he spotted an obscure footpath that climbed nearly two hundred vertical feet from the river to a grassy plateau known as the Plains of Abraham, just beyond the city's walls. The crucial bit of intelligence about the footpath apparently came from Robert Stobo, recently escaped from Quebec, who five years earlier had been turned over by Washington as a hostage to the French to ensure the terms of surrender of Fort Necessity.

General Wolfe entertained premonitions of death as he put his affairs in order the night before the battle aboard ship in the Saint Lawrence. As he was rowed in the predawn darkness toward the secret path, he recited over and over Thomas Gray's recently famous poem, "Elegy in a Country Churchyard."

> . . . all that beauty, all that wealth e'er gave,
> Awaits alike the inevitable hour.
> The paths of glory lead but to the grave.

When day broke on September 13, 1759, the Marquis de Montcalm looked out from the city walls and saw a silent British line of Redcoats and Highlanders two deep stretching a half mile across the plain. He ordered his French regulars in their white uniforms and his little-trained Canadian militia in their ordinary clothes to charge the red line. They ran forward, bayonets afixed, until about one hundred and fifty yards away, paused, and opened fire, but the British did not flinch. When the French had approached within forty paces, the British line fired in unison, putting the stunned French to rout. In the midst of this, General Wolfe took a French musket ball through the wrist, which he casually wrapped in a handkerchief, and then one through the intestines and another through the chest. Carried to the rear, he lay bleeding in a daze, when one of his officers shouted, "They run; see how they run!" General Wolfe roused himself briefly, gave orders to cut off the French retreat, turned on his side, and said, "Now, God be praised, I will die in peace!" And he did.

Suffering a parallel fate, the Marquis de Montcalm took grapeshot through torso and leg and was propped on his horse by two soldiers until he made it back inside the Quebec city walls. "It's nothing, it's nothing," he reassured the gathered and alarmed citizens as blood poured forth, according to one account. "[D]on't be troubled for me, my good friends." Then he, too, fell dead. The city of Quebec soon fell, too, and not long after that Montreal. Finally, all of Canada fell to the British. This evicted France, after nearly two centuries of occupation, from the eastern regions of North America and would radically change the political destiny of the continent.

When the smoke and dust finally cleared, and the shattered hulls sank beneath the waves, the British had vanquished the French in land and sea battles around the globe. The conflict would ultimately be called the Seven Years' War (in Europe) and the French and Indian War (in North America). In the Caribbean, the British took from the French their sugar-rich island colonies of Guadeloupe and Martinique. In India, the British took Calcutta and Pondicherry and ended French influence on the

Indian subcontinent. In West Africa, the British took Senegal and other French slave-trading ports. From Spain, the British took Havana in Cuba and Manila in the Philippines, while British ally Portugal took pieces of South America. In Europe itself, after tens of thousands of deaths, territory given and territory taken, alliances shifting, the map had not altered greatly.

Even before the Seven Years' War ended, the elderly King George II died one morning in October 1760 of a burst heart while exerting himself in his toilet closet. He was succeeded by his young and temperamental grandson, soon to gain infamy in the American colonies—King George III.

When the Treaty of Paris ended the hostilities and was finally signed in February 1763, George Washington was thirty-one years old. Nearly a decade had passed since that wet dawn at Jumonville Glen when he fired the opening shots of this global conflict. Despite his many threats to quit, he had served for the first five years. Once the war moved on from the Ohio wilderness and Virginia backcountry, however, he, too, had moved on—back to his plantation at Mount Vernon and marriage to Martha. As the signers of the peace affixed their names at Paris, Washington was busy shipping hogsheads of tobacco, seeking a gardener for his estate from within the ranks of the Virginia Regiment (they had just received their pay and were all drunk, reported his contact), and ordering goods from his London agent, Robert Cary & Co. He had arranged for a "pipe" of the best Madeira—126 gallons—to be shipped to him.

He emulated a British aristocratic lifestyle, as the Virginia planters did generally, but he—and they—needed the funds to support it. Land rich but cash poor, he ran up considerable debt with his London agents from his many purchases of imported necessities and luxury goods such as elegant shoes, silver serving plates, and mahogany dining sets. Never able to catch up, he found it frustrating to try to pay down his debts by growing tobacco and shipping it abroad to sell through an agent at uncertain prices. Endlessly energetic, he experimented with other possible sources of profits, such as growing wheat, manufacturing whiskey, and, in partnership with other entrepreneurs, promoting hugely ambitious schemes to drain the Great Dismal Swamp for farmland and build a shipping canal from the Potomac River into the Ohio Valley.

Always he focused on land, and especially the western lands beyond the mountains. He understood that the burgeoning population on the Eastern Seaboard would eventually spill over the Appalachians into the Ohio Valley, and he wanted to own as much of those lands as possible when the eager settlers arrived. The British Crown had offered "bounty lands" in the Ohio Valley to those who had joined to fight in the French and Indian War. Washington aggressively—some of his former officers said unfairly—took advantage of this offer and parlayed it into tremendous personal holdings in the Ohio. He helped in the survey of the lands, conducted by a close associate of his, and, not surprisingly, ended up with the best bottomlands for farming. He also used surrogates to contact other war veterans who were awarded land but might need money instead, eventually acquiring over thirty thousand acres.

London authorities attempted to slow settlement in the Ohio Valley, keeping the lands as an "Indian Reserve," by declaring in 1763 what became known as the "Proclamation Line." This imaginary line running down the crest of the Appalachians forbade British settlement beyond it. By 1768, however, as land-hungry whites pushed deeper into the continent's interior, the British had negotiated treaties with Indian tribes who found themselves under pressure, and opened much of the Ohio to settlement. Thus came to pass the worst of Delaware chief Shingas's fears, that the English had come to the Ohio ultimately to "drive us away and settle the Country. . . ."

Through all his entrepreneurial activities, Washington maintained a tranquil and by all appearances contented life with Martha and her children at Mount Vernon. Yet his passion and anger still erupted at times. He stormed about his British creditors, constantly complaining of shoddy goods and exorbitant prices. The British Proclamation Line infuriated him. He lashed out at a former officer, George Muse, who accused him of unfair treatment in Ohio bounty land dealings. Warning him never to make such accusations again, Washington called Muse an "ungrateful and dirty" fellow. His honor, his wealth, his western lands—Washington passionately defended them against every slight or incursion.

A profound threat to colonial American tranquility arose in the mid-1760s with the passage of the Stamp Act. Britain had spent enormous sums to vanquish the French on multiple continents during the prolonged Seven Years' War. At war's end, Parliament decided that the American colonies should help to pay for the crushing cost of their own defense and the continued expense of keeping British troops there. The first time the American colonies had been taxed directly by Parliament, the Stamp Act imposed a tax on newspapers, pamphlets, legal documents, and other items that had to be printed on special "stamped" paper. The act immediately set off an uproar in the American colonies in the spring of 1765. In Virginia, it inspired planter and lawyer Patrick Henry to storm in the House of Burgesses against the injustice of a British tax imposed on American colonies without their consent. It was, he claimed, a violation of "the distinguishing Characteristick of British Freedom and without which the ancient Constitution cannot subsist." When met by charges of "Treason!" from some outraged burgesses, Henry supposedly shouted, "If this be treason, make the most of it!"

Washington did not immediately jump into the fray, choosing to attend to that spring's wheat planting at Mount Vernon rather than hang around in Williamsburg debating a response to the Stamp Act. By that fall of 1765, however, his attitude against the British economic policies and attitudes began to harden, fueled in part by his own financial distress. He wrote a long letter to Robert Cary, his London agent and creditor, complaining about the low prices his tobacco fetched in London, suggesting that Cary needed to work harder to sell his tobacco, and lamenting the high price and shoddy quality of goods that Cary shipped to him. He then gave his take on the Stamp Act, calling it an "unconstitutional method of Taxation" and warning that his fellow citizens viewed it "as a direful attack upon their Liberties, & loudly exclaim against the violation. . . ."

At the same time, Washington consciously pushed toward personal economic independence from Britain and his creditors there. He largely abandoned tobacco in favor of other crops like wheat and flax, milling it in his own mills. He sold barrels of fish, packed in salt, that his slaves and workers netted in the Potomac during seasonal migration runs, and he tried to manufacture more of his own goods, such as ironwork and fabric for slave clothing, rather than import them from England.

Revoking the Stamp Act, the British Parliament quickly imposed other taxes on the American colonies, such as the Townshend duties, beginning in 1767, which taxed tea, paper, glass, paint, and other items. Opposition built anew in the colonies. Plans circulated for "nonimportation associations"—groups dedicated to boycotting British goods. Washington himself soon embraced this scheme, believing that Britain would likely respond most to economic sanctions. In a letter to his friend and neighbor planter George Mason, Washington, taking a radical stance for his time, invoked a lofty concept of "American freedom" and actually advocated taking up arms if nothing else worked.

> *At a time when our lordly Masters in Great Britain will be satisfied with nothing less than the deprivation of American freedom, it seems highly necessary that something shou'd be done to avert the stroke and maintain the liberty which we have derived from our Ancestors; but the manner of doing it . . . is the point in question.*
>
> *That no man shou'd . . . hesitate a moment to use a—ms [arms] in defence of so valuable a blessing, on which all the good and evil of life depends; is clearly my opinion; Yet A—ms [Arms] I wou'd beg leave to add, should be the last resource; the [last] resort.*

This was radical talk—taking up arms against the king in defense of American freedom. Historian John Ferling notes that at the time Washington wrote these words, in April of 1769, he displayed a deeper disaffection toward Britain than any other man who would come to be known as one of the Founding Fathers, with the exception of Sam Adams. He had recently turned thirty-seven years old. Ten years had passed since Washington left the Ohio wilderness. What had changed that would compel Washington to jeopardize everything he strived so hard to attain by advocating taking up arms against his own government, even as a last resort?

His personal resentments against the British ran deep. In his quest for recognition, he had approached British military superiors many times requesting a royal officer's commission. Each time they coldly rebuffed him. As a provincial officer he had suffered a thousand other slights from British royal officers, in whose dim view colonials amounted to ragtag, undisciplined, ill-clothed soldiers led by uncouth officers amid the dense thickets, fetid swamps, and towering forests of the American backwoods. Washington felt that chill rejection deep inside. They ignored him, dismissed him, pushed him subtly aside. He had never been one of them. He had explored every avenue he could imagine for advancement. He left his frontier post and rode to Philadelphia, Boston, New York, waiting for days and weeks to be seen by British generals who offered him nothing, only to return to find a Virginia frontier dark with the smoke of burning homesteads and strewn with tomahawked and animal-gnawed corpses. Frustration seethed into resentment.

He had first raised the issue of fair and equal treatment for American colonists twelve years earlier when he wrote to Governor Dinwiddie urging regular British Army status for himself and the Virginia Regiment. His list of complaints had only grown in the twelve years since then. Washington had spent five years of his life fighting for the lands that lay over the mountains, yet London possessed the power to stop American colonists from settling there simply by drawing an imaginary line on the map along the Appalachian crest. Myriad other resentments sprang from the insults contained in the Townshend Acts—the taxes on imported goods, the rights of customs inspectors to search and seize property, the powers given to a customs board. No representative from America sat in the Parliament that made these laws.

Washington's grasp of the scope of the North American continent may have fueled his resentments still further. Unlike most of his Virginia planter contemporaries, who confined themselves to the cultivated precincts of the Eastern Seaboard and whose gaze turned east toward Britain, Washington had glimpsed something far larger in his travels beyond the mountains—both geographically and metaphorically. On that first frigid journey as a twenty-one-year-old bearing a message for the French commandant, Washington had crossed the mountains and ridden for weeks nearly to the shores of Lake Erie, had seen the waters of the Ohio River flowing toward the distant Gulf of Mexico. On these early travels, he gained an inkling of the continent's vast and unsettled nature and a vision of what it might become. His own ancestors had helped push Virginia's bounds of cultivation ever farther west, like a wave, to the foot of the Appalachians. He knew the wave would inevitably spill over the tall ridges into the Ohio, felling the oaks and maples with powerful thuds onto the black soil, burning and digging out the gnarled stumps, overturning the moist bottomland to plant crops, building gristmills and villages, and one day digging the canals connecting the Ohio to the Atlantic, when an endless sequence of barges would export grain and import prosperity. Seeing a great potential for profit, Washington first acted on that vision as a teenage surveyor, buying land in the mountain valleys at the age of eighteen and acquiring much more in years to come. But London, on a whim, could obliterate that vision with a line inked on a map or a simple proclamation.

Initially he had found the frontier settlers crude and stupid, helpless, and sometimes pitiful, but that gradually changed. He had observed their enterprise in setting out, without the phalanxes of slaves he employed but with only their own bare hands and a sharpened ax to clear a life from a wild patch of forest. He understood in a way that few Virginia aristocrats could that over there, just over those mountains, lay a frontier inhabited by settlers who pursued a very different way of life, largely self-sufficient and sometimes beyond the reach of colonial authority, people who prized their independence, people who had escaped indentured servitude or debtors' prison or whatever ghosts from the past to make a life of their own devising, in their own control. In those five years on the frontier and wilderness, Washington had seen scores, hundreds, probably thousands of people who had carved out a new life. He had witnessed their self-sufficiency, independence of spirit, suspicion of outside authority, and passion for self-government. He had absorbed a distillation of that contrarian spirit.

Though he had dismissed the Native Americans he met in the Ohio wilderness as savage and barbaric, he admired their deftness in the forest, their incredible toughness, their devastating bush-fighting techniques. One wonders if he also admired their mobility and ability to adapt, their resourcefulness and ability to live off the land, unencumbered by heavy debt and the hoary institutions of Europe. In Braddock's Defeat, he had personally witnessed and barely survived the Indian attack in the Ohio wilderness that decimated what should have been a powerful European army. He had seen the vulnerability of the British Empire. He may have realized that one did not have to win the cities to win a war. In marked contrast to his head-on disaster at Fort Necessity, the Indians generally avoided making a last stand against daunting odds. They melted away to other places and lived to fight another day. He would not forget how they could disappear, and just as suddenly reappear.

The wilderness itself projected a power that Washington came to know: the endless forest with its massive hardwoods reaching up to the green canopy, the great humped mountains, the rivers surging down them, the frigid winters and deep snows, the steamy summers, the seasons of mud and rain. He had stumbled through the wilds nearly alone on that first wintry trip, half-frozen, with little food, almost perishing when he tumbled into an icy river. From that point forward he had treated it with respect. He knew that winter's cold or a season of mud could stop an army in its tracks.

All these strands may have fueled the passion with which, by April 1769, he was willing to stand up to British authorities. Other influences reached across the Atlantic from Enlightenment Europe, from Paris and London, and the thinkers and writers there. Books, pamphlets, and essays of the moment raised profound questions about the nature of government and about the divine right of kings. Here on the banks of the Potomac the two strands met—Enlightenment Europe and American wilderness, blending the values they both embodied—self-sufficiency and independence—with concepts of government by consensus and opposition to tyranny. It would prove a volatile mix.

Washington himself had changed since the early days of the French and Indian War. Powerfully ambitious and self-centered in his early twenties, with an iron will bent on personal advancement and a flaring temper when thwarted, he had grown beyond pure self-interest. Very quickly he found himself burdened with the responsibility of caring for others. Within a few months of that first, fateful journey into the frozen Ohio, the twenty-two-year-old Washington was commanding a body of men. Under his leadership, scores of them—men whom he knew—fell dead or wounded into the muddy, bloody trenches of Fort Necessity.

At age twenty-three, as a young aide-de-camp, he survived the horrible slaughter of Braddock's Defeat, delivering the dying general's requests for reinforcements as men groaned and screamed for help along the midnight trail. Before he turned twenty-four, he had taken charge of the entire Virginia Regiment. He had to feed it, clothe it, pay it, and, sometimes most difficult of all, discipline it. When he was twenty-five, trying to bring discipline, he swung two of his men from a forty-foot gallows and had to bear that weight. His cares evolved into something broader than his own advancement. His circle of responsibility ever spread and deepened. At twenty-six, he witnessed the nightmare spectacle of his own men shooting their comrades-in-arms in the foggy dusk, and tried to stop it. From a very early age, his choices, actions, and inclinations bore profound consequences for those around him.

Remarkable for his driving ambition and for bearing a great deal of responsibility so early in life, the young Washington was also remarkable for his acquisitiveness. He was highly protective of what he did possess. Utterly unafraid in the face of physical danger and apparently unafraid of losing his life, he seems to have feared above all else the loss of his reputation. Jealously guarding his property and money as closely as his reputation, he was on his guard against being cheated or taken advantage of while at the same time hungrily amassing land holdings and determinedly climbing his way into the intricate latticework of Virginia aristocracy. With his father dying when Washington was eleven and his older half-brother and mentor dying when he was twenty, a strained and distant relationship with his mother and her many needs and infatuated with a woman who was already married, the young Washington was, in many ways, on his own emotionally as well as financially and socially, an orphan with modest land holdings trying to find his way.

Following the ancestral Washington pattern, he married a wealthy young widow whose holdings launched him, as one historian put it, "to the top tier of Virginia's planter class." His marriage laid a rock-solid foundation that allowed him to move beyond his singular focus on establishing himself. In addition to her wealth and social standing, Martha's steadfast support gave him emotional security—a safety net of reassurance—that he had previously lacked. After scrambling upward with all his energy and all his focus for over a decade, he had arrived socially, financially, and emotionally at a place that once lay far beyond his reach.

He had undergone another transformation, too, during his five years in the wilderness and at war. Traveling between the known, settled world of the Tidewater coast and the unknown world of the Ohio wilderness, with its dark and powerful forces, young Washington, like the prototypical hero in Joseph Campbell's mythology, faced a series of trials. After great difficulty and many mistakes, he returned, ultimately, "the master of two worlds." The worlds he mastered were not only wilderness and civilization, but the disparate parts of his own self. Over and over, he listened to the pleadings of dying soldiers, or frontier women begging him for help, their husbands hacked outside their burning homesteads, their children captured. From the self-focused youth he had been, as his responsibilities broadened, so did his empathy. It did not happen all at once. Gradually he learned to control his drive, his passion, and his anger. Rather than lashing out at whatever obstacle blocked his self-advancement, he now channeled that drive and passion and anger for the greater good, shaping it into something powerful and useful.

Historian Joseph Ellis dates the moment of Washington's transformation quite precisely, at least in a metaphorical sense, to April of 1769, when he wrote his impassioned letter on nonimportation of British goods and pledged himself prepared as a last resort to take up arms against "our lordly Masters in Great Britain [who] will be satisfied with nothing less than the deprivation of American freedom."

Writes Ellis: "This was the moment when Washington first began to link the hard-earned lessons that shaped his own personality to the larger cause of American independence."

It was not only about him. At some point, or gradually over the years and challenges that he faced with varying degrees of success or failure, he came to realize that he could acquire far more honor and reputation through his selfless acts—his acts for the common good—than he could through the single-minded pursuit of self-interest. From his earliest years in the Ohio wilderness during his twenties, he seemed conscious of living on a historical stage with meticulously chosen dress to match. If he was initially playing the role, consciously shaping his image and his actions—wilderness messenger, warrior, planter, burgess—he became very good at it. Finally he became that person, that selfless leader, the one that is remembered as George Washington.

POSTSCRIPT

By the late 1760s, tensions between American colonists and British authorities had ratcheted still higher, over taxes imposed by Parliament without American consent, along with other measures considered repressive by colonists. In March 1770 in Boston, a hotbed of resistance, eight British troops opened fire on a crowd of angry American demonstrators in what became known as the "Boston Massacre." Rescinding some of the taxes, Parliament kept a tax on tea, partly to force Americans to bow to British power to tax them. Outraged Bostonians, some dressed as Mohawk Indian warriors, boarded ships in Boston Harbor in the dead of night on December 16, 1773, and dumped the tea chests overboard—the Boston Tea Party.

A little over a year later, in spring of 1775, seven hundred British troops sent to seize arms and gunpowder at Concord, outside Boston, were met by an armed militia of colonists. The two sides opened fire on each other, and as the British withdrew toward Boston, the colonists took aim from behind rocks, trees, and houses, exacting a high toll on the Redcoats and barricading the British troops inside the city under British general Thomas Gage, who had served alongside then aide-de-camp Washington and led the advance guard at Braddock's Defeat.

During the month following the Battle of Lexington and Concord, May 1775, representatives from the various colonies met in Philadelphia for the Second Continental Congress. One of the delegates from Virginia, forty-three-year-old French and Indian War veteran, wealthy planter, and longtime burgess Colonel George Washington, had worked at the forefront of the Virginia resistance, assembling and training independent county militias to defend Virginia's interests against the British. Traveling from Virginia to Philadelphia as spring flowers blossomed in the fine weather, Colonel Washington had ridden in aristocratic style in his luxurious four-wheeled chariot, imported from England, with coachmen and postilion. Along the way it was clear that a martial spirit had erupted among American colonists eager to defend their rights, and as he passed through cities like Baltimore, Colonel Washington was asked to review newly raised independent companies of troops.

As the Continental Congress began in the Pennsylvania State House, Colonel Washington showed up wearing the blue and buff uniform that he had designed for the Fairfax County Militia, which, for officers like himself, he had carefully specified be adorned with silk sashes, gorgets, and shoulder knots. He had far more military experience than any other delegate. As the congress considered what to do about the request from Massachusetts for military help, Virginia's Colonel Washington was named to head committees on ammunition supplies, on a budget for a twelve-month military campaign, and on other military matters. On June 14, the Continental Congress took a bold step and voted to recruit ten companies of troops from Pennsylvania, Maryland, and Virginia to help those in Massachusetts who were barricading British troops inside Boston by closing off its peninsula. These troops needed a commander. Eyes turned to Colonel Washington, wearing the resplendent officers' uniform, with his tall, stately bearing, his military expertise, his sound and steady judgment.

"He is a complete gentleman," Massachusetts delegate Thomas Cushing reported. "He is sensible, amiable, virtuous, modest, and brave."

"He possessed the gift of silence," John Adams would remember decades later. "He had great self-command."

"Dignity with ease," Adams's wife, Abigail, would remark of Washington.

"[H]e has so much martial dignity in his deportment that you would distinguish him to be a general and a soldier from among ten thousand people," the Philadelphia physician and Continental Congress delegate Benjamin Rush exclaimed. "There is not a king in Europe that would not look like a valet de chambre by his side."

On June 14, the same day that the congress authorized troops to be recruited, John Adams rose and spoke favorably of Colonel Washington, who, Adams later remembered, jumped up and with "his usual modesty darted into the library room." The following day, Thursday, June 15, the Continental Congress officially voted to appoint a general "to command all the continental forces raised, or to be raised, for the defense of American liberty."

Thomas Johnson of Maryland officially nominated George Washington of Virginia.

By a unanimous vote, Colonel Washington was elected "General and Commander in Chief of the army of the United Colonies."

Accepting the appointment the following day, Washington stood by his chair near the door. He acknowledged the "high honor" that the congress had given him. Despite his "great distress" that he lacked the abilities and military experience to do so, he would, as the congress desired it, nevertheless exert himself with his every power to carry out this "momentous duty" in support of "the glorious cause."

"But, lest some unlucky event should happen, unfavorable to my reputation," Washington qualified his acceptance, "I beg it may be remember[e]d by every Gent[lema]n in the room that I this day declare, with the utmost sincerity, I do not think myself equal to the command I [am] honoured with."

It would turn out that he was.

FATE OF THE CHARACTERS

Christopher Gist—Having explored part of the Ohio Valley for the British and the Ohio Company, and guided Washington and Braddock during the French and Indian War, Gist apparently died of smallpox soon thereafter in Georgia or South Carolina in 1759 at the age of about fifty-four.

Governor Robert Dinwiddie—In poor health and retiring from his post as lieutenant governor of Virginia in 1758, Dinwiddie returned to Britain and died in Bristol in 1770 at the age of seventy-eight, survived by his wife and two daughters.

Sally Fairfax and George William Fairfax—The Fairfaxes lived at Belvoir Manor and remained friends with their neighbors George and Martha Washington until 1773, when they returned to England and George William attempted to become the next Lord Fairfax after the death of his uncle. His effort failed, the American Revolution broke out, and Sally and George William remained in England, renting out Belvoir Manor. With the Cary family's Virginia estate confiscated in the Revolution and deprived of the Fairfax fortune, George William died at Bath in 1787, while Sally, not fully accepted by her aristocratic British in-laws, lived a quiet life for many years thereafter, dying in 1811, childless. Before her death, in a letter to a relative, she wrote, "I know now that the worthy man is to be preferred to the high-born."

Lieutenant Colonel Thomas Gage—The Royal Army officer who headed Braddock's advance party when it was attacked and who managed to survive, Gage went on to become commander of all British forces in North America in 1763 after the end of the French and Indian and Seven Years' Wars. As American colonists began to protest their treatment by the British Crown in the late 1760s and early 1770s, and Boston became a hotbed of resistance, Gage was named military governor of Massachusetts and moved his headquarters there with thousands of British troops. It was Gage who, under British orders to take action against the rebelling colonists, in April 1775 ordered British Redcoats to seize rebel gunpowder and arms at Concord, resulting in the Battle of Lexington and Concord and the start of the American Revolution. Soon thereafter, newly appointed General George Washington came to Boston at the head of the Continental Army. Gage, his old comrade in arms from Braddock's march and now opposing commander, was recalled to England, replaced by General William Howe.

Sir John St. Clair—Despite his obvious bravery, relentless energy, and devotion to his quartermaster's tasks, Sir John's volatile temper and lack of tact prompted his future commanders in the British Army to remove him from any active combat in North American operations and over his protests to relegate him to deskbound duties reviewing financial accounts. While posted in Philadelphia, Sir John met an attractive young woman from the city's society, Elizabeth "Betsey" Moland, who was delighted to bear the title "Lady St. Clair," and he in turn to partake of her family's wealth. Married in 1762, the couple moved to a country home outside Trenton, New Jersey, and kept a second residence just across the Hudson River from New York City, giving Sir John quick access to British Army headquarters there. In failing health due to his old wound from Braddock's Defeat and quarreling with his in-laws, Sir John died in 1767. His widow remarried a British officer, and his son, the next Sir John, fought on the Loyalist side during the Revolution, in which both the St. Clair and Moland families lost all of their American estates. "Sir John St. Clair's shortcomings were so severe," writes his biographer, Douglas Cubbison, "that, by the conclusion of the Forbes campaign of 1758, the man succeeded in writing himself out of history."

James Smith—James Smith escaped from Indian captivity in 1759 in Montreal and returned to Pennsylvania but found his "sweetheart" had married a few days before. He married Anne Wilson, and the couple became settlers on the Pennsylvania frontier, eventually having seven children. In 1765, Smith, in protest of British trade practices and treatment of Americans, led several hundred local settlers in armed raids and rebellions against British troops and frontier fortifications. Known as Smith's Rebellion or the Black Boys Rebellion because they painted their faces in red and black in the Indian style, this is sometimes called the first armed uprising against British rule in America and an early assertion of the American right to bear arms. A great proponent of Indian methods of wilderness warfare, Smith called on General George Washington at the commander's winter headquarters in Morristown, New Jersey, in 1777, and offered his services and those of his followers as bushfighters against the British troops. (Whether they discussed Braddock's Defeat—which each saw from opposite sides—remains unknown.) Smith eventually led a force into the Ohio wilderness in pursuit of British-allied Indians. After the Revolution, he moved to the newly opened Kentucky country, became a legislator, and died in 1813. Smith was portrayed by John Wayne in the 1939 Hollywood film *Allegheny Uprising* (aka *The First Rebel).*

Claude-Pierre Pécaudy de Contrecoeur—After the great French and Indian victory at the Battle of the Monogohela (Braddock's Defeat), during which he was commander of Fort Duquesne, Contrecoeur, having sustained an injury or illness during the campaign, asked the newly appointed governor-general of New France, Pierre de Rigaud de Vaudreuil de Cavagnial, to be removed from command. Contrecoeur officially retired in late 1758 to his lands near Montreal and after the British conquered Canada in 1763 decided to remain there. As seigneur of a feudal estate in 1765, he counted 371 persons; 6,640 cultivated acres; and 973 animals and was considered one of the most important people in Canada. He died in 1775.

Captain Louis Coulon de Villiers—Having defeated Washington's forces in 1754 at Fort Necessity and accepted the fort's capitulation, Villiers participated in French and Indian guerrilla warfare against the Pennsylvania frontier the following year, then went on to distinguish himself in the French capture of British forts at Oswego on Lake Erie and William Henry on Lake George. He was recommended for a cross of Saint Louis, with the commendation that "the family of the Sieur de Villiers has always distinguished itself in the service. There is not one of them who has not died in action against the enemy." He died of smallpox, however, a few days after receiving the award.

Half King—Despite his role in triggering the French and Indian War, the Half King hardly witnessed it unfold, falling ill in October 1754, a few months after the surrender of Fort Necessity, and dying a few days later—the victim, his followers said, of French witchcraft.

George Washington—He went on to become commander in chief of the Continental Army and first president of the United States. Retiring to Mount Vernon with Martha in 1797, he died in 1799 at age sixty-seven, apparently of a throat infection and severe bloodletting.

ACKNOWLEDGMENTS

This book had its origins in a trip I took to western Pennsylvania in 2006 while researching another book, about the "blank spots"—the unpopulated areas on the American map. To my surprise, I found that parts of western Pennsylvania today, in this age of suburban sprawl, are characterized by sparse population, heavy forest, and rugged mountains—a near wilderness not so different from the Ohio wilderness of centuries ago. I discovered that the region had a rich history and that a young George Washington's adventures and fate were deeply entwined with it. This was an entirely different Washington from the one I knew, a much less perfect and much more vulnerable one.

I wanted to know how that wilderness had shaped him as a young man. My explorations in pursuit of him have taken me to places I never would have discovered and to wonderful people whom I never would have met. I spent a chill autumn night camped on a forested mountain ridge along the old Braddock Road to sense the character of the woods that Washington might have known. I paddled French Creek—the Rivière Le Boeuf—to feel its swings and bends and riffles. I traced Washington's route into the Ohio wilderness and visited the scenes of his early skirmishes, battles, and surrender. To write the book, I also drew on my own experiences over many years in the wilderness, camping in deep snow, trekking through subzero forests with my belongings on my back. It occurred to me that, during my years in the outdoors, I've had a good deal of experience paddling canoes down half-frozen rivers and ice-rimmed lakes. It was visceral experiences like these—the sounds, the smells, the finger-numbing difficulties—that I drew on to help illuminate some of what Washington might have experienced on those icy forays into the Ohio wilderness.

Among the many generous individuals who helped me on this journey to discover the young Washington, I would like to thank John Robinson, my fellow researcher with a magic touch for putting his finger on the right information at the right time. The entire educational complex known generically as Mount Vernon, supported by the Mount Vernon Ladies' Association, helped me in countless ways, including the wonderful website the organization maintains, mountvernon.org, which offers a treasure trove of information and insight on all aspects of George Washington for anyone who would like to pursue it further. The staff and resident historians were very helpful at the beautiful new research library at Mount Vernon, the Fred W. Smith National Library for the Study of George Washington, especially Douglas Bradbury, Sarah Myers, Joseph Stolz, and Mary Thompson. Likewise, many thanks to the staff and organizations of Colonial Williamsburg; Historic Jamestowne; Fort Necessity National Battlefield; Carlyle House Historic Park; Winchester-Frederick County Historical Society; Braddock's Battlefield History Center and its director, Robert T. Messner; Fort LeBoeuf Historical Society and the Amos Judson House; and Fort Pitt Museum and the Heinz History Center. The museum and staff at the marvelously reconstructed Fort Ligonier, in Ligonier, Pennsylvania, have been a tremendous resource, and special thanks to Erica Nuckles, director of history and collections, and to Matt Gault and Jason Cherry.

The annual gathering known as the Jumonville Seminar, near the site of George Washington's first skirmish, sponsored by the Braddock Road Preservation Association, brings together scholars and history enthusiasts to learn in greater depth about this particular time and place in American history. I am grateful for the warm welcome the organization extended to me, and I encourage anyone who would like to know more about these events, the period, the clothing, and countless other details, to join it and attend the annual Jumonville Seminar. Special thanks to Walter L. Powell, president of Braddock Road Preservation Association; Jaye and Larry Beatty, who run the Jumonville Camp; David Preston, professor of history at the Citadel, speaker at the seminar, and author of the definitive *Braddock's Defeat;* Daniel J. Tortora; Douglas Cubbison, author of *On Campaign Against Fort Duquesne;* Carl Ekberg; Norman Baker; Bruce Egli; and Joan Mancuso.

Other historians and individuals who have helped me on this journey are J. William "Bill" Youngs, Andrew Laue, Scott Elrod, Tom and Sue Duffield, Fred Haefele and Caroline Patterson, David and Rosalie Cates, Martha Newell and Mike Kadas, and Christopher Preston. Patti Lazarus provided shelter from the storm in every way and Jim Mckenzie, writer and native of western Pennsylvania, offered engrossing discussions about history, poetry, life, and ice jams on the Allegheny River. Several individuals read parts of the manuscript in earlier drafts and gave valuable feedback, including Noel Ragsdale, Jane Ragsdale, Jim Mckenzie, Sean Sheehan, and Mike Kadas. I've also felt very fortunate to have had the response and insight of a Missoula book group that read the manuscript in draft form, as well as provided excellent food and drink during our discussions of the young Washington and his world: Ashby Kinch, John Bardsley, Carl McKay, Tom Webster, Jesse Johnson, and Dave Calkin. Thanks to Kent Madin and Linda Svendsen for our chilled horse-packing journey through the swamps and taiga forest of Manchuria, and to Fred Haefele, Matt and Steve Rinella, and Ian Frazier for our hunting experiences paddling the sometimes ice-choked Missouri Breaks—all of which helped inform this book. Greg Kiser got me launched on French Creek and entertained me with stories at dinner. Jim Ritter is an old and dear friend and longtime canoeing companion but also a journalist and an insatiable reader of history, with whom I originally discussed this idea while paddling Montana's Marias River in 2013. He has helped me in many ways in bringing it to a reality.

Kevin McCann, cartographer with an elegant style and passion for history, drew the wonderful maps for this volume. Suet Chong and the rest of the art and design team at Ecco/HarperCollins produced the beautiful design. Also at Ecco/HarperCollins, special thanks to master of many worlds Emma Janaskie, and to Miriam Parker, Martin Wilson, Meghan Deans, and Brenda Woodward. My editor Denise Oswald has brought her acute editorial insight, enthusiasm, and support to this project, and has been fun to work with besides, while publisher Daniel Halpern has thrown his generous support behind it. Hilary Redmon originally commissioned the book and backed the idea enthusiastically from the start. I can't count the many ways my agent Stuart Krichevsky helped generate the idea for this book and carry it to fruition with steady support and wisdom. Ross Harris, now an agent in his own right, was always there as backup, and Laura Usselman kept it all on track.

My deepest gratitude goes to my family, who have provided the emotional underpinnings, support, and understanding on which this and all my other writing projects rest: my mother, Judith Stark; late father, William Stark; siblings, Kate Stark Damsgaard, Ted Stark, and Sarah Stark; late father-in-law, Wilmott Ragsdale; and late mother-in-law, Jane Ragsdale. To our children, Molly Stark-Ragsdale and Skyler Stark-Ragsdale, I can only say, among a thousand other things, that I have the deepest pride in them. To my wife, Amy, goes my love beyond all depth and my inexpressible gratitude. Without her, none of this—this book, this life—would have been possible.

NOTES

CHAPTER ONE

3 *the old Indian town of Venango* Washington, "Journey to the French Commandant (1753)," 13.

log cabin See Freeman for description, 1:302. See also Washington, "Journey to the French Commandant (1753)," 13.

4 *Captain Phillipe Thomas Joncaire* For greater biographical details, see *Dictionary of Canadian Biography,* http://www.biographi.ca/en/bio/chabert_de_joncaire_philippe_thomas_3E.html, accessed June 28, 2016.

7 *"[T]hey were sensible the* English*"* Washington, "Journey to the French Commandant (1753)," 13.

10 *Plied by voyageur canoes* Preston, 135.

11 *"lived under the French colors"* Gist, 81.

13 *Andrew Montour, who costumed himself* Freeman, 1:383.

CHAPTER TWO

15 *leaving her at home with five young children* See Mary Ball Washington biography at http://www.mountvernon.org/digital-encyclopedia/article/mary-ball-washington/, accessed July 21, 2017.

16 *"War is horrid in fact"* Freeman, 1:69.

from the brother he called his "best friend" Ibid., 194.

"[S]he offers several trifling objections" Ibid., 195, citing William Fairfax to Lawrence Washington, September 9, 1746.

"cut him and staple him" Ibid., 198–99. "I think he had better be put apprentice to a tinker." Joseph Ball to Mary Washington, May 19, 1747.

fell well short Ibid., 269.

17 *Scholars have identified* Evans, 2.

In 1610, a new arrival *Encyclopedia Virginia,* http://www.encyclopediavirginia.org/Rolfe_John_d_1622, accessed October 16, 2017.

18 *Virginia tobacco, however* See, among other sources on tobacco cultivation in Jamestown, https://www.nps.gov/jame/learn/historyculture/tobacco-colonial-cultivation-methods.htm, accessed October 16, 2017.

As tobacco cultivation spread See "The Origins of Slavery in Virginia," http://www.virginiaplaces.org/population/slaveorigin.html, accessed October 16, 2017.

A new source of human energy See Lisa Rein, "Mystery of Virginia's First Slaves Unlocked 400 Years Later," *Washington Post,* September 3, 2006. Some of this original group of slaves may have already been Christianized and literate due to Portuguese contact in Africa. See also "The First Africans," http://historicjamestowne.org/history/the-first-africans/, accessed October 17, 2016. For institutionalization of slavery in Virginia, see "History of Slavery in America," https://www.ocf.berkeley.edu/~arihuang/academic/abg/slavery/history.html, accessed October 17, 2017, and "The Origins of Slavery in Virginia," http://www.virginiaplaces.org/population/slaveorigin.html, accessed October 17, 2017.

Some were merchants Evans, 6–7.

For a small "summe of money" Ibid., 9.

19 *Once in Virginia* Ibid., 9–13.

By some counts Freeman, 1:79.

The acreages expanded Evans, 15.

Families intermarried Ibid., 122.

"Virginia society" Boorstin, 101.

Voting rights steadily eroded Ibid., 101.

Along Tidewater rivers For Tidewater Virginia society and culture generally, see Boorstin and Evans.

20 *The Washingtons had occupied* See Fleming.

betrothing wealthy widows H. Clifford Smith, 34–36, 37.

this ancestral Lawrence Washington Ibid., 34–36.

When Henry VIII fought Ibid., 37–38.

The Washington family procreated Ibid., 40, 45–46. Robert Washington sold the estate, in two parts, to his nephew Lawrence Makepeace and one Simon Heynes in 1610.

A great-grandson of "Lawrence the Builder" Ibid., 46–47.

21 *"[He is] a common frequenter of alehouses"* Ibid., 47.

22 *Relatives may have* See "Samuel Argall," *Encyclopedia Virginia,* https://www.encyclopediavirginia.org/Argall_Samuel_bap_1580-1626, and "Edwin Sandys," https://www.encyclopediavirginia.org/sandys_sir_edwin_1561-1629, accessed October 17, 2017.

Her holds full See Hoppin, 151, and Freeman, 1:15.

Here Pope introduced Hoppin, 149.

The pleased father-in-law Freeman, 1:16–17.

23 *in scattered parcels* Ibid., 17.

His wife, Anne See "Anne Washington," https://www.geni.com/people/Anne-Washington/6000000003918387026, accessed October 17, 2017.

punish Indians for attacks Freeman, 1:23

Starting with simple techniques Ibid., 198.

24 *The equivalent of a tract* For English colonial land grants generally, see Osgood.

The Fairfaxes were neighbors Freeman, 1:200.

Lawrence had married For Lawrence and Anne generally, see Henriques. See also "The Blurred Line of Famous Families: The Fairfaxes and George Washington," PBS Frontline, http://www.pbs.org/wgbh/pages/frontline/shows/secret/famous/washington.html, accessed October 17, 2017. For Anne's nickname, see Freeman, 2:459. For acreage, Ibid., 1:76.

25 *Through Lawrence's marriage* For George William, see "George William Fairfax," *Dictionary of Virginia Biography,* http://www.lva.virginia.gov/public/dvb/bio.asp?b=Fairfax_George_William, accessed October 17, 2017.

"[B]edew no mans face" See "The Rules of Civility and Decent Behaviour," *George Washington Digital Encyclopedia,* http://www.mountvernon.org/digital-encyclopedia/article/the-rules-of-civility-and-decent-behaviour/, accessed October 17, 2017.

27 *the rolling, rising ground* Freeman, 1:210.

It was in the Shenandoah Washington, *Journal of My Journey Over the Mountains,* 21.

Sally's ancestors had been For Sally generally, see Cary, also "Sally Fairfax," *George Washington Digital Encyclopedia,* http://www.mountvernon.org/digital-encyclopedia/article/sally-fairfax/, accessed October 17, 2017. For *"the ruling oligarchy,"* Cary, 12. For *"the cleverest and far the most fascinating,"* ibid., 20, 28–29.

28 *Beyond the Blue Ridge* See Washington, *Journal of My Journey Over the Mountains*. For *"not being so good a Woodsman,"* ibid., 27. For *feather mattress,* ibid., 29. For *dining arrangements,* ibid., 34.

29 *"Saterday April 2d"* Ibid., 43–49.

30 *For a time* Ibid., 30–32.

"They clear a Large Circle" Ibid., 33.

"a great Company of People" Ibid., 45.

He surveyed nearly See "Surveying," *George Washington Digital Encyclopedia,* http://www.mountvernon.org/digital-encyclopedia/article/surveying/, accessed October 17, 2017.

31 *Becoming a certified county surveyor* Freeman, 1:243.

Using connections with Ibid., 236; see also Anderson, *Crucible of War,* 23, 27.

When that did not help Freeman, 1:247.

32 *George, now nineteen* See "Washington's Journey to Barbados," *George*

Washington Digital Encyclopedia, http://www.mountvernon.org/george-washington/washingtons-youth/journey-to-barbados, accessed October 17, 2017.

32 *They rented rooms* See "New Take on 'George Slept Here,'" *Boston Globe,* February 17, 2008.

Among other things Freeman, 1:264–65.

33 *Some historians point* See "Washington's Journey to Barbados," *George Washington Digital Encyclopedia,* http://www.mountvernon.org/george-washington/washingtons-youth/journey-to-barbados, accessed October 17, 2017.

34 *When he arrived* Freeman, 1:256.

CHAPTER THREE

35 *A cold December rain fell incessantly* Washington, "Journey to the French Commandant (1753)," 16, and Gist, 82.

38 *"all convenient and possible dispatch"* "Commission from Robert Dinwiddie, 30 October 1753," Founders Online, National Archives, http://founders.archives.gov/documents/Washington/02-01-02-0028, last modified November 26, 2017.

39 *In what's now Maine* "Indians of the Eastern Seaboard," Bureau of Indian Affairs (Dept. of the Interior), Washington, D.C., 1967, http://files.eric.ed.gov/fulltext/ED028871.pdf, retrieved October 19, 2017.

40 *maintained its neutrality* Anderson, *Crucible of War,* 16.

Whites settling in eastern Pennsylvania Ibid., 18.

41 *appointed a regent* Anderson, *The War That Made America,* 23.

not entirely clear Anderson, Crucible of War, 18.

boiled and eaten his father Freeman, 1:292.

42 *In an infamous land deal* See Eastern Delaware Nation website, http://www.easterndelawarenations.org/history.html, accessed September 28, 2016.

leaving many Delaware wandering in search of a home Anderson, *Crucible of War,* 22–23.

big bones, long limbs, and big hands and feet See "Personality," *George Washington Digital Encyclopedia,* http://www.mountvernon.org/digital-encyclopedia/article/personality/. Sources vary as to his exact height, between six feet tall and six feet, three inches.

44 *hooves spinning silently* See https://www.youtube.com/watch?v=HHPPOFbLb3c.

"The major and I went over first" Gist, 82.

45 *In the muddy low spots* Neither Gist nor Washington describes exactly the trail they might have followed along the Rivière Le Boeuf, but ten years later a traveler characterized it as "a narrow crooked Path, difficult Creeks, & several long Defiles." Colonel Henry Bouquet to Jeffrey Amherst, July 4, 1763, quoted in Wallace, *Indian Paths of Pennsylvania,* 170.

46 *red, black, yellow, white* For a precise painting representing Indian dress of a Delaware leader in 1735, see Merrell, 61. For other details of Indian food and clothing see Wallace, *Indians in Pennsylvania,* and Weslager, as well as contemporary descriptions in *Seaman's* (Morris) *Journal,* in Sargent.
winter capes sewn of layered turkey feathers Weslager, 54.
Both societies were matrilineal Fur, 27.

47 *river still too swift* Gist, 82.
"We passed over much good Land" Washington, "Journey to the French Commandant (1753)," 16.

48 *wanted to present himself* Ibid. Washington "prepar'd early to wait upon the Commander."
"a great deal of complaisance" Gist, 83.
"He is an elderly Gentleman" Washington, "Journey to the French Commandant (1753)," 16.

49 *"Whereas I have appointed"* "Passport from Robert Dinwiddie, 30 October 1753," *Founders Online,* National Archives, http://founders.archives.gov/documents/Washington/02-01-02-0030, last modified July 12, 2016.

51 *council of war* Washington, "Journey to the French Commandant (1753)," 16, and Gist, 83. Gist has a slightly different accounting of the dates on which these various events occurred.
late French commandant of Fort Le Boeuf Freeman, 1:292–93. Freeman identifies the commander as Sieur de Marin. See also Kent, *The French Invasion of Western Pennsylvania 1753,* 49.
"I am not afraid of Flies, or Musquitos" Washington, "Journey to the French Commandant (1753)," 8.

52 *"fair Promises of Love and Friendship"* Ibid., 18.
Working in July's heat J. Godfred, "Making a Birchbark Canoe," http://www.northwestjournal.ca/VIII4.htm, accessed September 9, 2016.

55 *"I went to the Half-King"* Washington, "Journey to the French Commandant (1753)," 19.

56 *"The Lands upon the River Ohio"* Ibid., 25–26.

CHAPTER FOUR

57 *crossed a small lake* See Gist, 83. Also see map in Albert, 566.

58 *thirty thousand or more paddle strokes* Nute, 27.
clambered over the gunwales Washington, "Journey to the French Commandant (1753)," 19–20.
paddled about sixteen miles Gist, 83.
a hollow tree where a bear lodged James Smith, 95.
some two million black bears https://www.nps.gov/shen/learn/nature/black-bear.htm, accessed September 14, 2016.

59 *four or five hundred yards across* Washington, "Journey to the French Commandant (1753)," 20.

61 *feed for the animals* See Freeman, 1:281 and 281n. On his journey to the French commandant, Washington's horses carried corn for feed. For caloric expenditure of a heavily working horse and daily feed requirements, see http://www.equinespot.com/horse-grain.html; http://horseandrider.com/article/how-much-should-you-feed-your-horse-18769.

"I told him I hoped he would guard against" Washington, "Journey to the French Commandant (1753)," 20.

It snowed all that day Gist, 84.

62 *"The Major desired me to set out on foot"* Ibid., 84.

Nearly seven hundred thousand people O'Connell, 9.

63 *"great Variety of other Dealers in Money"* Ibid., 13.

two pints a week consumed Ibid.

the opening of Hampton Court Bridge See *Gentleman's Magazine,* December 1753.

64 *Londoners celebrated Christmas* See "Christmas to Present(s): How the Great British Christmas Took Shape," *Daily Telegraph,* December 22, 2011. See also Besant, 436, for a Christmas recollection by an eighteenth-century traveler, and Robert Herrick's 1648 poem "Twelfth-Night."

they now took townhouses O'Connell, 10.

in that December of 1753 See *Internet Shakespeare Editions,* http://internetshakespeare.uvic.ca/Theater/production/stage/2994/, accessed October 3, 2016.

65 *One in every five women* See https://teainateacup.wordpress.com/2012/11/03/james-boswell-and-london-prostitutes/, also noted in *The Economist*'s review of *The Secret History of Georgian London: How the Wages of Sin Shaped the Capital,* by Dan Cruickshank, http://www.economist.com/node/14636924, accessed October 3, 2016.

the intellectual capital of Europe Stromberg, 100.

the darkness of ignorance See Jones, *Paris,* 178.

tolerant and liberal in its views See "Encyclopédie," *Encyclopaedia Britannica,* https://www.britannica.com/topic/Encyclopedie, accessed October 4, 2016. Also, the Historical Text Archive on "Diderot's Encyclopedia" notes that Diderot explicitly wrote against the Church in his articles: http://historicaltextarchive.com/sections.php?action=read&artid=780.

66 *"I prefer fame above all else"* See "Louis XIV, King of France," *Encyclopaedia Britannica,* https://www.britannica.com/biography/Louis-XIV-king-of-France, accessed October 4, 2016.

English political philosopher Thomas Hobbes See "social contract theory," *Internet Encyclopedia of Philosophy,* http://www.iep.utm.edu/soc-cont/#SH2a. See also Stromberg, 86–90.

67 *"equal and independent"* See http://press-pubs.uchicago.edu/founders/print _documents/v1ch2s1.html, accessed October 6, 2016.

68 *"the Restoration of the Arts and Sciences"* See https://www.files.ethz.ch /isn/125491/5018_Rousseau_Discourse_on_the_Arts_and_Sciences.pdf, accessed October 7, 2016.

an annual pension from King Louis XV See http://www.gutenberg.org /files/3913/3913-h/3913-h.htm#link2H_4_0008, accessed October 7, 2016.

"discovered nothing" Jean Starobinski, "Rousseau and Revolution," *New York Review of Books,* April 25, 2002, http://www.nybooks.com/articles/2002/04 /25/rousseau-and-revolution/, accessed October 8, 2016.

69 *"I . . . pulled off my Cloaths"* Washington, "Journey to the French Commandant (1753)," 21.

A matchcoat was an Indian robe For greater detail on matchcoats, see Becker.

70 *back on the trail at 2 A.M.* Gist, 84.

71 *"The Major's feet grew very sore"* Ibid., 85.

72 *"As you will not have him killed"* Ibid., 85–86.

73 *they would come to Shannopin's Town* Sipe, 121.

"[We] walked all the remaining Part of the Night" Washington, "Journey to the French Commandant (1753)," 21.

75 *"thought ourselves safe enough to sleep"* Gist, 86.

"[T]he Ice I suppose had broke up above" Washington, "Journey to the French Commandant (1753)," 21.

77 *seizing hold of one log of the raft* Ibid., 22.

CHAPTER FIVE

79 *"as fatiguing a Journey as it is possible to conceive"* Washington, "Journey to the French Commandant (1753)," 22.

January 16, 1754 Freeman, 1:324.

one hundred and four burgesses Ibid., 173.

80 *pistole fee* A Spanish gold coin then in circulation. See "Currency: What Is a Pistole?" in *The Geography of Slavery in Virginia,* http://www2.vcdh.virginia .edu/gos/currency.html.

an administrator For biographical information on Dinwiddie, see Freeman, 1:170–71, Lewis, 120–25, and *Encyclopedia of Virginia,* http://www.encyclope diavirginia.org/Dinwiddie_Robert_1692-1770.

Rebecca and Elizabeth See https://www.geni.com/people/Robert-Dinwiddie -of-Airdrie/6000000012357820712, accessed August 7, 2107.

81 *"in an hostile Manner"* From the *Maryland Gazette,* December 6, 1753: ". . . I have been alarmed by several Informations from our back Settlements, from the *Indians,* and from our neighbouring Governors, of a large body of *French* Regulars and *Indians* in their interest, having marched from Canada, to

the River Ohio, in an hostile Manner, to invade His Majesty's Territories, and having actually built a Fort on His Majesty's Lands. . . ."

81 *the pistole controversy still unresolved* Freeman, 1:327.

The governor was already in an ill humor Ibid.

governor-general of New France See *Dictionary of Canadian Biography,* http://www.biographi.ca/en/bio/duquesne_de_menneville_ange_4E.html, accessed October 14, 2016.

"the evidence and reality of the rights of the King" Freeman, 1:325.

82 *sent a party to build a "stronghouse"* Anderson, *Crucible of War,* 45–46.

83 *"With the Hope of doing it"* Washington, "Journey to the French Commandant (1753)," 23.

"a proper way of thinking" Robert Dinwiddie to James Abercromby, February 9, 1754, in Freeman, 1:328.

William Trent, the Indian trader Anderson, *Crucible of War,* 45.

84 *"Novices"* Freeman, 1:329.

"You may, with almost equal success" Ibid., 331, quoting letter to John Augustine Washington, May 28, 1755.

85 *an elegant, Palladio-inspired portico* See http://www.lva.virginia.gov/exhibits/capitol/colonial.htm, accessed October 17, 2016.

The Governor's Council, or Upper House See http://www.lva.virginia.gov/exhibits/capitol/colonial.htm.

86 *corpses of seven white settlers* Washington, "Journey to the French Commandant (1753)," 22.

". . . that poor unhappy Family!" Minutes for Thursday, February 14, 1754, *Journal of the Virginia House of Burgesses,* 1619–1776, 8:175–76.

87 *which lands in the Ohio Valley* Freeman, 1:332.

"[One Burgess] pretended to ascertain" Robert Dinwiddie to John Hanbury, March 12, 1754, in Dinwiddie, 1:102–3.

88 *"fiction and scheme"* George Washington to John Campbell, Earl of Loudoun, 10 January 1757, in Abbot, 4:79–93.

89 *Green Spring estate of Philip Ludwell* Abbot, 1:71 n1.

". . . [I]t is a charge too great" "From George Washington to Richard Corbin, February–March 1754," *Founders Online,* National Archives, http://founders.archives.gov/documents/Washington/02-01-02-0034, last modified October 5, 2016.

90 *"[the pay] was very much off"* "From George Washington to Robert Dinwiddie, 29 May 1754," *Founders Online,* National Archives, http://founders.archives.gov/documents/Washington/02-01-02-0054, last modified June 29, 2017.

91 *"This, with some other Reason's"* Ibid.

"Cutlasses, Halbards, Officer's half Pikes" "From George Washington to Robert Dinwiddie, 7 March 1754," *Founders Online,* National Archives, http://

founders.archives.gov/documents/Washington/02-01-02-0036, last modified October 5, 2016.

92 *John Carlyle, official commissary to the expedition* Freeman, 1:335–36.
Now he faced a "generl Clamour" "From George Washington to Robert Dinwiddie, 9 March 1754," *Founders Online,* National Archives, http://founders.archives.gov/documents/Washington/02-01-02-0037, last modified October 5, 2016.

93 *". . . which I think makes it necessary"* "To George Washington from Robert Dinwiddie, 15 March 1754," *Founders Online,* National Archives, http://founders.archives.gov/documents/Washington/02-01-02-0038, last modified June 29, 2017.

94 *Joshua Fry* For biographical information, see Freeman, 1:338–39. Also see *Encyclopedia of Virginia,* http://www.encyclopediavirginia.org/Fry_Joshua_ca_1700-May_31_1754#start_entry, accessed November 18, 2016.
Cherokee and Catawba Freeman gives the number as one thousand, quoting Dinwiddie to Lord Holderness, March 17, 1754, Freeman, 1:337. See also "To George Washington from Robert Dinwiddie, 15 March 1754," *Founders Online,* National Archives, http://founders.archives.gov/documents/Washington/02-01-02-0038, last modified October 5, 2016,

CHAPTER SIX

98 *Mounted on horseback* Freeman, 1:343.
one hundred and twenty men Ibid., 343–42.
notch in the ridgeline called Vestal's Gap Also known as Key's Gap, see Freeman, 1:345n. Key's Gap is located on Route 9 where it cuts through the Blue Ridge.

100 *Captain Adam Stephen* See brief biography in "George Washington to Robert Dinwiddie, 20 March 1754," *Founders Online,* National Archives, note 7, http://founders.archives.gov/documents/Washington/02-01-02-0039, last modified October 5, 2016, accessed October 24, 2016. Also see Freeman, 1:345.
"full power" Robert Dinwiddie to Major John Carlisle, January 26, 1754, in Dinwiddie, 1:54.

101 *"Out of Twenty four Waggons"* "From George Washington to Robert Dinwiddie, 25 April 1754," *Founders Online,* National Archives, http://founders.archives.gov/documents/Washington/02-01-02-0045, last modified October 5, 2016.
twelve wagons Freeman, 1:349.
"The difficulty of getting Waggons" "From George Washington to Thomas Cresap, 18 April 1754," *Founders Online,* National Archives, http://founders

.archives.gov/documents/Washington/02-01-02-0042, last modified October 5, 2016.

102 *"the Indians are very angry at our delay"* Kent, "Contrecoeur's Copy of George Washington's Journal for 1754," 10.

Norman gentry See *Dictionary of Canadian Biography,* http://www.biographi.ca/en/bio/duquesne_de_menneville_ange_4E.html, accessed October 31, 2016.

lordly and self-important demeanor Kent, *The French Invasion of Western Pennsylvania 1753,* 13.

103 *"the key of the Great West"* Parkman, 1:62.

"In the course of the month of May" Quoted in Kent, *The French Invasion of Western Pennsylvania 1753,* 18.

A crusty sixty-year-old veteran Ibid., 14.

104 *"The labor of our troops was excessive"* Ibid., 58.

105 *"sheer imagination"* Ibid., 79.

brother-in-law Ensign Edward Ward Freeman, 1:351.

106 *the Half King laid down the first timber* Ibid., 352.

107 *laying siege to castles* Wright, 629–44.

Le Mercier, chief of artillery See *Dictionary of Canadian Biography,* http://www.biographi.ca/en/bio/le_mercier_francois_marc_antoine_4E.html, accessed October 26, 2016.

108 *"[Contrecoeur] absolutely insisted"* Ensign Ward's Deposition, in Gist, 276.

109 *"[Contrecoeur] ask'd many Questions"* Ibid.

110 *"[T]he Half King stormed"* Ibid., 278.

hire enough packhorses Freeman, 1:350.

111 *clear a road over the mountains* For details of Washington's plan, see "From George Washington to Horatio Sharpe, 24 April 1754," *Founders Online,* National Archives, http://founders.archives.gov/documents/Washington/02-01-02-0044, last modified October 5, 2016.

112 *"steadfast adherence to us"* Ibid.

113 *"You are to act on the Difensive"* "To George Washington from Robert Dinwiddie, January 1754," *Founders Online,* National Archives, http://founders.archives.gov/documents/Washington/02-01-02-0031, last modified June 29, 2017. Original source: Abbot, 1:63–67. Dinwiddie's instructions echoed those given to him by the Earl of Holderness, when Dinwiddie warned the Crown about the French in the Ohio: "[I]f You shall find, that any Number of Persons, whether Indians, or Europeans, shall presume to erect any Fort or Forts within the Limits of Our Province of Virginia, . . . You are to require of Them peaceably to depart, and not to persist in such unlawfull Proceedings, & if, notwithstanding Your Admonitions, They do still endeavour to carry on any such unlawfull and unjustifiable Designs, We do hereby strictly charge, & command You, to drive them off by Force of Arms." See Editorial Note preceding "Com-

mission from Robert Dinwiddie, 30 October 1753," *Founders Online,* National Archives, http://founders.archives.gov/documents/Washington/02-01-02-0028, last modified June 29, 2017. Original source: Abbot, 1:56–60.

113 *They "slaved" their way along* "From George Washington to Robert Dinwiddie, 9 May 1754," *Founders Online,* National Archives, http://founders.archives.gov/documents/Washington/02-01-02-0047, last modified October 5, 2016.
"to amend and alter the Roads" Ibid.
brought by British traders Freeman, 1:358.

114 *"400 more French"* Kent, "Contrecoeur's Copy of George Washington's Journal for 1754," 14–15.
The men who had served Freeman, 1:357. Also see "From George Washington to Robert Dinwiddie, 9 May 1754," *Founders Online,* National Archives, http://founders.archives.gov/documents/Washington/02-01-02-0047, last modified October 5, 2016.
Sally Fairfax's sister-in-law "To George Washington from Sarah Carlyle, 17 June 1754," *Founders Online,* National Archives, http://founders.archives.gov/documents/Washington/02-01-02-0068, last modified December 28, 2016. This letter refers to Washington's letter to Sarah Fairfax Carlyle of May 15, 1754, which does not survive.

115 *"Giving up my commission"* "From George Washington to Robert Dinwiddie, 18 May 1754," *Founders Online,* National Archives, http://founders.archives.gov/documents/Washington/02-01-02-0050, last modified October 5, 2016.
list of complaints Freeman, 1:362n suggests that this list can be reconstructed from Dinwiddie's letter of May 25.

116 *the impossibility of clearing a road* Kent, "Contrecoeur's Copy of George Washington's Journal for 1754," 16–17.
Washington, a lieutenant, an Indian guide "Expedition to the Ohio, 1754: Narrative," *Founders Online,* National Archives, http://founders.archives.gov/documents/Washington/01-01-02-0004-0002, last modified October 5, 2016. Original source: Jackson, 1:174–210.
"rough, rocky, and scarcely passable" "From George Washington to Joshua Fry, 23 May 1754," *Founders Online,* National Archives, http://founders.archives.gov/documents/Washington/02-01-02-0051, last modified October 5, 2016.
"nearly forty feet perpendicular" Ibid.

117 *building earthwork defenses* Kent, "Contrecoeur's Copy of George Washington's Journal for 1754," 17.
all Joshua Fry's reinforcements Ibid., 16n, 17, entry for May 12–15. See note for his expectations about other troops arriving.
"[T]herfor my Brotheres" Half King's letter is transcribed in "From George Washington to Robert Dinwiddie, 27 May 1754," *Founders Online,* National Ar-

chives, http://founders.archives.gov/documents/Washington/02-01-02-0053, last modified October 5, 2016.

117 *reached a crossing of the Youghiogheny* Ibid. Also see map of trails and camps of this area in Kent, "Contrecoeur's Copy of George Washington's Journal for 1754," 10–11.

118 *"prepar'd a charming field for an Encounter"* "From George Washington to Robert Dinwiddie, 27 May 1754," *Founders Online,* National Archives, http://founders.archives.gov/documents/Washington/02-01-02-0053, last modified October 5, 2016.

sentries around his camp Ibid.

119 *camped about six miles away* "From George Washington to Robert Dinwiddie, 29 May 1754," *Founders Online,* National Archives, http://founders.archives.gov/documents/Washington/02-01-02-0054, last modified June 29, 2017.

"[H]e imagined the whole party" Kent, "Contrecoeur's Copy of George Washington's Journal for 1754," 20.

"black as pitch" Ibid., 21.

Iroquois sachem named Monacatoocha Both Freeman, 1:290, and Preston, 28–29, refer to Monacatoocha's presence at the glen. I take the spelling from Preston's version. Monacatoocha is also known as Scaroudy.

120 *"disposition to attack on all sides"* "From George Washington to Robert Dinwiddie, 29 May 1754," *Founders Online,* National Archives, http://founders.archives.gov/documents/Washington/02-01-02-0054, last modified June 29, 2017.

Captain Thomas Waggoner Kent, "Contrecoeur's Copy of George Washington's Journal for 1754," 103n.

French were just waking from sleep See "John Shaw's Affidavit" in Baker-Crothers and Hudnut for a detailed description of what occurred at the glen, as he heard it told to him by several soldiers who were there. Some French soldiers were still asleep, testified Shaw, and others eating breakfast.

Some had gone off in the woods See Monceau's account, as told by Bonin, 97.

CHAPTER SEVEN

123 *They were the French* For a summary of various accounts, see n59 to Washington's "Expedition to the Ohio 1754," *Founders Online,* National Archives, http://founders.archives.gov/documents/Washington/01-01-02-0004-0002, last modified November 26, 2017. This includes his account of the Jumonville Glen affair and the French version of events as reported by a survivor named Monceau and contained in a letter from Contrecoeur to Governor of New France Duquesne. Other accounts include one from Adam Stephen appearing in the *Maryland Gazette,* August 29, 1754, and the *Pennsyl-*

vania Gazette, September 19, 1754, and in Stephen's autobiography. See also Baker-Crothers and Hudnut. Contemporary historians have given their own interpretations of the affair based on multiple sources, including Fred Anderson, who cites an account by Kaninguen in *Crucible of War,* 57. Some of the most recent documentary evidence has come to light through the research of David Preston and an eyewitness account of an Iroquois warrior that he discovered in a French archive (Preston, Appendix E).

123 *they were the enemy* Freeman makes this same point. Freeman, 1:375.

Washington himself fired the first shot Preston, Appendix E. Eyewitness testimony given by an Iroquois warrior.

Major Washington and his men on the rock rim Washington took most of the French fire, according to his account in Kent, "Contrecoeur's Copy of George Washington's Journal for 1754."

dampened the gunpowder The Iroquois warrior testimony cites wet gunpowder.

124 *fallen to the ground* See Kaninguen testimony in Anderson, *Crucible of War,* 57.

Joseph Coulon de Villiers de Jumonville See *Dictionary of Canadian Biography,* http://www.biographi.ca/en/bio/coulon_de_villiers_de_jumonville_joseph_3F.html, accessed November 8, 2016.

It warned that Contrecoeur had heard Contrecoeur Summons, in Moreau, 119–20.

the mountains "in Peace" Italics are in original summons, as translated in Moreau, 119.

125 *The Half King raised his tomahawk* See "From George Washington to Robert Dinwiddie, 3 June 1754," in which he describes how the Indians knocked the French wounded in the head and took scalps, and also that seven men got lost in the dark. *Founders Online,* National Archives, http://founders.archives.gov/documents/Washington/02-01-02-0062, last modified June 29, 2017.

The blows split open Ensign Jumonville's skull John Shaw in his Affidavit describes this quite closely, including that the Half King first asked Jumonville if he were English and was told that he was French. See Baker-Crothers and Hudnut for "John Shaw's Affidavit."

dozen or so Ibid. Shaw puts the number killed and corpses he saw at thirteen or fourteen.

126 *The eighteenth-century British commentator* Chernow, 45. Francis Parkman also writes, "Judge it as we may, this obscure skirmish began the war that set the world on fire." Parkman, 1:150.

Some historians have speculated "If he cherished any hope of reestablishing his (or the Six Nations') authority on the Ohio, [the Half King] would have known that he could do it only with British support. . . . [H]e had good reason to believe that only severe provocation to the French—enough to cause them to retaliate militarily—would galvanize [the colonies of Virginia and Pennsylvania] into action." Anderson, *Crucible of War,* 56–57. See also 5–7, 52–59. Killing Jumonville in so grisly a manner might have been a conscious decision on the Half King's part to cause maximum outrage among the French.

126 *because he wished to protect* Preston, 27–28.

remaining thirty-nine men Ibid., 26. Preston puts Washington's force at forty men, which minus one dead would equal thirty-nine. He puts the French force at thirty-five to start. Thirty-five to start minus one escapee, Morceau, and twenty-one captured would make thirteen French dead.

twenty-one surviving French Kent, "Contrecoeur's Copy of George Washington's Journal for 1754."

127 *one killed and one or two wounded* This is Washington's casualty count from his letter to his brother John Augustine, often called Jack. "From George Washington to John Augustine Washington, 31 May 1754," *Founders Online,* National Archives, http://founders.archives.gov/documents/Washington/02-01-02-0058, last modified June 29, 2017.

portable writing case For a similar writing case used by Washington in the Revolutionary War, see http://georgewashington.si.edu/portrait/09a_1.html.

in hopes of a meeting there "From George Washington to Robert Dinwiddie, 27 May 1754," *Founders Online,* National Archives, http://founders.archives.gov/documents/Washington/02-01-02-0053, last modified October 5, 2016.

128 *"Now Colo. W"* "To George Washington from Robert Dinwiddie, 25 May 1754," *Founders Online,* National Archives, http://founders.archives.gov/documents/Washington/02-01-02-0052, last modified December 28, 2016.

"to charge me with ingratitude" "From George Washington to Robert Dinwiddie, 29 May 1754," *Founders Online,* National Archives, http://founders.archives.gov/documents/Washington/02-01-02-0054, last modified October 5, 2016.

"We are debarr'd the pleasure of good Living" Ibid., 107–15.

129 *"in conjuncton with the Half King"* Ibid.

That was it for an actual account Ibid. Note 14 to this letter explains possible Indian roles through different accounts, such as John Shaw's.

Washington had grown convinced Ibid. "[T]hese Enterpriseing Men were purposely choose out to get intelligence, which they were to send Back by some brisk dispatches with mention of the Day that they were to serve the Summon's; which could be through no other view, than to get sufficient Reinforcements to fall upon us imediately after." See also Kent, "Contrecoeur's Copy of George Washington's Journal for 1754," 20: "ordered our munitions put in a secure place, for fear that this was a stratagem of the French to attack our camp."

prepared to attack first Preston calls the Jumonville incident an "ambush." Preston, 27.

130 *as far north as Lake Erie* "From George Washington to Robert Dinwiddie, 10 June 1754," *Founders Online,* National Archives, http://founders.archives

.gov/documents/Washington/02-01-02-0066, last modified December 28, 2016.

130 *"[B]ut that I know to be False"* "From George Washington to Robert Dinwiddie, 29 May 1754," *Founders Online,* National Archives, http://founders.archives.gov/documents/Washington/02-01-02-0054, last modified June 29, 2017.

131 *Writing to his younger brother Jack* "From George Washington to John Augustine Washington, 31 May 1754," *Founders Online,* National Archives, http://founders.archives.gov/documents/Washington/02-01-02-0058, last modified December 28, 2016.

"If the whole detachment of French" "From George Washington to Robert Dinwiddie, 3 June 1754," *Founders Online,* National Archives, http://founders.archives.gov/documents/Washington/02-01-02-0062, last modified June 29, 2017. Original source: Abbot, 1:122–26.

"driving them to the [damned] Montreal" Ibid. His actual words were "driving them to the (D)—Montreal."

a French attack might be imminent Kent, "Contrecoeur's Copy of George Washington's Journal for 1754," 23. "Fearing that as soon as the news of this defeat should reach the French we might be attacked by considerable forces, I began to raise a fort with a little palisade."

132 *"one Inch of what we have gaind"* "From George Washington to Joshua Fry, 29 May 1754," *Founders Online,* National Archives, http://founders.archives.gov/documents/Washington/02-01-02-0057, last modified June 29, 2017.

At its center stood a shack "From George Washington to Robert Dinwiddie, 3 June 1754," *Founders Online,* National Archives, http://founders.archives.gov/documents/Washington/02-01-02-0062, last modified June 29, 2017.

133 *Governor Dinwiddie, at Winchester* Ibid. See mention of Dinwiddie's arrival at Winchester and wish to meet with Half King and other chiefs of Six Nations.

This day the flour ran out Freeman, 1:281.

134 *He was to replace Colonel Fry* For a detailed description of Dinwiddie's orders for changes in leadership of the Virginia Regiment, see "To George Washington from Robert Dinwiddie, 4 June 1754," *Founders Online,* National Archives, http://founders.archives.gov/documents/Washington/02-01-02-0063, last modified October 5, 2016.

whom Dinwiddie appointed Ibid. Also see Freeman, 1:384.

find Colonel Innes's leadership "agreeable" "To George Washington from Robert Dinwiddie, 4 June 1754," *Founders Online,* National Archives, http://founders.archives.gov/documents/Washington/02-01-02-0063. See also footnotes to letter that explain Innes's command of the overall expedition and Washington's command of the Virginia Regiment.

"[I] hope the good Spirits of Yr Soldiers" "To George Washington from Robert Dinwiddie, 1 June 1754," *Founders Online,* National Archives, http://founders

.archives.gov/documents/Washington/02-01-02-0059, last modified October 5, 2016.

135 *and in his recent threats to quit* See Washington's letter to Governor Dinwiddie, referring to acquainting Colonel Fairfax with his intention of resigning. "From George Washington to Robert Dinwiddie, 29 May 1754," *Founders Online,* National Archives, http://founders.archives.gov/documents/Washington/02-01-02-0054, last modified June 29, 2017.

Dinwiddie may have sensed an uncertainty See *Virginia Encyclopedia of Biography,* http://www.encyclopediavirginia.org/Dinwiddie_Robert_1692-1770#start_entry, accessed November 19, 2016.

a gift of rum from the governor's private stock "From George Washington to Robert Dinwiddie, 10 June 1754," *Founders Online,* National Archives, http://founders.archives.gov/documents/Washington/02-01-02-0066, last modified October 5, 2016.

the commemorative medal For medal struck to commemorate success, see "To George Washington from Robert Dinwiddie, 2 June 1754," *Founders Online,* National Archives, http://founders.archives.gov/documents/Washington/02-01-02-0061, last modified June 29, 2017.

"[N]ow I shall not have it in my power" "From George Washington to Robert Dinwiddie, 10 June 1754," *Founders Online,* National Archives, http://founders.archives.gov/documents/Washington/02-01-02-0066, last modified December 28, 2016.

136 *Robert Stobo* For brief biography see Freeman, 1:382.

two experts See ibid., 391 and note, for account of Croghan's arrival and his and Montour's expertise. For Zinzendorf description of Montour, see Freeman, 1:383.

137 *Croghan was commissioned to provide flour* See Anderson, *Crucible of War,* 60. About sending George Croghan to Washington, see "To George Washington from Robert Dinwiddie, 1 June 1754," *Founders Online,* National Archives, http://founders.archives.gov/documents/Washington/02-01-02-0060, last modified June 29, 2017.

the French had offered a bounty Anderson, *Crucible of War,* 25.

two days after the arrival of Montour Kent, "Contrecoeur's Copy of George Washington's Journal for 1754," 24 and n125.

"I was as sensibly disappointed" "From George Washington to Robert Dinwiddie, 10 June 1754," *Founders Online,* National Archives, http://founders.archives.gov/documents/Washington/02-01-02-0066, last modified December 28, 2016.

139 *"an Officer of some Experience & Importance"* "To George Washington from Robert Dinwiddie, 4 May 1754," *Founders Online,* National Archives, http://founders.archives.gov/documents/Washington/02-01-02-0046, last modified December 28, 2016.

140 *an "Evil tendancy" would result* "From George Washington to Robert Dinwiddie, 10 June 1754," *Founders Online,* National Archives, http://founders.archives.gov/documents/Washington/02-01-02-0066, last modified December 28, 2016.

Dinwiddie also insisted that for morale See Harrington, 27.

141 *"Can not you spare a few Men"* "To George Washington from John Carlyle, 17 June 1754," *Founders Online,* National Archives, http://founders.archives.gov/documents/Washington/02-01-02-0067, last modified June 29, 2017.

142 *nine swivel guns* Ibid.

143 *A soft light suffused* Kent, "Contrecoeur's Copy of George Washington's Journal for 1754," 27. That they are inside a cabin at Gist's is clear from Washington's entry of June 20. "Thereupon the spokesman spread out his blanket on the floor, and upon this blanket placed various belts and strings of wampum in the order he had received them from the French."

Delaware chief Shingas Freeman, 1:392.

"My brothers, we your brothers" Kent, "Contrecoeur's Copy of George Washington's Journal for 1754," 26.

144 *The trader and negotiator Croghan* As Freeman notes, Washington undoubtedly had Croghan and Montour's advice in protocol and in shaping his speeches to the masterful Indian orators. Freeman, I:392

refer to the governor of New France Anderson, *Crucible of War,* 15, 26.

"have a beautiful speech" Kent, "Contrecoeur's Copy of George Washington's Journal for 1754," 26.

146 *he wrote to Washington to proceed cautiously* "To George Washington from Robert Dinwiddie, 27 June 1754," *Founders Online,* National Archives, http://founders.archives.gov/documents/Washington/02-01-02-0072, last modified June 29, 2017.

"the French act with great Secrecy" Ibid.

CHAPTER EIGHT

147 *After a series of confusing rumors* See "Journey to the French Commandant: Narrative," *Founders Online,* National Archives, https://founders.archives.gov/documents/Washington/01-01-02-0003-0002, accessed October 20, 2017. See also, Freeman, 1:396.

The Forks now teemed Freeman, 1:396.

148 *In the meantime* Ibid., 397.

The Indians who remained Ibid., 398.

Made up of his officers See "Minutes of a Council of War, 28 June 1754," *Founders Online,* National Archives, https://founders.archives.gov/documents/Washington/02-01-02-0075, accessed October 20, 2017.

149 *Taking the lead* "Adam Stephen's Letter," *Maryland Gazette,* August 29, 1754.

149 *Washington's personal servant* "From George Washington to Carter Burwell, 20 April 1755," *Founders Online,* National Archives, https://founders.archives.gov/documents/Washington/02-01-02-0125, accessed October 20, 2017.

They dined on better food "From George Washington to William Byrd, 20 April 1755," *Founders Online,* National Archives, http://founders.archives.gov/documents/Washington/02-01-02-0124, accessed October 20, 2017.

150 *"the roughest and most hilly"* "Adam Stephen's Letter," *Maryland Gazette,* August 29, 1754.

Struggling uphill and down Ibid.

151 *He resolved to stay* Freeman, 1:401–2.

"[T]hose pleasing reflections" "To George Washington from Sarah Carlyle, 17 June 1754," *Founders Online,* National Archives, https://founders.archives.gov/documents/Washington/02-01-02-0068, accessed October 20, 2017.

152 *Sensing impending disaster* Freeman, 1:402.

"After [the Jumonville incident]" Preston, 352.

"Perhaps the strongest feature" "George Washington," *Colonial Williamsburg,* http://www.history.org/almanack/people/bios/biowash2.cfm, accessed October 21, 2017.

As Washington's troops Anderson, *Crucible of War,* 57–58.

Since mid-April For Fort Duquesne generally, see Hunter.

153 *When news of Jumonville's* For Captain Coulon de Villiers, see *Dictionary of Canadian Biography,* http://www.biographi.ca/en/bio/coulon_de_villiers_louis_3E.html, accessed October 21, 2017.

"The English have murdered my children" Parkman, 1:154.

154 *"Both hatchets and wine"* Ibid.

The French burnished muskets Ibid., 154–55.

"The path was so rough" Ibid., 155.

155 *In the downpour* Moreau, 177.

Early that morning "Adam Stephen's Letter," *Maryland Gazette,* August 29, 1754.

156 *At about the same time* For accounts of the battle, see "I. 19 July 1754," *Founders Online,* National Archives, https://founders.archives.gov/documents/Washington/02-01-02-0076-0002, accessed October 21, 2017; Stephen, "The Ohio Expedition of 1754," 47–48; Moreau, 178-81.

As the Virginia Abbot, 1:163.

157 *Washington ordered the British* See Washington, *George Washington Remembers,* 17.

Their fire "extinguished" Moreau, 179.

As if shaken Washington, *George Washington Remembers,* 17.

The British dead slumped Stephen, "The Ohio Expedition of 1754," 49.

158 *Startlingly, they possessed* "Adam Stephen's Letter," *Maryland Gazette,* August 29, 1754.

With so many guns jammed Washington, *George Washington Remembers,* 17.

158 *The British struggled* "Adam Stephen's Letter," *Maryland Gazette,* August 29, 1754.

and fell, lethally wounded "From George Washington to Carter Burwell, 20 April 1755," *Founders Online,* National Archives, https://founders.archives.gov/documents/Washington/02-01-02-0125, accessed October 20, 2017.

A shout came Freeman, 1:405.

159 *As France and Britain* Moreau, 179–80.

160 *He agreed to negotiate* "II., 3 July 1754," *Founders Online,* National Archives, http://founders.archives.gov/documents/Washington/02-01-02-0076-0003, accessed October 21, 2017.

Word was sent back For documents relating to Fort Necessity, see Cleland.

"As our Intention have never" Abbot, 1:166.

161 *They did not realize* For French circumstances generally, see Villiers account in Moreau.

The proud and outspoken For his personal account, see Stephen, "The Ohio Expedition of 1754."

162 *For the rest of his life* "From George Washington to Colonel Adam Stephen, 20 July 1776," *Founders Online,* National Archives, https://founders.archives.gov/documents/Washington/03-05-02-0298, accessed October 21, 2017.

CHAPTER NINE

163 *sandy Duke of Gloucester Street* Burnaby, *Travels in North America,* from Parkman, I:162–63n.

appeared in hard print *Virginia Gazette,* July 19, 1754.

a joint report on the battle "I., 19 July 1754," *Founders Online,* National Archives, http://founders.archives.gov/documents/Washington/02-01-02-0076-0002, last modified February 21, 2017.

164 *Villiers listed only three dead* *Villiers Journal* in Moreau, 181.

greeted the retreating British officers Colonel James Innes's account, in *Maryland Gazette,* August 1, 1754.

Not a single Indian had fought with the British John Shaw's account, in Baker-Crothers and Hudnut, 28.

they left their wounded at this camp Ibid., 26.

A group of sixteen Virginia volunteers Ibid.

165 *"Not an English flag"* Parkman, 1:161.

Governor Dinwiddie reacted to the news Robert Dinwiddie to James Abercromby, July 24, 1754, (postscript) and Robert Dinwiddie to Colonel Innes, July 20, 1754, both in Dinwiddie.

166 *sheaf of letters* Robert Dinwiddie's correspondence with the Lords of Trade, Secretary of War, Earl of Albemarle, and others, July 24, 1754, in Dinwiddie. Also Anderson, *Crucible of War,* 66.

166 *carefully noted to the Lords of Trade* Robert Dinwiddie to the Lords of Trade, July 24, 1754, in Dinwiddie.

"Thus have a few brave Men been exposed" *Virginia Gazette,* July 19, 1754. Also Freeman, 1:423.

the widow Christiana Campbell See "Christiana Campbell," http://www.history.org/almanack/people/bios/christiana_campbell.cfm?showSite=mobile, accessed December 10, 2016.

167 *"He was a good-natured man"* *Journal of Conrad Weiser* in Colonial Records of Pennsylvania, 6:151–52, https://archive.org/stream/colonialrecordsov6harr/colonialrecordsov6harr_djvu.txt], accessed February 22, 2017.

"I wish Washington had acted with prudence" Freeman, 1:415, quoting William Johnson to Goldsbrow Banyar, July 29, 1754. [NOTE: The author has changed the word *doubt* to *question,* as that appears to be its modern meaning.]

"Washington and many such may have courage" Freeman 1:423–24, quoting letter of September 11, 1754.

"All North America will be lost" Quoted in Anderson, *Crucible of War,* 67.

168 *simply could not comprehend* Preston, 30.

"I can with truth assure you" "From George Washington to John Augustine Washington, 31 May 1754," *Founders Online,* National Archives, http://founders.archives.gov/documents/Washington/02-01-02-0058, last modified June 29, 2017.

"He would not say so" Freeman, 423.

in a letter to the Maryland Gazette "Adam Stephen's Letter," *Maryland Gazette,* August 29, 1754.

"Let any of these brave Gentlemen" Ibid.

the local view in Virginia Freeman, 1:423.

169 *He now ordered Washington* "To George Washington from Robert Dinwiddie, 1 August 1754," *Founders Online,* National Archives, http://founders.archives.gov/documents/Washington/02-01-02-0082, last modified December 6, 2016.

"Consider, I pray you, Sir" "From George Washington to William Fairfax, 11 August 1754," *Founders Online,* National Archives, http://founders.archives.gov/documents/Washington/02-01-02-0085, last modified February 21, 2017.

170 *paddled the upper reaches of the Potomac* "From George Washington to Charles Carter, August 1754," *Founders Online,* National Archives, http://founders.archives.gov/documents/Washington/02-01-02-0094, last modified December 6, 2016.

171 *The great Mingo Indian ally* Freeman, 1:433.

He socialized at the Raleigh Tavern Freeman, 1:436 n168. This is clearly the Raleigh Tavern, as Alexander Finnie operated it for a number of years, including in the 1750s. For *Raleigh Tavern Architectural Report,* see Macomber.

171 *Washington had ordered from London* "Invoice, 23 October 1754," *Founders Online,* National Archives, http://founders.archives.gov/documents/Washington/02-01-02-0109, last modified February 21, 2017.

172 *He traveled to Williamsburg* See "To George Washington from William Fitzhugh, 4 November 1754," *Founders Online,* National Archives, http://founders.archives.gov/documents/Washington/02-01-02-0113, last modified February 21, 2017.

"disputes between the regulars" Lewis, 161.

ten independent companies See explanation of events and reorganization plan in "To George Washington from William Fitzhugh, 4 November 1754," *Founders Online,* National Archives, http://founders.archives.gov/documents/Washington/02-01-02-0113, last modified February 21, 2017. In Dinwiddie's plan, the captains were to be put on the "regular" establishment, by having London send blank commissions that he would fill out with the captains' names. See also Freeman, 1:439–40.

173 *Governor Horatio Sharpe of Maryland* See "To George Washington from William Fitzhugh, 4 November 1754," *Founders Online,* National Archives, http://founders.archives.gov/documents/Washington/02-01-02-0113, last modified February 21, 2017.

would not have to take orders Freeman, 1:443.

"[F]or my Part," Fitzhugh wrote "To George Washington from William Fitzhugh, 4 November 1754," *Founders Online,* National Archives, http://founders.archives.gov/documents/Washington/02-01-02-0113, last modified February 21, 2017.

view over the Potomac For description and artist's rendering of Belvoir Manor in the 1700s and rental notice description in 1774, after the Fairfaxes left for England, never to return, see http://www.belvoir.army.mil/history/18C.asp, accessed December 19, 2016.

"a woman of unusually fine mind" Cary, 28–29.

174 *"[I]f you think me capable of holding"* "From George Washington to William Fitzhugh, 15 November 1754," *Founders Online,* National Archives, http://founders.archives.gov/documents/Washington/02-01-02-0114, last modified December 6, 2016.

CHAPTER TEN

177 *The Duke of Newcastle's problem* For Cumberland and Newcastle, see Preston, 30–31, and Anderson, *Crucible of War,* chapter 6.

When news of the British For arrival of the news in London, see Preston, 29; for *"all North America will be lost,"* see Anderson, *Crucible of War,* 67.

178 *"The key to success"* Anderson, *Crucible of War,* 67.

178 *He agreed with Newcastle* Ibid. and Preston, 31.
Instead of one campaign Preston, 31.
serving as the king's representative Ibid., 53–54; for further description, 44–55.
His uncle by marriage Ibid., 45. This uncle's name was John Blow.
179 *As historian David Preston* Ibid., 44–55.
His younger sister Ibid., 46.
As a young man Ibid., 48.
180 *In early January* For an expanded list of supplies, see ibid., 67–68.
181 *In an ironic turn* For documentation of inheritance, see Freeman, 2:2–3.
she had quickly remarried For Anne generally, see "Anne Lee," https://www.geni.com/people/Anne-Lee/6000000003914650750. For slave, see "Editorial Note," http://founders.archives.gov/documents/Washington/02-01-02-0115-0001, and "II., 10 December 1754," *Founders Online,* National Archives, http://founders.archives.gov/documents/Washington/02-01-02-0115-0003, last modified December 6, 2016.
182 *Washington spent that Christmas* For Christmas traditions, see "Christmas in Colonial Virginia," http://www.history.org/almanack/life/christmas/hist_inva.cfm; for importance of dancing, see "Dance during the Colonial Period," http://www.EncyclopediaVirginia.org/Dance_During_the_Colonial_Period. Both accessed October 22, 2017. For Washington's winnings, see Freeman, 2:4.
Christmas of 1754 Freeman, 2:5.
183 *He had just leased* See "Ten Facts about the Mansion," http://www.mountvernon.org/the-estate-gardens/the-mansion/ten-facts-about-the-mansion/, and "Lease of Mount Vernon, 17 December 1754," http://founders.archives.gov/documents/Washington/02-01-02-0116. Both accessed October 22, 2017.
"Reviewing, Recruiting, and other" See *Maryland Gazette,* January 2, 1755.
184 *He busied himself* For domestic duties, see Freeman, 2:4–5; for ship's arrival, see Preston, 69.
On February 22 See McCardell, 139, and Preston, 39.
"Under the drawing room candelabra" McCardell, 139–40.
185 *this assistance included* Ibid., 141.
The governor had advertised Freeman, 2:12.
His deputy quartermaster general Ibid, 2:10.
186 *"The worst road I ever traveled"* Pargellis, 61.
In his forties when he arrived For St. Clair, see Cubbison, 9.
"[A] mad sort of fool" Ibid.
"[They are] totally ignorant" Preston, 71–72.
Hilaritus Sapentise et Bense Macomber.
187 *How could he watch* Freeman, 2:11.
188 *"[General Braddock] has ordered me"* See "To George Washington from

Robert Orme, 2 March 1755," *Founders Online,* National Archives, http://founders.archives.gov/documents/Washington/02-01-02-0120, last modified June 29, 2017. Accessed October 22, 2017.

188 *"To be plain, Sir"* "From George Washington to Robert Orme, 2 April 1755," *Founders Online,* National Archives, http://founders.archives.gov/documents/Washington/02-01-02-0122, last modified November 26, 2017.

To his brother "From George Washington to Augustine Washington, 14 May 1755," *Founders Online,* National Archives, http://founders.archives.gov/documents/Washington/02-01-02-0134, accessed October 22, 2017.

189 *Working their way up* Hamilton, 9.

His Majesty's warships For arrival, see McCardell, 154; for descriptions of British soldiers, see Preston, 56.

Alexandria's most prosperous merchant Preston, 73.

190 *Sitting on a low bluff* Ibid., 73–74.

"[B]y Sum means or another" Ibid., 75.

At the moment, however For relationship between Washington and his mother, see Abbott, 1:246–48.

191 *"I find myself"* "From George Washington to Robert Orme, 2 April 1755," *Founders Online,* National Archives, http://founders.archives.gov/documents/Washington/02-01-02-0122, last modified November 26, 2017.

"[W]henever you find it" Ibid., 249.

". . . I can very truly say" "From George Washington to William Byrd, 20 April 1755," *Founders Online,* National Archives, http://founders.archives.gov/documents/Washington/02-01-02-0124, accessed October 22, 2017.

192 *"Every man"* See Cohen et al., "Insult, Aggression, and the Southern Culture of Honor: An 'Experimental Ethnography,'" https://deepblue.lib.umich.edu/bitstream/handle/2027.42/92155/InsultAggressionAndTheSouthernCulture.pdf, accessed October 22, 2017.

"Mine honour is" Quoted in "Hein: Learning Responsibility and Honor," *Washington Times,* July 3, 2008.

193 *In his later years* "The Letter Book for the Braddock Campaign," *Founders Online,* National Archives, http://founders.archives.gov/documents/Washington/02-01-02-0119, accessed October 22, 2017.

194 *To curtail the rowdiness* Sargent, 291–96.

195 *On April 15* Preston, 90.

Alexandria had quieted Ibid., 77.

196 *Some interpretations assert* Ibid., 79–80.

Seeing an opportunity Ibid., 82.

When St. Clair found See Colonial Records of Pennsylvania, 368–69.

197 *"That instead of marching"* Ibid., 368.

called Pennsylvania's behavior Freeman, 2:21.

198 *True to his hidden mission* Houston, "Benjamin Franklin and the 'Wagon Affair' of 1755."
Just before returning Preston, 93.
"The general and all" Franklin, 154.
199 *"[He] said the damned"* Ben Franklin to Deborah, April 26, 1755, in Houston.
Toting £800 in gold Preston, 93.
200 *General Braddock, his aide Orme reported* Ibid., 95.
But Franklin harbored Franklin, Chapter 8, and Preston, 93.
201 *Ben Franklin would* For British views of the colonists, see Franklin, and Preston, 54. For *"He smiled at my ignorance . . . ,"* see Franklin, 159.

CHAPTER ELEVEN

203 *"I have no expectation of reward"* "From George Washington to William Byrd, 20 April 1755," *Founders Online,* National Archives, http://founders.archives.gov/documents/Washington/02-01-02-0124, last modified November 26, 2017.
wealthy planter, Virginia burgess "From George Washington to Carter Burwell, 20 April 1755," *Founders Online,* National Archives, http://founders.archives.gov/documents/Washington/02-01-02-0125, last modified December 28, 2016. See also *Encyclopedia Virginia* under Carter Burwell.
requesting £50 to cover valuables "From George Washington to Carter Burwell, 20 April 1755," *Founders Online,* National Archives, http://founders.archives.gov/documents/Washington/02-01-02-0125, last modified December 28, 2016.
"I am just ready to embark" Ibid.
204 *"[T]he sole motive wch envites me"* "From George Washington to John Robinson, 20 April 1755," *Founders Online,* National Archives, http://founders.archives.gov/documents/Washington/02-01-02-0126, last modified December 28, 2016.
"I think will be tedious in advancing" "From George Washington to William Fairfax, 23 April 1755," *Founders Online,* National Archives, http://founders.archives.gov/documents/Washington/02-01-02-0127, last modified December 28, 2016.
his white personal servant, John Alton Ibid. and Freeman, 2:27 and 55.
three or four days' hard riding Freeman, 2:27.
205 *"It will be needless"* "From George Washington to Sarah Cary Fairfax, 30 April 1755," *Founders Online,* National Archives, http://founders.archives.gov/documents/Washington/02-01-02-0128, last modified June 29, 2017.
Here the general and his staff Freeman, 2:31.
206 *as Governor Dinwiddie had promised* Ibid., 2:32. See also *Orme Journal,* in Sargent, 287.

206 *There were none* Freeman, 2:32.
seven hundred men *Orme Journal,* in Sargent, 380.
207 *"I am very happy in the Generals Family"* "From George Washington to Mary Ball Washington, 6 May 1755," *Founders Online,* National Archives, http://founders.archives.gov/documents/Washington/02-01-02-0132, last modified December 28, 2016.
mounted guard of light horses McCardell, 181.
208 *"no describing the badness of the roads"* Ibid., 180. See Mrs. Browne's journal entries at http://www.loudounhistory.org/history/loudoun-braddock-march-1755.htm#t1.
The path crossed one stream *Seaman's (Morris) Journal,* in Sargent, 371–72.
"There is nothing round us but trees" McCardell, 180; Wahll, 181.
"Their Cattle are near as wild as Deer" Letter from anonymous British officer writing home, at http://www.loudounhistory.org/history/loudoun-braddock-march-1755.htm#t1, accessed January 5, 2017.
209 *"[T]he Wound itches"* Wahll, 181.
he had been assigned, by Commodore Keppel Ibid., 10.
210 *"Here lives one Colonel Cressop"* *Seaman's (Morris) Journal,* Sargent, 372.
"and a D____d Rascal" Ibid., 372n.
General Braddock and his entourage Ibid., 373.
211 *It had to be buried* *Orme Journal,* in Sargent, 313–15.
"Indeed the Officers are as ill" Wahll, 181–82.
"the folly of Mr. Dinwiddie" Preston, 124.
Washington wrote to Sally Fairfax "From George Washington to Sarah Cary Fairfax, 14 May 1755," *Founders Online,* National Archives, http://founders.archives.gov/documents/Washington/02-01-02-0138, last modified June 29, 2017.
The fort itself consisted of a log stockade Preston, 107.
212 *amazing the Indians* *Orme Journal,* in Sargent, 374.
Washington boasted to his younger brother Jack "From George Washington to John Augustine Washington, 14 May 1755," *Founders Online,* National Archives, http://founders.archives.gov/documents/Washington/02-01-02-0137, last modified December 28, 2016. [NOTE: The letter was not sent, apparently because Washington soon saw Jack and others to whom he wrote.]
to pay farmers and contractors Freeman, 2:39.
He avoided a visit to his mother Ibid., 42.
213 *buy a slave named Harry for £45* See "From George Washington to John Augustine Washington, 28 May 1755," *Founders Online,* National Archives, note 9, http://founders.archives.gov/documents/Washington/02-01-02-0146, last modified February 21, 2017.
he impatiently waited two days "Memorandum, 15–30 May 1755," *Founders Online,* National Archives, http://founders.archives.gov/documents/Washington/02-01-02-0140, last modified December 28, 2016.

213 *twenty-five hundred strong* Preston, 121.
three independent companies See Freeman, 2:44, for list of companies.
Doctors checked the latter Hamilton, 16.
one thousand lashes each Ibid., 17.
"fish out," as he expressed it "From George Washington to John Augustine Washington, 28 May 1755," *Founders Online,* National Archives, http://founders.archives.gov/documents/Washington/02-01-02-0146, last modified February 21, 2017.

214 *as Governor Dinwiddie had promised* See *Orme Journal,* in Sargent, 287, on sending Christopher Gist's son, Nathaniel, to summon the Cherokee and Catawba. See also Preston, 118, on some Cherokee starting north but soon running into problems; also on the outbreak of fighting between Iroquois and Cherokee.
"let them want for nothing" Kopperman, 100–101.
what did Braddock intend to do Ibid., 101.

215 *"To wch Genl Braddock answered"* Ibid.

CHAPTER TWELVE

217 *"almost a perpendicular rock"* *Orme Journal,* in Sargent, 323.

218 *he wrote a brief, almost curt, letter* "From George Washington to Mary Ball Washington, 7 June 1755," *Founders Online,* National Archives, http://founders.archives.gov/documents/Washington/02-01-02-0150, last modified June 29, 2017.
in far more affectionate terms "From George Washington to Sarah Cary Fairfax, 7 June 1755," *Founders Online,* National Archives, http://founders.archives.gov/documents/Washington/02-01-02-0153, last modified December 28, 2016.
To his brother Jack he gave an update "From George Washington to John Augustine Washington, 7 June 1755," *Founders Online,* National Archives, http://founders.archives.gov/documents/Washington/02-01-02-0154, last modified December 28, 2016.

220 *the complete silver service* Kopperman, 171.
"by frequent breaches of Contracts" "From George Washington to William Fairfax, 7 June 1755," *Founders Online,* National Archives, http://founders.archives.gov/documents/Washington/02-01-02-0148, last modified June 29, 2017.
symphonic accompaniment of thunder Preston, 173. *Orme Journal,* in Sargent, 326, discusses the departure and dates. He gives a list of the various companies on 327–29.

221 *Colonels Halkett and Dunbar* See *Orme Journal,* in Sargent, 331, for a listing of officers attending the council of war.
governor of the Colony of Massachusetts Bay https://www.britannica.com/biography/Thomas-Gage.

221 *sixteen large and heavily reinforced wagons* McCardell, 220; for other details see *Orme Journal,* in Sargent, 331–32.

222 *several four-inch mortars* Cubbison, 88. He notes that they actually measure four and two-fifths inches.

hauling twelve cannon, four howitzers Hamilton, 41.

Washington's contribution of his best horse "From George Washington to John Augustine Washington, 28 June–2 July 1755," *Founders Online,* National Archives, http://founders.archives.gov/documents/Washington/02-01-02-0160, last modified December 28, 2016.

personal servant Ibid.

two thousand pounds Cubbison, 88.

an extra sixty thousand pounds Ibid.

223 *attempting to avoid streams* Ibid., 83–85.

With axes and long military blades Ibid., 83.

The miners call the blasters Hamilton, 18.

224 *they chopped twigs and branches* Cubbison, 81–83.

always chose to ford streams Wahll, 278–79. See also Cubbison, 81–85.

225 *The old soldiers worried* Wahll, 182.

"formal attacks & Platoon-firing" Adam Stephen to John Hunter, July 18, 1755, in Kopperman, 225–28.

Washington, writing many years later "Remarks" in Washington, *George Washington Remembers,* 18–19.

"We this day passed the Aligany Mountain" *Orme Journal,* in Sargent, 334–35.

226 *a gloomy, low-lying pine forest* Preston, 180; McCardell, 222.

one of the hunters brought in a bear Hamilton, 18–19.

The baggage train stretched out "From George Washington to John Augustine Washington, 28 June–2 July 1755," *Founders Online,* National Archives, http://founders.archives.gov/documents/Washington/02-01-02-0160, last modified December 28, 2016.

level every mole hill Ibid.

"I urgd it in the warmest terms" Ibid.

227 *French troops moving from Montreal* Preston, 181–82.

The "flying column" under Sir John Ibid., 182; Wahll, 284.

each pulled by a team of nine horses Preston, 176.

"marched out the Knight" Ibid., 184, citing John Rutherford to Richard Peters, August 15, 1755.

228 *the inflammation of the intestines* http://www.medicalnewstoday.com/articles/171193.php, accessed March 10, 2017.

"It is the Desire of [everyone] in the Family" "To George Washington from Roger Morris, 23 June 1755," *Founders Online,* National Archives, http://founders.archives.gov/documents/Washington/02-01-02-0158, last modified June 29, 2017.

229 *Monacatoocha was captured* *Orme Journal,* in Sargent, 336–37.
"They had stripped and painted some trees" Ibid., 341.
shot him four times in the belly Hamilton, 44–45.
symbols depicting the scalps they had taken Ibid., 45.
French officers had written their names Ibid.
The scattered human bones Ibid.
230 *discovered a child of seven* Sargent, 383.
the flying column approached Ibid., 345; for map, see Preston, 167.
would blow up Fort Duquesne Adam Stephen to John Hunter, July 18, 1755, in Kopperman, 225–28.
twenty-five miles "From George Washington to John Augustine Washington, 28 June–2 July 1755," *Founders Online,* National Archives, http://founders.archives.gov/documents/Washington/02-01-02-0160, last modified June 29, 2017.
an opportunity he would not miss "From George Washington to Robert Orme, 30 June 1755," *Founders Online,* National Archives, http://founders.archives.gov/documents/Washington/02-01-02-0161, last modified December 28, 2016.

CHAPTER THIRTEEN

231 *an eighteen-year-old named James Smith* For biographical information about James Smith and his family homestead area, see http://mhs.mercersburg.org/; http://smithrebellion1765.com/?page_id=9; http://www.ohiohistorycentral.org/w/James_Smith; and http://www.fortloudounpa.com/, accessed January 16, 2017.
had "fallen violently in love" James Smith, 5.
233 *"a general rejoicing around me"* Ibid., 8–9.
234 *"What news from Braddock's army?"* Ibid., 10–11.
within fifteen miles of Fort Duquesne Freeman, 2:61.
"The British Gentlemen" Adam Stephen to John Hunter, July 18, 1755, in Kopperman, 225–28; also, attributed to unnamed British officer, in Freeman, 2:64.
on July 6 Preston, 204.
While the father grieved Hamilton, 46–47.
235 *A torrential rain soaked* Ibid.
a covered wagon Washington, *George Washington Remembers,* 19.
"You may thank my Fds" "From George Washington to John Augustine Washington, 28 June–2 July 1755," *Founders Online,* National Archives, http://founders.archives.gov/documents/Washington/02-01-02-0160, last modified June 29, 2017.
commander back at Fort Cumberland Preston, 136.

235 *". . . I have been greatly surprisd"* "From George Washington to James Innes, 2 July 1755," *Founders Online,* National Archives, http://founders.archives.gov/documents/Washington/02-01-02-0163, last modified June 29, 2017.

236 *situated on a wooded plateau* Preston, 211.

scores and scores of white tents See Orme diagram in Preston, 203.

237 *The first party roused at 2 A.M.* See orders for the day in Wahll, 338–39. See also Freeman, 2:64.

distinguished themselves by their size See "British Grenadiers—Soldier Profile," https://www.military-history.org/articles/early-modern/british-grenadiers-soldier-profile.htm, accessed March 15, 2017.

brought the party to about three hundred men See Freeman, 2:64n, for various accounts of the strength of this advance party.

yellow silk regimental banners Carter, 9.

Christopher Gist, George Croghan Preston, 119.

the two hundred axmen See Robert Orme to Robert Napier, July 18, 1755, in Pargellis, 98–100.

flushing out a party of Indians Sargent, 384.

lead a lightning night raid on Fort Duquesne Sargent, 352. See also Freeman, 2:62.

238 *They marched downward* Sargent, 352n, puts the distance from the Monongahela Camp to the river on the night of July 8 as two miles.

tie cushions to his saddle Washington, *George Washington Remembers,* 17.

the main body of the great procession Sargent, 353.

the two miles For map, see Preston, 227.

two or three hundred yards across Hamilton, 27.

twelve feet high Freeman, 2:65. See also Pargellis, 106, for height of bank.

239 *remains of a trading post* Preston, 221.

higher plateau some four hundred feet above From measurements on Google Earth, the difference between the river elevation and the tops of the high bluffs.

grenadiers, wearing their tall wizard-like hats For illustrations of grenadiers, see Washington, *George Washington Remembers,* 44–45. Also see http://www.militaryheritage.com/7ywrepca.htm; http://www.militaryheritage.com/mitre.htm.

ordered the troops to ford the river Sargent, 384.

"Some talk of Alexander" http://lyricsplayground.com/alpha/songs/b/britishgrenadierguards.shtml, accessed October 21, 2017.

240 *red coats with white lace* See 44th and 48th regiment uniform colors at http://www.warof1812.ca/charts/7warchtb.htm, accessed January 21, 2017. See also http://www.kronoskaf.com/syw/index.php?title=44th_Foot.

the yellow silk banners of the 44th http://www.kronoskaf.com/syw/index.php?title=44th_Foot.

240 *stretched out over a mile* Preston, 229, gives the dimensions of the column as a mile in length as it moved up from the far bank, before the attack, and two or three hundred yards across.

the grenadiers' march For grenadiers' march, see https://www.youtube.com/watch?v=1zSowOS4Wyg.

their hundreds of sharpened bayonets Sargent, 384.

It was the most beautiful sight Witnesses of that river crossing, including George Washington, recalled it years later. Sargent 217, 218, 218n. In Sargent, 218n, a veteran of the battle revisited the site of the crossing many years later, in 1799, and described it to a companion who recorded it as follows: "'A finer sight could not have been beheld,' according to an account based on an eyewitness, 'the shining barrels of the muskets, the excellent order of the men, the cleanliness of their apparel, the joy depicted on every face at being so near Fort Du Quesne—the highest object of their wishes. The music reechoed through the mountains. How brilliant the morning. . . .'" The Jared Sparks biography of Washington includes this passage: "Washington was often heard to say during his lifetime, that the most beautiful spectacle he had ever beheld was the display of the British troops on this eventful morning. Every man was neatly dressed in full uniform, the soldiers were arranged in columns and marched in exact order, the sun gleamed from their burnished arms, the river flowed tranquilly on their right, and the deep forest over shadowed them with solemn grandeur on their left. Officers and men were equally inspirited with cheering hopes and confident anticipations." Sparks, 62.

If they had not attacked by now Sargent, 385.

242 *served as commandant of Michilimackinac* Ibid., 132.

as mayor of Quebec McCardell, 141.

Daniel-Hyacinthe-Marie Liénard de Beaujeu See *Dictionary of Canadian Biography,* http://www.biographi.ca/en/bio/lienard_de_beaujeu_daniel_hyacinthe_marie_3E.html, accessed January 24, 2017.

Troupes de la Marine See Presto, Appendix, under French Armed Forces.

"goes toward the musket shots" Ibid., 134.

send reinforcements to the Ohio Valley Ibid., 131. See also "Duquesne," *Dictionary of Canadian Biography,* http://www.biographi.ca/en/bio/duquesne_de_menneville_ange_4E.html, accessed January 22, 2017.

about thirty-six feet long Preston, 137–38.

244 *Up and down ravines* Ibid., 142.

the porters would haul 2,378 barrels Ibid., 143.

pining for his wife and children Ibid., 130, 163. Preston quotes letters from Contrecoeur saying that he had lost everything from his household, and his family was scattered.

245 *six to seven hundred Indian warriors* Ibid., 149.

Shawnee, Miami, Hurons See ibid., 149–50, for listing of tribes.

245 *their land in eastern Pennsylvania* See http://delawaretribe.org/blog/2013/06/27/the-walking-purchase/, accessed January 25, 2017.

247 *Captain Beaujeu then sang their war song* See Kopperman, Appendix F, 267.

to within six leagues of the fort Preston, 210.

248 *"a great stir in the fort"* James Smith, 11.

Captain Beaujeu set off For Le Courtois's account of the battle, see Preston, Appendix F, 355.

numbered about two hundred and fifty Ibid., 222, lists one hundred and eight from the *Troupes de la Marine* and one hundred and eighty-six from the Canadian militia.

red, black, brown, and blue James Smith, 8.

followed a path a few miles north Preston, 224; for other details of terrain and movement, see Preston maps on 227, 236.

249 *scouts had carefully noted the formation* See Le Courtois's report in Preston, Appendix F, 355.

250 *overseen by French officers* Ibid.

CHAPTER FOURTEEN

251 *who tended his master's seven horses* "The Journal of Captain Robert Cholmley's Batman," Hamilton, 31.

"joy at our Good Luck" "Gordon's Account," Pargellis, 106.

252 *"for I do not think"* "The Journal of Captain Robert Cholmley's Batman," Hamilton, 27–28.

two hundred axmen Robert Orme to Robert Napier, July 18, 1755, in Pargellis, 98–100.

Wearing his father's long silk sash Preston, 256. For details on the sash, see http://www.mountvernon.org/preservation/collections-holdings/browse-the-museum-collections/object/w-86/.

on his cushioned saddle "Remarks" in Washington, *George Washington Remembers,* 18.

bloody flux Freeman, v. ii, calls it bloody flux in the caption to an illustration of Washington's accounts book, where he buys milk.

"Those heroes of antiquity" https://militarymusic.com/blogs/military-music/13515769-the-british-grenadiers-march.

253 *an advance battalion* Hamilton, 6.

"the Indiens was upon us" "The Journal of Captain Robert Cholmley's Batman," Hamilton, 28.

Three hundred, he estimated "Gordon's Account," Pargellis, 106.

254 *British officers on horseback* Pouchot account, Kopperman, Appendix F, 263.

Make ready! Commands for firing from British Royal Army's *Manual Exercise as Ordered by His Majesty, in the year 1764.*

254 *the British have the advantage* Godefroy account, Kopperman, 259.
velocity of 1,000 feet per second See chart of cannon ballistics and muzzle velocities: http://www.arc.id.au/CannonBallistics.html, accessed March 18, 2017, from discussion on website https://thefiringline.com/forums/archive/index.php?t-16135.html, "Tests conducted with French Model 1769 muskets in the 19th century showed a muzzle velocity of 320 meters per second which is approximately 1,130 feet per second," accessed March 18, 2017.
one hundred French-Canadian militia Dumas account, Kopperman, 251–52. The Contrecoeur account also describes Canadians running from the action, ibid., 250.
255 *"Every man for himself!"* Dumas account, ibid., 251.
The tremendous blast Ibid.
"God save the King!" Ibid. [NOTE: Or "long live the king," from French Officer A account, ibid., 254.]
naked torso ripped by grapeshot That grapeshot was shot from the cannon is clear from the Godefroy account, ibid., 259. Also, in French Officer C account, that Beaujeu was killed by a third volley of British artillery, this after the French had stepped back twice with the first two volleys, ibid., 257. Dumas says that Beaujeu was killed by the third volley of musketry. The Contrecoeur account matches the above, with the French falling back.
tumbling into the bottom of the ravine Dumas account, ibid., 252.
a British flanking party Gordon account, ibid., 199.
"Fix your bayonets!" Preston, 231; Gage account, Kopperman, 59. "Fix your bayonets" is the correct command from the British Army's Manual Exercise, 8.
256 *the death cry* Eastman, 26. https://www.youtube.com/watch?v=HdJRBEYwyXs, accessed February 3, 2017.
sound resembling the rumble of distant thunder See survivors' accounts in Peter Stark, *Astoria: John Jacob Astor and Thomas Jefferson's Lost Pacific Empire* (New York: HarperCollins Publishers, 2014).
the original line of march Gage account, Kopperman, 192.
not aiming, simply firing up Ibid., 192.
"Don't throw away your fire!" Ibid.
can't see more than two Indians Ibid.
Gage gives the order to fall back Gordon account, ibid., 199.
behind tree trunks Adam Stephen quote, ibid., 226.
fifteen of the eighteen officers Gates account, ibid., 196.
257 *He orders Colonel Burton* Orme account, ibid., 214.
five hundred men taken Preston, 241.
make an assault on the hillside Orme account, Kopperman, 214.
"[I]f we saw of them" "The Journal of Captain Robert Cholmley's Batman," Hamilton, 29.
258 *baggage train is being attacked* Gordon account, Kopperman, 199–200.

258 *in red and blue uniforms* McCardell, 250.
making a half-moon shape "The Journal of Captain Robert Cholmley's Batman," Hamilton, 28.
"The whole now were got together" Orme account, Kopperman, 214.
"Confusion and Panick" Gordon account, ibid., 200.
"storys they had heard of the Indians" "The Journal of a British Officer," Hamilton, 50.
"will haunt me" Matthew Leslie to an unknown recipient, July 30, 1755, Kopperman, 204.
"[T]he whole army was" Gage account, ibid., 192.

259 *"Men dropped like leaves"* British Officer D, ibid., 176.
"[I] beg'd of him for God-Sake" St. Clair, ibid., 224.
all is ruin and he should retreat Cubbison, 107, quoting letter from Dr. Alexander Hamilton, who later treated St. Clair's wound and heard a detailed account of the battle from him.
His servant ties Ibid., 107.
His son, Lieutenant James Halkett Preston, 243–44, who recounts the story. He also relates the discovery of the bodies during the Forbes Campaign in 1758, ibid., 312. See also the full story quoted from Benjamin West's account, as West witnessed it in 1758, told in *Lives of Painters,* v. ii, by Allan Cunningham. Quoted in Carter, *Historical Record of the Forty-Fourth, or the East Essex Regiment of Foot,* 12n–14n.
the two regimental colors Preston, 247.

260 *the American colonial soldiers* "The Journal of a British Officer," Hamilton, 50. See also British Officer A, Kopperman, 164.
They cut loose their teams of horses "The Journal of a British Officer," Hamilton, 52.
striking them with the flat of his sword Sargent, 230 n2, quoting Burd letter to Morris. See also Preston, 253.
"Cowards!" he shouts Sargent, 230.

262 *offer to lead the men* "Remarks" in Washington, *George Washington Remembers,* 19.
Royal Army artillerymen Preston, 245.
orders a detachment Gordon account, Kopperman, 200.
Shirley dead with a bullet Preston, 254.

263 *out of the milling, panicky masses* Orme account, Kopperman, 215. See also Preston, 254. Preston puts the number at one hundred and fifty.
musket ball in the hip Preston, 254, citing Orme to Keppel, July 18, 1755.
The entire body suddenly turns "The Journal of a British Officer," Hamilton, 51.
These are the general's last orders The Gordon account, Kopperman, 200, states that the general's last order was protecting twelve-pounders and sending an advance up the hill. See also Gage account, ibid., 192.

263 *his right arm and chest* Preston, 256.

puffs of smoke now come faster Gordon account, Kopperman, 200.

Officers stagger back Ibid.

Fear surges in the surviving troops Gage in his account blames the fear on all the stories the soldiers have heard from "country people."

Few officers stand to lead the men See British Officer C, Kopperman, 174, for a brief description of Waggoner, with one hundred and seventy Virginians, and Polson, another brave Virginian, leading groups of brave men, one cut down by friendly fire, the other cut to pieces by Indians.

The horrible shrieks of the Indians Gordon account, ibid., 200.

264 *"[M]any of the officers called out"* Croghan account, ibid., 185.

about two hundred men Gordon account, ibid., 200.

Two American companies Preston, 259, who also quotes an anonymous British officer, whose long letter is found in Pargellis, 112–24.

stays by his side For George Washington's account, see "Remarks" in Washington, *George Washington Remembers,* 20. See also Croghan account, Kopperman, 185: "[T]he unfortunate General was intirely abandoned in his waggons with only his servants and a person or two more . . ."; and Gage account, ibid., 193: "The General was saved by the dexterity of his servants. . . . [T]he General was defended by very few but officers in his repassing of the river. . . ."

Washington directing "Remarks" in Washington, *George Washington Remembers,* 20. This seems accurate, as it would make sense that Washington would be at his side as the last aide standing and that it would fall to him to direct the care of the general. Whether he directed the whole retreat, as it's been stated, is somewhat more questionable.

with the help of other officers Gordon account, Kopperman, 200.

There is a pause For Gage recalling this, see Preston, 257.

265 *"fell upon them"* Le Courtois account, Preston, 257 (quoted), and also in full in Appendix F. See also Pouchot account, Kopperman, 263–64: "The Indians taking this movement of the column from the front toward the rear, as a tendency to retreat, rushed uponn them with their tomhawks, as did the French also, when they disbanded, and a great massacre followed."

"The war cries of the Indians" French Officer A, Kopperman, 254.

"threw away their arms and ammunition" Sargent, 356. See also W. Dunbar, Wahll, 350, who reports that most of them threw their arms away.

Everyone now is running for the ford Gage account, Kopperman, 193: "the whole were put to flight." See also *Orme Journal,* in Sargent, 256.

the wounded are abandoned "Journal of Cholmley's Batman," Kopperman, 183. See also Pouchot account, ibid., 264.

blood dyes the summer-warm water "The Journal of a British Officer," Hamilton, 52: "during our crossing, they Shot many in ye Water both Men & Women, & dyed ye stream with their blood, scalping & cutting them in a most barbarous manner."

266 *Road engineer Harry Gordon* "Gordon's Account," Pargellis, 107–8. See also Preston, 261.

"Coll: Burton thho' very much Wounded" Kopperman, 89–90.

CHAPTER FIFTEEN

269 *"I heard a number of scalp halloo's"* James Smith, 12.

270 *Nec Aspera Terrent* "G.R." (Georgis Rex translated King George) surmounted by a crown be embroidered on the front. A small red flap, embroidered with "NEC ASPERA TERRENT" See http://www.militaryheritage.com/mitre.htm, accessed March 27, 2017.

"these ravenous Hellhounds" Cameron, quoted in Kopperman, 178–79.

271 *Halt the retreat* Washington, *George Washington Remembers,* 20.

with wagons to evacuate Sargent, 256.

272 *"The shocking Scenes"* Washington, *George Washington Remembers,* 20–21.

Deserters began to stream out See Sargent, 235–37, for scene at Dunbar's Camp, destruction of stores, and Braddock's last statements.

273 *"[T]he miraculous care of Providence"* "From George Washington to John Augustine Washington, 18 July 1755," *Founders Online,* National Archives, http://founders.archives.gov/documents/Washington/02-01-02-0169, last modified June 29, 2017.

unsubstantiated rumors would persist One Thomas Fausett of Pennsylvania would later claim to have shot Braddock. Sargent, 244–53. See also Kopperman, Appendix A, 138, for account of the original report and assessment of it.

274 *"[T]he Confusion, hurry and Conflagration"* Preston, 273, quoting unnamed officer.

ate nothing but the luxurious hams Hamilton, 32.

Braddock asked for his pistols Kopperman, 87, quoting Croghan. See also Croghan account in Kopperman, 185, quoting Charles Swaine to Richard Peters, August 5, 1755.

"Who would have thought it?" Franklin, *Autobiography,* 133.

275 *wrapped in two blankets* Hamilton, 32.

he may have ordered *Seaman's (Morris) Journal* says he was buried "decently, though privately," in Sargent, 388.

a Royal Army officer of Braddock's rank Freeman, 2:83 n135.

Out of 1,469 officers and soldiers Preston, 277, and Appendix D, 347.

sixty-three killed Sargent, 237–38.

276 *"Now we must employ Americans"* Kopperman, 302 n12.

276 *"Yor Name is more talked off"* "To George Washington from Christopher Gist, 15 October 1755," *Founders Online,* National Archives, http://founders.archives.gov/documents/Washington/02-02-02-0109, last modified February 21, 2017.

". . . I cannot but hope that Providence" Kopperman, 130.

"Who is Mr. Washington?" Ibid., 130.

277 *"were struck with such a panick"* "From George Washington to Mary Ball Washington, 18 July 1755," *Founders Online,* National Archives, http://founders.archives.gov/documents/Washington/02-01-02-0167, last modified June 29, 2017.

"The Virginians behavd like men" "From George Washington to Robert Dinwiddie, 18 July 1755," *Founders Online,* National Archives, http://founders.archives.gov/documents/Washington/02-01-02-0168, last modified June 29, 2017.

278 *"I doubt not but you have heard"* "From George Washington to Augustine Washington, 2 August 1755," *Founders Online,* National Archives, http://founders.archives.gov/documents/Washington/02-01-02-0176, last modified June 29, 2017.

"very unexpectedly I must own" "From George Washington to Robert Dinwiddie, 18 July 1755," *Founders Online,* National Archives, http://founders.archives.gov/documents/Washington/02-01-02-0168, last modified February 21, 2017.

"every proper occasion" Washington, *George Washington Remembers,* 18–19.

279 *stay at taverns along the way* Freeman, 87 and 87n.

"After thanking Heaven" "To George Washington from Sarah Cary Fairfax, Ann Spearing, and Elizabeth Dent, 26 July 1755," *Founders Online,* National Archives, http://founders.archives.gov/documents/Washington/02-01-02-0172, last modified June 29, 2017.

281 *"I was employ'd to go a journey"* "From George Washington to Augustine Washington, 2 August 1755," *Founders Online,* National Archives, http://founders.archives.gov/documents/Washington/02-01-02-0176, last modified June 29, 2017.

an alarmed Governor Dinwiddie Dinwiddie, 2:357n.

"no Man can gain any Honour" "From George Washington to Warner Lewis, 14 August 1755," *Founders Online,* National Archives, http://founders.archives.gov/documents/Washington/02-01-02-0184, last modified March 30, 2017.

282 *"Let them not be deceiv'd,"* "From George Washington to Charles Lewis, 14 August 1755," *Founders Online,* National Archives, http://founders.archives.gov/documents/Washington/02-01-02-0185, last modified June 29, 2017.

His mother opposed the idea Freeman, 2:107.

282 *it would "dishonour" him* "From George Washington to Mary Ball Washington, 14 August 1755," *Founders Online,* National Archives, http://founders.archives.gov/documents/Washington/02-01-02-0183, last modified June 29, 2017.

283 *carried him by canoe* James Smith, 13.

"as if he had been plucking a turkey" Ibid., 14.

"My son," said the old chief Ibid., 16.

284 *"On this the warriors began to advance"* Ibid., 18–19.

noteworthy Virginians urged him "From George Washington to John Campbell, Earl of Loudoun, 10 January 1757," *Founders Online,* National Archives, http://founders.archives.gov/documents/Washington/02-04-02-0045, last modified June 29, 2017. Washington wrote that he had "long since been satisfied of the impossibility of continueing in this Service without loss of Honour: Nay, was fully convinced of it, before I accepted the Command the Second time (seeing the gloom that sat hovering over us) and did, for this Reason, reject the offer till I was ashamed to deny; not careing to expose my Character to Publick Censure: But the Sollicitations of the Country overcame my objections."

touch-up at the town's "French Barber" "Editorial Introduction," *Founders Online,* National Archives, note 6, http://founders.archives.gov/documents/Washington/02-02-02-0001-0001, last modified March 30, 2017.

to add another one thousand men *Virginia Gazette,* September 5, 1755.

to bring its total force to twelve hundred Freeman, 2:111. Freeman says these could be recruited either as volunteers or by drafting unmarried militiamen.

285 *"& for repellg the unjust & hostile Invasions"* "I. Commission, 14 August 1755," *Founders Online,* National Archives, http://founders.archives.gov/documents/Washington/02-02-02-0001-0002, last modified March 30, 2017.

each of the Virginia Regiment's captains "General Instructions for Recruiting, 1–3 September 1755," *Founders Online,* National Archives, http://founders.archives.gov/documents/Washington/02-02-02-0005, last modified March 30, 2017.

But at Fredericksburg Freeman, 2:115.

"No men, no discipline, no clothing" Ibid., 117.

the new platoon way of exercising "From George Washington to Adam Stephen, 11 September 1755," *Founders Online,* National Archives, http://founders.archives.gov/documents/Washington/02-02-02-0021, last modified June 29, 2017.

286 *"Have the Arms all clean'd"* "Memorandum, 17 September 1755," *Founders Online,* National Archives, http://founders.archives.gov/documents/Washington/02-02-02-0035, last modified June 29, 2017.

287 *now made a lieutenant colonel* Ibid., note 6.

288 *"the most deplorable Situation"* "To George Washington from Adam Stephen, 4 October 1755," *Founders Online,* National Archives, http://founders.archives.gov/documents/Washington/02-02-02-0068, last modified March 30, 2017.

soldiers discovered three bodies Freeman, 2:132.

they captured young Marie Le Roy Le Roy and Leininger, 407–20.

289 *"blow out my brains"* "From George Washington to Robert Dinwiddie, 11–14 October 1755," *Founders Online,* National Archives, http://founders.archives.gov/documents/Washington/02-02-02-0099, last modified March 30, 2017.

"I must with great regret decline the honor" Ibid.

290 *"I desire you will bring some Indians"* "From George Washington to Andrew Montour, 19 September 1755," *Founders Online,* National Archives, http://founders.archives.gov/documents/Washington/02-02-02-0048, last modified March 30, 2017.

"I was greatly enraptur'd" "From George Washington to Andrew Montour, 10 October 1755," *Founders Online,* National Archives, http://founders.archives.gov/documents/Washington/02-02-02-0095, last modified March 30, 2017.

291 *Captain Dagworthy, while a Marylander* "To George Washington from Adam Stephen, 4 October 1755," *Founders Online,* National Archives, http://founders.archives.gov/documents/Washington/02-02-02-0068, last modified June 29, 2017. Note 6 explains the ranking situation.

Washington arrived in Williamsburg Freeman, 2:135. See also balls, gambling, and drinking in the capital, ibid., 141.

eight hundred Virginians "From George Washington to Adam Stephen, 18 November 1755," *Founders Online,* National Archives, http://founders.archives.gov/documents/Washington/02-02-02-0178, last modified March 30, 2017.

292 *"The Accounts of Our Behaviour"* "To George Washington from Adam Stephen, 22 November 1755," *Founders Online,* National Archives, http://founders.archives.gov/documents/Washington/02-02-02-0182, last modified March 30, 2017.

"We have heard of General Braddock's Defeat" "To George Washington from Joseph Ball, 5 September 1755," *Founders Online,* National Archives, http://founders.archives.gov/documents/Washington/02-02-02-0007, last modified June 29, 2017.

"[A]ll the Publick Prints, private Letters" "From George Washington to Adam Stephen, 18 November 1755," *Founders Online,* National Archives, http://founders.archives.gov/documents/Washington/02-02-02-0178, last modified March 30, 2017.

"I think Mount Vernon would afford" "To George Washington from Robert Orme, 10 November 1755," *Founders Online,* National Archives, http://founders.archives.gov/documents/Washington/02-02-02-0168, last modified June 29, 2017.

293 *including the death penalty for deserters* Freeman, 2:136–37.

293 *"the command of Captain Dagworthy"* "From George Washington to Robert Dinwiddie, 5 December 1755," *Founders Online,* National Archives, http://founders.archives.gov/documents/Washington/02-02-02-0203, last modified March 30, 2017.

ten officers brought in only twenty new recruits Ibid.

294 *"riding from place to place"* Ibid.

He set up an office This structure still exists, at the corner of West Cork and Braddock streets, and is maintained as the George Washington Office Museum by the Winchester-Frederick County Historical Society. https://winchesterhistory.org/george-washingtons-office/.

"Any Soldier who shall desert" "George Mercer's Orders, 25 December 1755," *Founders Online,* National Archives, http://founders.archives.gov/documents/Washington/02-02-02-0231, last modified June 29, 2017.

"We had an extreamly good dinner" "To George Washington from Adam Stephen, 26 December 1755," *Founders Online,* National Archives, http://founders.archives.gov/documents/Washington/02-02-02-0237, last modified March 30, 2017.

295 *one of the four prongs* Anderson, *The War That Made America,* 74.

"Then 'tis decreed" *Gentleman's Magazine,* October 1755, 496.

CHAPTER SIXTEEN

297 *instructing him to sound out Dagworthy* "From George Washington to Adam Stephen, 28 December 1755," *Founders Online,* National Archives, http://founders.archives.gov/documents/Washington/02-02-02-0242, last modified June 29, 2017.

thus command over Fort Cumberland Freeman, 2:133–35.

"[Lieutenant Bacon] then told the Gentlemen" "Memorandum, 7 January 1756," *Founders Online,* National Archives, http://founders.archives.gov/documents/Washington/02-02-02-0269, last modified March 30, 2017.

298 *Washington addressed his officers* "Address, 8 January 1756," *Founders Online,* National Archives, http://founders.archives.gov/documents/Washington/02-02-02-0271, last modified June 29, 2017.

". . . I have determined to resign a commission" "From George Washington to Robert Dinwiddie, 14 January 1756," *Founders Online,* National Archives, http://founders.archives.gov/documents/Washington/02-02-02-0296, last modified June 29, 2017.

299 *accompanied by two servants* Freeman, 2:156.

founded by Benjamin Franklin For history and founding in 1743, see https://www.amphilsoc.org/about, accessed April 6, 2017.

fifteen blocks northward See 1755 map of Manhattan in Library of Congress online, https://www.loc.gov/resource/g3804n.ar110100/, accessed April 6, 2017.

300 *Susannah had a younger sister* Freeman, 2:159–60.

passing the Block House https://www.loc.gov/resource/g3804n.ar110100/, accessed April 6, 2017.

Cromwell's Head Tavern https://drinkingboston.oncell.com/en/cromwells-head-19837.html, accessed April 7, 2017.

301 *Despite admonishing his Virginia Regiment* Freeman, 2:175.

"a gentleman who has a deservedly high reputation" Ibid., 2:164, quoting the edition of March 1, 1756.

"It is my orders that Colonel Washington" "To George Washington from William Shirley, 5 March 1756," *Founders Online,* National Archives, http://founders.archives.gov/documents/Washington/02-02-02-0321, last modified March 30, 2017.

could well give advantage to his protégé Freeman, 2:167.

302 *asked that they make him second-in-command* "From George Washington to Robert Hunter Morris, 9 April 1756," *Founders Online,* National Archives, note 4, http://founders.archives.gov/documents/Washington/02-02-02-0343, last modified June 29, 2017.

"All my Ideal hopes" "From George Washington to Robert Dinwiddie, 16 April 1756," *Founders Online,* National Archives, http://founders.archives.gov/documents/Washington/02-03-02-0001-0001, last modified June 29, 2017.

"The timidity of the inhabitants" "From George Washington to John Robinson, 16 April 1756," *Founders Online,* National Archives, http://founders.archives.gov/documents/Washington/02-03-02-0002, last modified June 29, 2017.

303 *"Indians are the only match for Indians"* "From George Washington to John Robinson, 7 April 1756," *Founders Online,* National Archives, http://founders.archives.gov/documents/Washington/02-02-02-0333, last modified June 29, 2017.

"It will give me the greatest pleasure" "From George Washington to John Robinson, 18 April 1756," *Founders Online,* National Archives, http://founders.archives.gov/documents/Washington/02-03-02-0011, last modified June 29, 2017.

304 *an estimated one hundred Indians and French* "To George Washington from William Stark, 18 April 1756," *Founders Online,* National Archives, http://founders.archives.gov/documents/Washington/02-03-02-0012, last modified March 30, 2017.

chopped down trees and bushes "Orders, 20 April 1756," *Founders Online,* National Archives, http://founders.archives.gov/documents/Washington/02-03-02-0019, last modified March 30, 2017.

305 *"[N]othing but a large and speedy reinforcement"* "From George Washington to Henry Lee and William Fairfax, 21 April 1756," *Founders Online,* National Archives, http://founders.archives.gov/documents/Washington/02-03-02-0026, last modified March 30, 2017.

305 *But within a mere two weeks* Governor Sharpe writes and agrees to Washington as second-in-command of the Virginia Regiment on April 10, 1756. Freeman, 2:171.

men's brains beaten out *Maryland Gazette,* March 11, 1756.

306 *"But what can I do?"* "From George Washington to Robert Dinwiddie, 22 April 1756," *Founders Online,* National Archives, http://founders.archives.gov/documents/Washington/02-03-02-0033, last modified June 29, 2017.

CHAPTER SEVENTEEN

309 *"rather let Braddocks bed be your aim"* "To George Washington from Landon Carter, 21 April 1756," *Founders Online,* National Archives, http://founders.archives.gov/documents/Washington/02-03-02-0031, last modified March 30, 2017. "[H]ow are we grievd? to hear Colo. George Washington hinting to his Country he is willing to retire[.] Sir[,] Merit begets Envy, and should such A thing happen at this hour it must Glut the malice of those who wish you ill. will they not then say[,] see Yr4 darling cloaking fear under the Colour of disgust? Give me leave then as your intimate Friend to persuade you to forget that if any thing has been said to your dishonr . . . If I expostulate with you too warmly tis only to save my self and your other friends from much difficulty that must attend our endeavours to Justifye yr conduct should yo. decline No Sir rather let Braddocks bed be your aim than anything that might discolour those Laurels that I promise my self are kept in store for you."

"When ever you are mentiond" "To George Washington from Charles Carter, 22 April 1756," *Founders Online,* National Archives, http://founders.archives.gov/documents/Washington/02-03-02-0037, last modified March 30, 2017.

310 *"Your good Health and Fortune"* "To George Washington from William Fairfax, 26–27 April 1756," *Founders Online,* National Archives, http://founders.archives.gov/documents/Washington/02-03-02-0053, last modified March 30, 2017.

"nay, scarcely a minute passes" "From George Washington to Robert Dinwiddie, 24 April 1756," *Founders Online,* National Archives, http://founders.archives.gov/documents/Washington/02-03-02-0044, last modified December 30, 2015.

311 *"The Indians were then in great hopes"* James Smith, 47–48.

312 *"Desolation and murder still increase"* "From George Washington to Robert Dinwiddie, 27 April 1756," *Founders Online,* National Archives, http://founders.archives.gov/documents/Washington/02-03-02-0055, last modified December 30, 2015.

313 *close to nine hundred men in Winchester* Freeman, 2:188–89.

he wanted to make an example "From George Washington to Robert Dinwiddie, 23 May 1756," *Founders Online,* National Archives, http://founders

.archives.gov/documents/Washington/02-03-02-0169, last modified December 30, 2015.

313 *He chastised one officer* "From George Washington to Henry Woodward, 5 May 1756," *Founders Online,* National Archives, http://founders.archives.gov/documents/Washington/02-03-02-0090, last modified December 30, 2015.

an ailing and worn-out Governor Dinwiddie *Encyclopedia Virginia,* http://www.encyclopediavirginia.org/Dinwiddie_Robert_1692-1770#start_entry, accessed April 12, 2017. See also Freeman, 2:198.

In mid-July, back at Winchester "To George Washington from William Fairfax, 10 July 1756," *Founders Online,* National Archives, http://founders.archives.gov/documents/Washington/02-03-02-0236, last modified March 30, 2017.

315 *Lord Loudoun to replace* Freeman, 2:196.

this French action against Minorca Anderson, *Crucible of War,* 170. France declared war three weeks later. Borneman, 66.

to write a letter of recommendation "To George Washington from Robert Dinwiddie, 27 May 1756," *Founders Online,* National Archives, http://founders.archives.gov/documents/Washington/02-03-02-0180, last modified December 30, 2015.

316 *"We humbly represent to your Lordship"* "From George Washington to John Campbell, Earl of Loudoun, 25 July 1756," *Founders Online,* National Archives, http://founders.archives.gov/documents/Washington/02-03-02-0266, last modified March 30, 2017.

Washington took up fencing lessons Freeman, 2:204.

"make the best appearance possible" "From George Washington to Adam Stephen, 6 September 1756," *Founders Online,* National Archives, http://founders.archives.gov/documents/Washington/02-03-02-0345, last modified December 30, 2015.

"The Profession of Soldier" *Virginia Gazette,* September 3, 1756. See also Freeman, 2:210–11.

317 *"What useless lumber"* "To George Washington from John Kirkpatrick, 22 September 1756," *Founders Online,* National Archives, note 2, http://founders.archives.gov/documents/Washington/02-03-02-0355, last modified December 30, 2015.

He would quit Freeman, 2:211.

"In the Short time I have had" "To George Washington from John Kirkpatrick, 22 September 1756," *Founders Online,* National Archives, http://founders.archives.gov/documents/Washington/02-03-02-0355, last modified December 30, 2015.

318 *"courage & bravery both"* "To George Washington from William Ramsay, 22 September 1756," *Founders Online,* National Archives, http://founders.archives.gov/documents/Washington/02-03-02-0356, last modified December 30, 2015.

318 *"I am sensible you will be blamed"* "To George Washington from Augustine Washington, 16 October 1756," *Founders Online,* National Archives, http://founders.archives.gov/documents/Washington/02-03-02-0370, last modified December 30, 2015.

319 *attended by his servant* See references to Thomas Bishop in Freeman 2:156, 199, 234.

he assembled the Virginia Regiment Baker, 30, citing letter in *Virginia Gazette,* August 27, 1756.

"for divers good causes" Freeman, 2:205.

hard-headed, extremely vain See "Montcalm," Dictionary of Canadian Biography, http://www.biographi.ca/en/bio/montcalm_louis_joseph_de_3E.html, accessed April 12, 2017.

In a fierce exchange of cannon fire Anderson, *Crucible of War,* 162–63.

320 *narrowly missed an ambush* Freeman, 2:217–18.

"I passed & escaped almost certain destruction" Washington, *George Washington Remembers,* 22.

"otherwise we must have fallen a Sacrifice" "From George Washington to Robert Dinwiddie, 9 November 1756," *Founders Online,* National Archives, http://founders.archives.gov/documents/Washington/02-04-02-0001-0001, last modified December 30, 2015.

seven men and three women Freeman, 2:218.

"This Jaunt afforded me great opportunity" "From George Washington to Robert Dinwiddie, 9 November 1756," *Founders Online,* National Archives, http://founders.archives.gov/documents/Washington/02-04-02-0001-0001, last modified December 30, 2015.

321 *insulted that Washington seemed to blame* "To George Washington from Robert Dinwiddie, 16 November 1756," *Founders Online,* National Archives, http://founders.archives.gov/documents/Washington/02-04-02-0006-0001, last modified March 30, 2017.

"my open & disinterested way of writing" "From George Washington to Robert Dinwiddie, 24 November 1756," *Founders Online,* National Archives, http://founders.archives.gov/documents/Washington/02-04-02-0008, last modified March 30, 2017.

CHAPTER EIGHTEEN

323 *"I cannot conceive what service"* "To George Washington from Robert Dinwiddie, 2 February 1757," *Founders Online,* National Archives, http://founders.archives.gov/documents/Washington/02-04-02-0056, last modified December 30, 2015.

late February and early March 1757 Freeman, 2:237 and 237n.

324 *he had written Lord Loudoun* "From George Washington to John Campbell,

Earl of Loudoun, 10 January 1757," *Founders Online,* National Archives, http://founders.archives.gov/documents/Washington/02-04-02-0045, last modified December 30, 2015.

324 *"[D]on't think My Lord I am going to flatter"* Ibid.

325 *cover letter to Lord Loudoun's aide-de-camp* "From George Washington to James Cuninghame, 28 January 1757," *Founders Online,* National Archives, http://founders.archives.gov/documents/Washington/02-04-02-0055, last modified December 30, 2015.

326 *a stocky, fifty-two-year-old Scottish bachelor* Freeman, 2:236–37.

Lord Loudoun brusquely dismissed See Ferling, 38–39. See also Freeman, 2:241, for Loudoun thanking Washington.

"the burning fury and humiliation" Ferling, 39.

"that being Americans shoud deprive us" "From George Washington to Robert Dinwiddie, 10 March 1757," *Founders Online,* National Archives, http://founders.archives.gov/documents/Washington/02-04-02-0062, last modified December 30, 2015.

328 *"highly disagreeable to your [correspondent]"* "Memorial to John Campbell, Earl of Loudoun, 23 March 1757," *Founders Online,* National Archives, http://founders.archives.gov/documents/Washington/02-04-02-0066, last modified December 30, 2015.

329 *having moved from his rented house* Washington had moved to Fort Loudoun by December 2, 1756. Baker, 34.

an elegant mahogany dining set "Enclosure: Invoice to Richard Washington, 15 April 1757," *Founders Online,* National Archives, http://founders.archives.gov/documents/Washington/02-04-02-0075-0002, last modified March 30, 2017.

"I have been posted" "From George Washington to Richard Washington, 15 April 1757," *Founders Online,* National Archives, http://founders.archives.gov/documents/Washington/02-04-02-0075-0001, last modified March 30, 2017.

due to the strains of the office *Virginia Gazette,* April 22, 1757.

330 *to maintain the Virginia Regiment* Freeman, 2:246.

"I . . . hope, that your Honor" "From George Washington to Robert Dinwiddie, 29 April 1757," *Founders Online,* National Archives, http://founders.archives.gov/documents/Washington/02-04-02-0080, last modified December 30, 2015.

". . . I am tired of the place" "From George Washington to Robert Dinwiddie, 2 December 1756," *Founders Online,* National Archives, http://founders.archives.gov/documents/Washington/02-04-02-0012, last modified June 29, 2017.

331 *Parties of Indian allies* Freeman, 2:246.

nor money to pay "To George Washington from John Robinson, 21 June 1757," *Founders Online,* National Archives, http://founders.archives.gov/documents/Washington/02-04-02-0157, last modified June 29, 2017. "Mr Atkins wrote it seems very pressingly to the Govr to send up Money to pay the Indians for Scalps, and the Govr accordingly issued his Warrant to me to pay £240 for that Purpose, which I have paid tho' not justified for it by the Act of Assembly, for the Act directs that only Ten Pounds for every Scalp should be paid by me, and the remaining thirty to be paid at the next Assembly. . . ."

331 *his aide Captain George Mercer* Freeman, 2:130.

"the Govr knew not how to treat Indians" "To George Washington from George Mercer, 24 April 1757," *Founders Online,* National Archives, http://founders.archives.gov/documents/Washington/02-04-02-0078, last modified December 30, 2015.

A Cherokee chief warrior, Youghtanno Ibid., note 6.

332 *the arrival of Edmond Atkin* Freeman, 2:247. Also spelled Edmund.

"[T]hey are the most insolent" "From George Washington to Robert Dinwiddie, 24 May 1757," *Founders Online,* National Archives, http://founders.archives.gov/documents/Washington/02-04-02-0091, last modified December 30, 2015.

A large party of seventy "From George Washington to John Stanwix, 28 May 1757," *Founders Online,* National Archives, http://founders.archives.gov/documents/Washington/02-04-02-0096, last modified December 30, 2015. That Lewis is Andrew Lewis. Freeman, 2:249.

"An Indian will never forget a promise" "From George Washington to Robert Dinwiddie, 30 May 1757," *Founders Online,* National Archives, http://founders.archives.gov/documents/Washington/02-04-02-0098, last modified December 30, 2015.

333 *He considered himself lucky* "From George Washington to John Stanwix, 28 May 1757," *Founders Online,* National Archives, http://founders.archives.gov/documents/Washington/02-04-02-0096, last modified December 30, 2015. Also, Stanwix identification in Freeman, 2:249.

Colonel Stanwix had deep roots Chip Twellman Haley, "Fort namesake Stanwix had unusual life, death," *Rome (NY) Sentinel,* June 11, 2017.

334 *four-thousand-man Royal American Regiment* See http://www.militaryheritage.com/60thregt.htm.

Another of Washington's frustrations "To George Washington from Robert Dinwiddie, 1 June 1757," *Founders Online,* National Archives, http://founders.archives.gov/documents/Washington/02-04-02-0100, last modified December 30, 2015.

335 *"[W]hen the Frenchmen came"* "To George Washington from James Baker, 10 June 1757," *Founders Online,* National Archives, http://founders.archives.gov/documents/Washington/02-04-02-0117, last modified December 30, 2015.

"loiterers," "deserters" "To George Washington from Robert Dinwiddie, 6 June 1757," *Founders Online,* National Archives, http://founders.archives.gov/documents/Washington/02-04-02-0107, last modified December 30, 2015.

336 *"what is fit for the Service"* "Memoranda, 13 June 1757," *Founders Online,* National Archives, http://founders.archives.gov/documents/Washington/02-04-02-0124, last modified December 30, 2015.

"[A] Safe Delivarence" "To George Washington from James Livingston, 14 June 1757," *Founders Online,* National Archives, http://founders.archives.gov/documents/Washington/02-04-02-0127, last modified December 30, 2015.

338 *"If the Enemy is coming down"* "From George Washington to Horatio Sharpe, 16 June 1757," *Founders Online,* National Archives, http://founders.archives.gov/documents/Washington/02-04-02-0135, last modified December 30, 2015.

"a very considerable Force" "From George Washington to John Dagworthy, 16 June 1757," *Founders Online,* National Archives, http://founders.archives.gov/documents/Washington/02-04-02-0133, last modified December 30, 2015.

Colonel Stanwix arranged for wagons "To George Washington from John Stanwix, 18 June 1757," *Founders Online,* National Archives, http://founders.archives.gov/documents/Washington/02-04-02-0142, last modified December 30, 2015.

339 *Fairfax rode hard* "To George Washington from William Fairfax, 17 June 1757," *Founders Online,* National Archives, http://founders.archives.gov/documents/Washington/02-04-02-0140, last modified December 30, 2015.

the militia from five other counties "To George Washington from Robert Dinwiddie, 20 June 1757," *Founders Online,* National Archives, http://founders.archives.gov/documents/Washington/02-04-02-0152, last modified December 30, 2015.

"your Guns are but muskets" "From George Washington to John Stanwix, 20 June 1757," *Founders Online,* National Archives, http://founders.archives.gov/documents/Washington/02-04-02-0150, last modified December 30, 2015.

"The storm which threatned us" "From George Washington to William Fairfax, 25 June 1757," *Founders Online,* National Archives, http://founders.archives.gov/documents/Washington/02-04-02-0169, last modified December 30, 2015.

340 *"surprising" in an officer* "From George Washington to John Robinson, 10 July 1757," *Founders Online,* National Archives, http://founders.archives.gov/documents/Washington/02-04-02-0190, last modified December 30, 2015.

"obstinate and perverse" "From George Washington to John Stanwix, 15 July 1757," *Founders Online,* National Archives, http://founders.archives.gov/documents/Washington/02-04-02-0200, last modified December 30, 2015.

Cherokee Indians wrongly jailed Freeman, 2:256 for both Atkin jailing Indians and Washington removing Peter Hog. For bad food sickening Cherokees in jail, see "To George Washington from Edmond Atkin, 20 July 1757," *Founders Online,* National Archives.

340 *The burgesses had enacted a draft* Freeman, 2:258.

renewed a law punishing desertion with death Ibid.

"I am greatly at a loss how to proceed" "From George Washington to John Robinson, 10 July 1757," *Founders Online,* National Archives, http://founders.archives.gov/documents/Washington/02-04-02-0190, last modified December 30, 2015.

341 *"I have a Gallows near 40 feet high"* "From George Washington to John Stanwix, 15 July 1757," *Founders Online,* National Archives, http://founders.archives.gov/documents/Washington/02-04-02-0200, last modified December 30, 2015.

342 *a carpenter and a good dancer and fiddler* "General Court-Martial, 25–26 July 1757," *Founders Online,* National Archives, http://founders.archives.gov/documents/Washington/02-04-02-0218, last modified December 30, 2015.

"The Prisoner Ignatius Edwards" Ibid.

twenty-two prisoners held for desertion "From George Washington to John Stanwix, 15 July 1757," *Founders Online,* National Archives, note 2, http://founders.archives.gov/documents/Washington/02-04-02-0200, last modified March 30, 2017.

343 *William Smith, a twenty-year-old saddler* "From George Washington to William Crawford, 20 July 1757," *Founders Online,* National Archives, note 1, http://founders.archives.gov/documents/Washington/02-04-02-0211 last modified December 30, 2015.

344 *"I hope excuse my hanging"* "From George Washington to Robert Dinwiddie, 3 August 1757," *Founders Online,* National Archives, http://founders.archives.gov/documents/Washington/02-04-02-0232, last modified March 30, 2017.

Washington issued orders to his officers "Instructions to Company Captains, 29 July 1757," *Founders Online,* National Archives, http://founders.archives.gov/documents/Washington/02-04-02-0223, last modified December 30, 2015.

brought the Virginia Regiment "From George Washington to Robert Dinwiddie, 3 August 1757," *Founders Online,* National Archives, http://founders.archives.gov/documents/Washington/02-04-02-0232, last modified March 30, 2017.

His detailed instructions "Instructions to Company Captains, 29 July 1757," *Founders Online,* National Archives, http://founders.archives.gov/documents/Washington/02-04-02-0223, last modified March 30, 2017.

345 *"Discipline," he wrote prophetically* Ibid.

CHAPTER NINETEEN

347 *At the same time as the hangings* Freeman, 2:264, 274. See also Robert Steward's letter explaining the dire situation in early November to Governor Dinwiddie: "For upwards of three Months past Colo. Washington has labour'd under a Bloudy Flux, about a week ago his Disorder greatly increas'd attended with bad Fevers, the day before yesterday he was seiz'd with Stitches & violent Pleuretick Pains," "Robert Stewart to Robert Dinwiddie, 9 November 1757," *Founders Online,* National Archives, http://founders.archives.gov/documents/Washington/02-05-02-0029, last modified March 30, 2017.

broadaxes and socket chisels "Invoice from Richard Washington, 20 August 1757," *Founders Online,* National Archives, http://founders.archives.gov/documents/Washington/02-04-02-0244, last modified March 30, 2017.

348 *"[N]o man that ever was employed"* "From George Washington to Robert Dinwiddie, 17 September 1757," *Founders Online,* National Archives, http://founders.archives.gov/documents/Washington/02-04-02-0260, last modified December 30, 2015.

"My Conduct to Yo." "To George Washington from Robert Dinwiddie, 24 September 1757," *Founders Online,* National Archives, http://founders.archives.gov/documents/Washington/02-04-02-0268, last modified December 30, 2015.

"I do not know that I ever gave your Honor cause" "From George Washington to Robert Dinwiddie, 5 October 1757," *Founders Online,* National Archives, http://founders.archives.gov/documents/Washington/02-05-02-0001, last modified December 30, 2015.

349 *"I exert every means in my power"* "From George Washington to John Stanwix, 8 October 1757," *Founders Online,* National Archives, http://founders.archives.gov/documents/Washington/02-05-02-0003, last modified December 30, 2015.

"No troops in the universe" "From George Washington to John Robinson, 25 October 1757," *Founders Online,* National Archives, http://founders.archives.gov/documents/Washington/02-05-02-0017, last modified December 30, 2015.

From Beverley Robinson "To George Washington from Beverley Robinson, 8 August 1757," *Founders Online,* National Archives, http://founders.archives.gov/documents/Washington/02-04-02-0239, last modified December 30, 2015. Robinson updated Washington on the status of the fleet before the attack began.

With five thousand French soldiers Ibid., note 1.

dug trenches to approach Anderson, *The War That Made America,* 109–15. See also Anderson, *Crucible of War,* 194–99.

350 *the full honors of war* Anderson, *The War that Made America,* 111.

Washington's friend George Mercer "To George Washington from George Mercer, 17 August 1757," *Founders Online,* National Archives, http://founders.archives.gov/documents/Washington/02-04-02-0242, last modified December 30, 2015.

pleaded desperately to his ally Anderson, *Crucible of War,* 176–77.

351 *"[Y]ou have been frequently indulg'd"* "To George Washington from Robert Dinwiddie, 19 October 1757," *Founders Online,* National Archives, http://founders.archives.gov/documents/Washington/02-05-02-0010, last modified December 30, 2015.

it was not for a "party of pleasure" "From George Washington to Robert Dinwiddie, 24 October 1757," *Founders Online,* National Archives, http://founders.archives.gov/documents/Washington/02-05-02-0014, last modified December 30, 2015.

"But His Honor was pleased" "From George Washington to John Robinson, 25 October 1757," *Founders Online,* National Archives, http://founders.archives.gov/documents/Washington/02-05-02-0017, last modified December 30, 2015.

wrote Governor Dinwiddie on November 5 "From George Washington to Robert Dinwiddie, 5 November 1757," *Founders Online,* National Archives, http://founders.archives.gov/documents/Washington/02-05-02-0028, last modified December 30, 2015.

353 *recipes for comforting dishes* "From George Washington to Sarah Cary Fairfax, 15 November 1757," *Founders Online,* National Archives, http://founders.archives.gov/documents/Washington/02-05-02-0035, last modified March 30, 2017.

354 *"I have been reduced"* "From George Washington to John Stanwix, 4 March 1758," *Founders Online,* National Archives, http://founders.archives.gov/documents/Washington/02-05-02-0073, last modified March 30, 2017.

of what he called "Decay" Ibid.

I believe has receivd great Injury Ibid., last modified December 30, 2015. Also see footnotes to this letter for events leading to his quick turnaround.

CHAPTER TWENTY

355 *After a brief stay at the plantation* Freeman, 2:278.

Martha Dandridge Custis http://www.mountvernon.org/digital-encyclopedia/article/martha-washington/, accessed April 26, 2017.

made her the wealthiest widow Ibid.

356 *He almost surely knew of her* Freeman, 2:278.

probably had made her acquaintance See Freeman, 2: Appendix II-1, on the possible circumstances of the meeting of Washington and Martha Custis.

356 *George William, had sailed off* "To George Washington from George William Fairfax, 6 December 1757," *Founders Online,* National Archives, http://founders.archives.gov/documents/Washington/02-05-02-0042, last modified June 29, 2017.

seventeen thousand acres Freeman, 2:297.

aide-de-camp to the Duke of Marlborough For these and other details about Daniel Parke, see http://www.encyclopediavirginia.org/Parke_Daniel_1669-1710, accessed April 25, 2017.

delivered the news to Queen Anne Freeman, 2:280–82.

357 *far more than his Virginia estate* Ibid., 283.

But the lawsuits continued Ibid., 298–99.

358 *had an "agreeable" personality* http://www.mountvernon.org/george-washington/martha-washington/george-and-martha-washingtons-relationship/.

and dark hair From description based on her portrait. Chernow, 80.

news came across the Atlantic Freeman, 2:304–5. According to Freeman, the first news of the change in command reached New York on March 4 and was quickly carried along the coast.

359 *his post on the frontier* Ibid., 2:301. See also this letter, written "on the road to Winchester": "From George Washington to John Blair, 2 April 1758," *Founders Online,* National Archives, http://founders.archives.gov/documents/Washington/02-05-02-0080, last modified December 30, 2015.

a few days after this second visit Freeman, 2:302.

361 *two thousand British Army regulars* Anderson, *Crucible of War,* 236.

the powerful British Royal Navy Ibid., 212–13.

". . . I should be extreamly sorry" "From George Washington to John Forbes, 23 April 1758," *Founders Online,* National Archives, note 1, http://founders.archives.gov/documents/Washington/02-05-02-0102, last modified December 30, 2015.

John Blair, serving as acting governor Freeman, 2:308. Also see http://www.encyclopediavirginia.org/Blair_John_ca_1687-1771, accessed April 26, 2017.

362 *wrote to . . . John Stanwix* "From George Washington to John Stanwix, 10 April 1758," *Founders Online,* National Archives, http://founders.archives.gov/documents/Washington/02-05-02-0087, last modified December 30, 2015.

Crown would reimburse the colonies Anderson, *Crucible of War,* 226, quoting William Pitt's letter to colonial authorities.

Connecticut authorizing five thousand men Ibid., 227.

These, the burgesses decided, would be split "From George Washington to John Blair, 9 April 1758," *Founders Online,* National Archives, note 3, http://founders.archives.gov/documents/Washington/02-05-02-0085, last modified December 30, 2015.

William Byrd III, an aristocratic member Anderson, *Crucible of War,* 230.

362 *Washington feared they would drift away* "From George Washington to John Blair, 9 April 1758," *Founders Online,* National Archives, http://founders.archives.gov/documents/Washington/02-05-02-0085, last modified December 30, 2015.

enough presents with which to reward them "From George Washington to John Stanwix, 10 April 1758," *Founders Online,* National Archives, http://founders.archives.gov/documents/Washington/02-05-02-0087, last modified December 30, 2015.

363 *who served as secretary to the general* Freeman, 2:327.

"All is lost by Heavens!" "From George Washington to Francis Halkett, 2 August 1758," *Founders Online,* National Archives, http://founders.archives.gov/documents/Washington/02-05-02-0284, last modified December 30, 2015. See also note 2 to this letter for brief overview of Washington's attempts to convince Forbes to take Braddock Road.

General Forbes's field commander Anderson, *Crucible of War,* 268.

the seventy-mile-long ridge known as Laurel Hill "From George Washington to Francis Halkett, 2 August 1758," *Founders Online,* National Archives, http://founders.archives.gov/documents/Washington/02-05-02-0284, last modified December 30, 2015.

364 *he made certain to announce* Freeman, 2:317.

"show your face" Ibid., 318.

365 *sought the private advice of those* Ibid., 319.

"Col. Bouquet was at first in a great dillemma" "To George Washington from Adam Stephen, 19 July 1758," *Founders Online,* National Archives, http://founders.archives.gov/documents/Washington/02-05-02-0244, last modified March 30, 2017.

"I hope no exception were taken" "From George Washington to James Wood, 28 July 1758," *Founders Online,* National Archives, http://founders.archives.gov/documents/Washington/02-05-02-0278, last modified December 30, 2015.

366 *a schoolteacher's annual salary* For money and price comparisons then and now, see http://www.history.org/foundation/journal/summer02money2.cfm, accessed April 28, 2017.

"I am now at the bottom of their scheme" John Forbes to Henry Bouquet, August 9, 1758. Freeman, 2:328–29.

367 *known to the British as Ticonderoga* Anderson, *Crucible of War,* 240–41.

General James Abercromby Ibid., 233.

Known as "Granny" to his men Ibid., 241.

"an aged gentleman" Parkman, 89.

bush warfare techniques Ibid., 90.

former counterfeiter, Robert Rogers See http://www.newenglandhistoricalsociety.com/george-howe-british-general-might-changed-history/.

"No women follow the camp" Parkman, 90.

367 *entirely covered the lake's surface* Ibid., 92.

368 *"the soul of General Abercromby's army"* Ibid., 97.

like a forest flattened by a hurricane Ibid., 101.

about thirty-five hundred French defenders Anderson, *Crucible of War,* 241.

the sunny, warm afternoon of July 8 Ibid., 243.

369 *"Cut Down Like Grass"* Ibid., 244.

". . . I could hear the men screaming" Ibid., 244–45, quoting David Perry, "Recollections of an Old Soldier . . . Written by Himself," *Magazine of History* 137, 1928, 9–10.

by an admiring Montcalm's counting Parkman, 106–7.

1,944 officers and men killed Ibid., 110.

370 *They jumped into their boats* Anderson, *Crucible of War,* 246.

"to any blockhead" Ibid., 247–48.

Dunfermline, home of Scottish kings http://publications.fifedirect.org.uk/c64_PP_History.pdf, accessed April 30, 2017.

371 *claimed it ruled over the tribes* See Anderson, *Crucible of War,* 15–18, for explanation of dynamics of Iroquois, Shawnee, Delaware, Mingoes, and the Ohio Valley.

In return, he received promises Ibid., 207.

He returned to Philadelphia Ibid., 270.

372 *Christian Frederick Post* http://explorepahistory.com/hmarker.php?markerId=1-A-21B, accessed May 1, 2017.

message to follow Christ's living example http://www.moravian.org/the-moravian-church/history/, accessed May 2, 2017.

"[Post] now accepted his terrible mission" Parkman, 144.

Ohio wilderness toward the end of July Freeman, 2:354.

373 *"We are still Incampd here"* "From George Washington to John Robinson, 1 September 1758," *Founders Online,* National Archives, http://founders.archives.gov/documents/Washington/02-05-02-0351, last modified December 30, 2015.

374 *General Forbes had told the two* Forbes, 52.

kept up his complaints "From George Washington to Francis Fauquier, 25 September 1758," *Founders Online,* National Archives, http://founders.archives.gov/documents/Washington/02-06-02-0034-0001, last modified December 30, 2015.

earned him the description "obstinate" Freeman, 2:335 and 335n.

a shameful scheme Freeman, 2:328–29, quoting John Forbes to Henry Bouquet, August 9, 1758.

376 *"headlong to grasp at a name"* Quoted in ibid., 347.

admitted to a "repugnance" Ibid., 342.

establishing a forward base Anderson, *Crucible of War,* 272.

talked Colonel Bouquet into his plan Freeman, 2:340–49.

great faith in the superiority Ibid., 342.

377 *vanquish the enemy Indians* Ibid., 340.
Highlanders carrying shortened muskets For detailed description of 78th Highlander dress (assuming 77nd is similar), see http://www.militaryheritage.com/78thregt.htm, accessed May 3, 2017.
Major Grant set it in motion Most details of Major Grant's attack are from Freeman, 2:340–49.
378 *"I thought we had nothing to fear"* Ibid., 345.
"My heart is broke" Forbes, 50.

CHAPTER TWENTY-ONE

379 *now went by the name William* Freeman, 2:335.
"In silence I now express my Joy" "From George Washington to Sarah Cary Fairfax, 12 September 1758," *Founders Online,* National Archives, http://founders.archives.gov/documents/Washington/02-06-02-0013, last modified March 30, 2017.
380 *twice in the spring* Ibid., note 3.
382 *"Do we still misunderstand"* "From George Washington to Sarah Cary Fairfax, 25 September 1758," *Founders Online,* National Archives, http://founders.archives.gov/documents/Washington/02-06-02-0033, last modified December 30, 2015.
his seven hundred and fifty troops Anderson, *Crucible of War,* 272.
"that does not rather Envy" "From George Washington to Sarah Cary Fairfax, 25 September 1758," *Founders Online,* National Archives, http://founders.archives.gov/documents/Washington/02-06-02-0033, last modified December 30, 2015.
383 *"O Marcia, let me hope"* Ibid., note 5.
384 *"If the weather does not favor"* Freeman, 2:353. Forbes wrote of the need for some dry days, "God grant them soon."
385 *when their pay ran out* Ibid., 2:355–56.
"[Time is] a thing at present" Ibid., 2:355, quoting John Forbes to William Pitt, October 20–27, 1758.
a body of two hundred French and Indians "Orderly Book, 12 November 1758," *Founders Online,* National Archives, note 1, http://founders.archives.gov/documents/Washington/02-06-02-0106, last modified December 30, 2015. This gives several accounts of the event. Also see Freeman, 2:357–58.
party of five hundred Virginian troops Freeman, 2:357. Washington's memory, many years later, was that he led the second group to relieve the first.
Washington's party advanced For an account of the friendly fire incident, also see *Maryland Gazette,* December 7, 1758.
386 *Then a figure ran into the open* This is the recollection of Washington's fellow officer Thomas Bullitt. For Washington's full account and other

accounts of the "friendly fire" incident, see "Orderly Book, 12 November 1758," *Founders Online,* National Archives, note 1, http://founders.archives .gov/documents/Washington/02-06-02-0106, last modified March 30, 2017.

386 *With the blade of his sword* Washington, *George Washington Remembers,* 23.

387 *It appears in none* Freeman, 2:358.

"to the Ground where the Skirmish was" "Orderly Book, 12 November 1758," *Founders Online,* National Archives, http://founders.archives.gov/documents /Washington/02-06-02-0106, last modified March 30, 2017.

388 *The tally of the dead* "Orderly Book, 13 November 1758," *Founders Online,* National Archives, http://founders.archives.gov/documents/Washington/02-06 -02-0107, last modified March 30, 2017.

Under threat of torture and death Freeman, 2:358–59.

389 *divided them into three brigades* Ibid., 359.

"quite as feeble as a child" Anderson, *Crucible of War,* 271–72.

banned women from the march "Orderly Book, 15 November 1758," *Founders Online,* National Archives, http://founders.archives.gov/documents/Wash ington/02-06-02-0110, last modified March 28, 2016.

390 *Washington's men cleared six miles of road* "From George Washington to John Forbes, 15 November 1758," *Founders Online,* National Archives, http:// founders.archives.gov/documents/Washington/02-06-02-0111, last modified March 28, 2016.

with one man injured by a falling tree "From George Washington to John Forbes, 16 November 1758," *Founders Online,* National Archives, http:// founders.archives.gov/documents/Washington/02-06-02-0112, last modified June 29, 2017.

making another six miles before nightfall "From George Washington to John Forbes, 17 November 1758," *Founders Online,* National Archives, http:// founders.archives.gov/documents/Washington/02-06-02-0116, last modified March 28, 2016.

They paused to reconnoiter See n196 in Freeman, 2:363. Letter from Bouquet describes stopping twelve miles from Fort Duquesne on the 23rd and halting on the 24th to gain intelligence.

The dogs had to be tied up "Orderly Book, 24 November 1758," *Founders Online,* National Archives, http://founders.archives.gov/documents/Wash ington/02-06-02-0132, last modified March 28, 2016.

If the silent, stalking enemy "Orderly Book, 23 November 1758," *Founders Online,* National Archives, http://founders.archives.gov/documents/Wash ington/02-06-02-0131, last modified March 28, 2016.

391 *tall red-painted poles* *Post Journal,* in Thomson, 133.

"The Heavens alone were our Covering" Ibid., 133.

"Therefore I was resolved to go forward" Ibid., 131–32.

"I prayed the Lord to blind them" Ibid., 134.

392 *his two marriages to Indian women* For a brief biography of Christian Frederick Post, see http://explorepahistory.com/hmarker.php?markerId=1-A-21B, accessed May 10, 2017.

393 *three hundred French officers* *Post Journal,* in Thomson, 149.

394 *"And by this belt I take you by the Hand"* Ibid., 148.

395 *"[W]e have great Reason to believe"* Ibid., 153.

396 *A voice shouted from the woods* Password for scouts arriving November 24 in camp. "Orderly Book, 24 November 1758," *Founders Online,* National Archives, http://founders.archives.gov/documents/Washington/02-06-02-0132, last modified March 28, 2016.

397 *thump of an explosion* *Maryland Gazette,* December 21, 1758.
now reasserted their dominance Anderson, *Crucible of War,* 276.

398 *"I sit there as a Bird on a Bough"* Ibid., 277.
the Penn family Ibid., 278.
"Thus the form of Iroquois dominance" Ibid., 278.
"[R]eturn forthwith to your Towns" Ibid., 280.

399 *arrived at the Forks* *Maryland Gazette,* December 21, 1758.
camped on a midriver island Freeman, 2:365.
another paddled canoes up the Allegheny Anderson, *Crucible of War,* 283.

400 *twenty-six-year-old Colonel Washington* Freeman, 2:364–65, suggests that there may have been some disappointment among the troops and Washington.

EPILOGUE

403 *General Forbes assembled a procession* Freeman, 2:367.

404 *arrived at Belvoir Manor in time* "From George Washington to Francis Fauquier, 9 December 1758," *Founders Online,* National Archives, note 1, http://founders.archives.gov/documents/Washington/02-06-02-0138, last modified March 30, 2017.
"In you we place the most implicit confidence" See: "Address from the Officers of the Virginia Regiment, 31 December 1758," *Founders Online,* National Archives, http://founders.archives.gov/documents/Washington/02-06-02-0147, last modified March 28, 2016.
"I must esteem [your approval]" "From George Washington to the Officers of the Virginia Regiment, 10 January 1759," *Founders Online,* National Archives, http://founders.archives.gov/documents/Washington/02-06-02-0152, last modified March 28, 2016.

405 *immediately achieved membership* Ellis, 40.
Probably sterile Ibid., 42.
as a vestryman in the local parish Longmore, 86–87.

406 *The crucial bit of intelligence* Anderson, *Crucible of War,* 352, citing Stobo's own memoir.

"The paths of glory lead but to the grave" Parkman, 2:285–86. Parkman quotes a professor of natural philosophy at the University of Edinburgh, who was a young midshipman at the time, in the same boat as Wolfe.

407 *"They run; see how they run!"* Ibid., 297.

"It's nothing, it's nothing" Ibid., 297, quoting an account handed down from an eyewitness.

408 *exerting himself in his toilet closet* Anderson, *Crucible of War,* 419–20, citing Horace Walpole account.

He had arranged for a "pipe" "From George Washington to Robert Cary & Company, 30 September 1762," *Founders Online,* National Archives, http://founders.archives.gov/documents/Washington/02-07-02-0099, last modified March 30, 2017.

409 *such as growing wheat* Ferling, 76.

"drive us away and settle the Country" *Post Journal,* in Thomson, 153.

constantly complaining of shoddy goods Chernow, 138.

410 *an "ungrateful and dirty" fellow* "From George Washington to George Muse, 29 January 1774," *Founders Online,* National Archives, http://founders.archives.gov/documents/Washington/02-09-02-0344, last modified June 29, 2017.

"the distinguishing Characteristick" From Patrick Henry's Virginia Resolves, in May 1765, Third Resolve, http://www.encyclopediavirginia.org/Virginia_Resolves_on_the_Stamp_Act_1765, accessed May 18, 2017.

that spring's wheat planting Ferling, 67.

his attitude against the British See Longmore, 76–81, for detailed analysis of Washington's letter to Robert Cary, September 20, 1765, and what it meant in terms of his evolving thought.

"unconstitutional method of Taxation" "From George Washington to Robert Cary & Company, 20 September 1765," *Founders Online,* National Archives, http://founders.archives.gov/documents/Washington/02-07-02-0252-0001, last modified March 28, 2016.

personal economic independence Longmore, 82–85.

411 *"At a time when our lordly Masters"* "From George Washington to George Mason, 5 April 1769," *Founders Online,* National Archives, http://founders.archives.gov/documents/Washington/02-08-02-0132, last modified March 28, 2016.

he displayed a deeper disaffection Ferling, 68.

412 *No representative from America* Ibid., 70.

415 *highly protective of what he did possess* See Ellis, 46–47, for examples of how "he was excessively and conspicuously assiduous in the defense of his own interests, especially when he suspected he was being cheated out of money or land."

416 *"to the top tier of Virginia's planter class"* Ibid., 40.

"our lordly Masters in Great Britain" "From George Washington to George Mason, 5 April 1769," *Founders Online,* National Archives, http://founders.archives.gov/documents/Washington/02-08-02-0132, last modified March 30, 2017.

417 *"This was the moment when Washington"* Ibid.

POSTSCRIPT

420 *his luxurious four-wheeled chariot* For more on Washington's chariot, see http://www.history.org/history/teaching/enewsletter/april03/iotm.cfm, accessed September 14, 2017. See also Chernow, 182, and Longmore, 158.

a martial spirit had erupted Longmore, 158.

Washington was named to head committees Ibid., 161–62.

421 *"He is a complete gentleman"* Chernow, 185–86.

"He possessed the gift of silence" Ibid., 185.

"Dignity with ease" Longmore, 182.

"[H]e has so much martial dignity" Ibid., 182.

Washington stood by his chair Chernow, 186–87.

BIBLIOGRAPHY

Abbot, W. W., ed. *The Papers of George Washington.* 10 vols. Colonial Series. Charlottesville, VA: University Press of Virginia, 1983–1995. (Volumes 7–10, Abbot, W. W., and Dorothy Twohig, eds.)

Albert, George Dallas. *The Frontier Forts of Western Pennsylvania.* Harrisburg, PA: Wm. Stanley Ray, State Printer, 1916.

Alberts, Robert. *A Charming Field for an Encounter: The Story of George Washington's Fort Necessity.* Washington, D.C.: National Park Service, 1975.

Amber, Charles. *George Washington and the West.* Chapel Hill, NC: University of North Carolina Press, 1936.

Anderson, Fred. *Crucible of War: The Seven Years' War and the Fate of Empire in British North America, 1754–1766.* New York: Vintage Books, 2001.

———. *The War That Made America: A Short History of the French and Indian War.* New York: Penguin, 2006.

Axelrod, Alan. *Blooding at Great Meadows: Young George Washington and the Battle that Shaped the Man.* Philadelphia: Running Press, 2007.

Bailey, Kenneth P. *The Ohio Company of Virginia and the Westward Movement 1748–1792: A Chapter in the History of the Colonial Frontier.* Glendale, CA: Arthur H. Clark, 1939.

Baker, Norman L. *Fort Loudoun: Washington's Fort in Virginia.* Winchester, VA: French and Indian War Foundation, 2006.

Baker-Crothers, Hayes, and Ruth Allison Hudnut. "A Private Soldier's Account of Washington's First Battles in the West: A Study in Historical Criticism." *Journal of Southern History* 8, no. 1 (February 1942): 23–62.

Beauchamp, William M. "Wampum and Shell Articles Used by the New York Indians." *Bulletin of the New York State Museum* 8, no. 41 (February 1941).

Beaujeu, Monongahela de. *The Hero of the Monongahela: Historical Sketch.* Translated by the Reverend G. E. Hawes. New York: William Post, 1913.

Becker, Marshall Joseph. "Matchcoats: Cultural Conservatism and Change." *Ethnohistory* 52, no. 4 (Fall 2005): 727–87.

Besant, Sir Walter. *London in the Eighteenth Century.* London: Adam & Charles Black, 1902.

Bonin, Jolicoeur Charles. *Memoir of a French and Indian War Soldier.* Edited by Andrew Gallup. Bowie, MD: Heritage Books, 1993.

Boorstin, Daniel. *The Americans: The Colonial Experience.* New York: Vintage Books, 1958.

Borneman, Walter R. *The French and Indian War: Deciding the Fate of North America.* New York: Harper Perennial, 2007.

Bouton, Terry. *Taming Democracy: "The People," the Founders, and the Troubled Ending of the American Revolution.* Oxford: Oxford University Press, 2009.

Brumwell, Stephen. *Redcoats: The British Soldier and War in the Americas, 1755–1763.* Cambridge: Cambridge University Press, 2006.

Buck, Solon J., and Elizabeth Hawthorn Buck. *The Planting of Civilization in Western Pennsylvania.* Pittsburgh: University of Pittsburgh Press, 1939.

Bullitt, Thomas Walker. *My Life at Oxmoor: Life on a Farm in Kentucky before the War.* Louisville, KY: John P. Morton, 1911.

Burnaby, Andrew. *Burnaby's Travels Through North America.* Reprinted from the third edition of 1798. New York: A. Wessels, 1904.

Carter, Thomas. *Historical Record of the Forty-Fourth, or the East Essex Regiment of Foot.* London: W. O. Mitchell, 1864.

Cary, Wilson Miles. *Sally Cary: A Long Hidden Romance of Washington's Life.* New York: De Vinne Press, 1916.

Cavendish, Richard. "Coronation of Queen Anne." *History Today* 52, no. 4 (April 2002).

Chernow, Ron. *Washington: A Life.* New York: Penguin Books, 2010.

Clary, David. *George Washington's First War: His Early Military Adventures.* New York: Simon & Schuster, 2011.

Cleland, Hugh. *George Washington in the Ohio Valley.* Pittsburgh: University of Pittsburgh Press, 1955.

Colonial Records of Pennsylvania, VI, April 2, 1754–January 29, 1756, https://archive.org/stream/colonialrecordsov6harr/colonialrecordsov6harr_djvu.txt

Contrecoeur, Dumas, and other French accounts. *Memorial Containing a Summary View of the Facts with Their Authorities, in Answer to the Observations Sent by the English Ministry to the Courts of Europe.* Translated from the French. Philadelphia: James Chattin, 1757.

Cruickshank, Dan. *The Secret History of Georgian London: How the Wages of Sin Shaped the Capital.* New York: Random House, 2009.

Cubbison, Douglas R. *On Campaign Against Fort Duquesne: The Braddock and Forbes Expeditions, 1755–1758, Through the Experiences of Quartermaster Sir John St. Clair.* Jefferson, NC: McFarland, 2015.

Dictionary of Canadian Biography. http://www.biographi.ca/en/index.php.

Dinwiddie, Robert. *The Official Records of Robert Dinwiddie, Lieutenant-Governor of the Colony of Virginia, 1751–1758.* Richmond, VA: Virginia Historical Society, 1884.

Dixon, David. "A High Wind Rising: George Washington, Fort Necessity, and the Ohio Country Indians." *Pennsylvania History* 74, no. 3 (Summer 2007): 333–53.

Eastern Delaware Nations. http://www.easterndelawarenations.org/history.html.

Eastman, Mary H. *American Aboriginal Portfolio*. Philadelphia: Lippincott, Grambo, 1853.

Ellis, Joseph J. *His Excellency: George Washington*. New York: Knopf, 2003.

Encyclopaedia Britannica. https://www.britannica.com/.

Encyclopedia Virginia. https://www.encyclopediavirginia.org/.

Evans, Emory G. *A "Topping People": The Rise and Decline of Virginia's Old Political Elite, 1680–1790*. Charlottesville, VA: University of Virginia Press, 2009.

Ferling, John. *The Ascent of George Washington*. New York: Bloomsbury Press, 2009.

Fitzpatrick, John C. *The George Washington Scandals*. Alexandria, VA: Washington Society of Alexandria, 1929.

———, ed. *The Writings of George Washington from the Original Manuscript Sources*. 39 vols. Washington, D.C.: Government Printing Office, 1931–1944.

Fleming, Thomas. "George Washington in Love." *American Heritage*, Fall 2009.

Flexner, James Thomas. *George Washington: The Forge of Experience (1732–1775)*. Boston: Little, Brown, 1965.

Forbes, John. *Letters of General John Forbes Relating to the Expedition Against Fort Duquesne in 1758*. Compiled by Irene Stewart. Pittsburgh: Allegheny County Committee, 1927.

Founders Online. National Archives. https://founders.archives.gov/.

Franklin, Benjamin. *The Autobiography of Benjamin Franklin*. Edited by Jared Sparks. London: George Bell & Sons, 1884.

Freeman, Douglas Southall. *George Washington: A Biography*. 7 vols. New York: Scribner's, 1948.

Fur, Gunlög. *A Nation of Women: Gender and Colonial Encounters Among the Delaware Indians*. Philadelphia: University of Pennsylvania Press, 2009.

George Washington's Mount Vernon. http://www.mountvernon.org/.

Gist, Christopher. *Christopher Gist's Journals, with Historical, Geographical and Ethnological Notes and Biographies of His Contemporaries*. Edited by William M. Darlington. Pittsburgh: J. R. Weldin, 1893.

Glover, Lori. *Founders as Fathers: The Private Lives and Politics of the American Revolutionaries*. New Haven, CT: Yale University Press, 2014.

Grant, James. *British Battles on Land and Sea*. London: Cassell, [n.d.].

Hamilton, Charles, ed. *Braddock's Defeat: The Journal of Captain Robert Cholmley's Batman, The Journal of a British Officer, Halkett's Orderly Book*. Norman, OK: University of Oklahoma Press, 1959.

Harden, William. "James Mackay, of Strathy Hall, Comrade in Arms of George Washington." *Georgia Historical Quarterly* 1, no. 2 (June 1917): 77–98.

Harrington, J. C. "Fort Necessity: Scene of George Washington's First Battle." *Journal of the Society of Architectural Historians* 13, no. 2 (May 1954): 25–27.

Henriques, Peter R. "Major Lawrence Washington versus the Reverend Charles Green: A Case Study of the Squire and the Parson." *Virginia Magazine of History and Biography* 100, no. 2 (April 1992): 233–64.

Hildeburn, Charles. "Sir John St. Clair, Baronet: Quarter-Master General in America, 1755–1767." *Pennsylvania Magazine of History and Biography* 9, no. 1 (April 1885): 1–14.

Hoppin, Charles Arthur. *The Washington Ancestry, and Records of the McClain, Johnson, and Forty Other Colonial American Families: Prepared for Edward Lee McClain.* Greenfield, OH: privately printed, 1932.

Houston, Alan. "Benjamin Franklin and the 'Wagon Affair' of 1755." *The William and Mary Quarterly,* 3rd series, 66, no. 2 (April 2009): 235–86.

Hunter, William A. *Forts on the Pennsylvania Frontier.* Harrisburg, PA: Pennsylvania Historical and Museum Commission, 1960.

"Indians of the Eastern Seaboard." Washington, D.C.: Bureau of Indian Affairs, 1967. http://files.eric.ed.gov/fulltext/ED028871.pdf, retrieved October 19, 2017.

Innes, Colonel James. "Account of the Battle of Fort Necessity." *Maryland Gazette,* August 1, 1754.

Internet Encyclopedia of Philosophy. http://www.iep.utm.edu/.

Jackson, Donald, ed. *The Diaries of George Washington.* Vol. 1, *11 March 1748—13 November 1765.* Charlottesville, VA: University Press of Virginia, 1976.

Jennings, Francis. *Empire of Fortune: Crowns, Colonies & Tribes in the Seven Years War in America.* New York: W. W. Norton, 1988.

Jones, Colin. *Paris: Biography of a City.* New York: Viking, 2005.

Kent, Donald H., ed. "Contrecoeur's Copy of George Washington's Journal for 1754." *Pennsylvania History: A Journal of Mid-Atlantic Studies* 19, no. 1 (January 1952): 1–32.

———. *The French Invasion of Western Pennsylvania 1753.* Harrisburg, PA: Pennsylvania Historical and Museum Commission, 1954.

Knollenberg, Bernhard. *George Washington: The Virginia Period, 1732–1755.* Durham, NC: Duke University Press, 1964.

Kopperman, Paul E. *Braddock at the Monongahela.* Pittsburgh: University of Pittsburgh Press, 1977.

La Force, Melanie, and Norman La Force. "Michel Pepin *dit* La Force and John Carlyle: A Link to the Seven Years' War." *Carlyle House Docent Dispatch* (April 2007).

Le Roy, Marie, and Barbara Leininger. "Three Years Captives Among the Indians." *Pennsylvania Magazine of History and Biography* 29, no. 4 (1905): 407–20.

Lengel, Edward G. *Inventing George Washington: America's Founder in Myth and Memory.* New York: Harper, 2011.

Lewis, Thomas A. *For King and Country: The Maturing of George Washington, 1748–1760.* New York: HarperCollins, 1993.

Locke, John. *Two Treatises on Government.* Edited by Peter Laslett. New York: Mentor Books, 1965.

Longmore, Paul K. *The Invention of George Washington.* Charlottesville, VA: University Press of Virginia, 1999.

Macomber, Walter M. *Raleigh Tavern Architectural Report, Block 17 Building 6A Lot 54.* Originally titled *Raleigh Tavern.* Colonial Williamsburg Foundation Library Research Report Series 1348. Williamsburg, VA: Colonial Williamsburg Foundation Library, 1990.

The Manual Exercise as Ordered by His Majesty, in the Year 1764. Together with Plans and Explanations of the Method generally Practiced at Reviews and Field-Days. Philadelphia: J. Humphreys, R. Bell, and R. Aitken, 1776.

Marshal, Henry, F.R.S.E. *Military Miscellany; Comprehending a History of the Recruiting of the Army, Military Punishments, etc. etc.* London: John Murray, 1846.

Marston, Daniel P. "Swift and Bold: The 60th Regiment and Warfare in North America, 1755–1765," master's thesis, history department, McGill University, Montreal, March 1997.

McCardell, Lee. *Ill-Starred General: Braddock of the Coldstream Guards.* Pittsburgh: University of Pittsburgh Press, 1986.

Merrell, James H. *Into the American Woods: Negotiations on the Pennsylvania Frontier.* New York: W. W. Norton, 2000.

Middlekauff, Robert. *Washington's Revolution: The Making of America's First Leader.* New York: Knopf, 2015.

Moreau, Jacob-Nicolas. *Memorial Containing a Summary View of Facts with Their Authorities, in Answer to the Observations Sent by the English Ministry to the Courts of Europe.* Philadelphia: James Chattin, 1757.

Mulkearn, Lois. "Half King, Seneca Diplomat of the Ohio Valley." *Western Pennsylvania Historical Magazine* 37, no. 2 (Summer 1954): 65–66.

Neely, Sylvia. "Mason Locke Weems's 'Life of George Washington' and the Myth of Braddock's Defeat." *Virginia Magazine of History and Biography* 107, no. 1 (Winter 1999): 45–72.

Neill, Edward D., et al. "The Ancestry and Early Life of George Washington." *Pennsylvania Magazine of History and Biography* 16, no. 3 (October 1892): 261–68.

Nute, Grace Lee. *The Voyageur.* New York: D. Appleton, 1931.

O'Connell, Sheila, et. al. *London 1753.* Boston: David R. Godine, 2003.

Osgood, Herbert L. "The Proprietary Province as a Form of Colonial Government." *American Historical Review* 2, no. 4 (July 1897): 644–64.

Pargellis, Stanley, ed. *Military Affairs in North America 1748–1765: Selected Documents from the Cumberland Papers in Windsor Castle.* New York: D. Appleton-Century, 1936.

Parkman, Francis. *Montcalm and Wolfe: France and England in North America, Part Seventh.* 2 vols. Boston: Little, Brown, 1885.

Perles, Stephanie J., et al. *Condition of Vegetation Communities in Fort Necessity National Battlefield and Friendship Hill National Historic Site: Eastern Rivers and Mountains Network Summary Report 2007–2009.* Natural Resource Data Series NPS/ERMN/NRDS—2010/035. University Park, PA: National Park Service, Northeast Region, 2010. https://irma.nps.gov/DataStore/DownloadFile/154152, retrieved October 15, 2017.

Preston, David L. *Braddock's Defeat: The Battle of the Monongahela and the Road to Revolution.* Oxford: Oxford University Press, 2015.

Rousseau, Jean-Jacques. *The Confessions of Jean Jacques Rousseau: Complete.* London: Aldus Society, 1903.

———. "Discourse on the Arts and Sciences. A Discourse Which Won the Prize at the Academy of Dijon in 1750 on This Question Proposed by the Academy: Has the Restoration of the Arts and Sciences had a Purifying Effect upon Morals?" *International Relations and Security Network: Primary Sources in International Affairs.* https://www.files.ethz.ch/isn/125491/5018_Rousseau_Discourse_on_the_Arts_and_Sciences.pdf. Accessed October 19, 2017.

Sargent, Winthrop, ed. *The History of an Expedition Against Fort Du Quesne, in 1755; under Major-General Edward Braddock.* Philadelphia: Historical Society of Pennsylvania, 1856.

Schoolcraft, Henry R. *The Red Race of America.* New York: Wm. H. Graham, 1847.

Schwartz, Barry. *George Washington: The Making of an American Symbol.* New York: Free Press, 1987.

Sipe, C. Hale. "The Principal Indian Towns of Western Pennsylvania." *Western Pennsylvania Historical Magazine* 13, no. 2 (April 1930): 104–22.

Smith, H. Clifford. *Sulgrave Manor and the Washingtons: A History and Guide to the Tudor Home of George Washington's Ancestors.* New York: MacMillan, 1933.

Smith, James. *An Account of the Remarkable Occurrences in the Life and Travels of Col. James Smith, (Now a Citizen of Bourbon County, Kentucky,) During His Captivity with the Indians, in the years 1755, '56, '57, '58, & '59.* Lexington, KY: John Bradford, 1799.

Sparks, Jared. *The Life of George Washington.* Boston: Little, Brown, 1853.

Starobinksi, Jean. "Rousseau and Revolution." *New York Review of Books,* April 25, 2002. http://www.nybooks.com/articles/2002/04/25/rousseau-and-revolution/. Accessed October 19, 2017.

Stephen, Adam. "The Ohio Expedition of 1754." *Pennsylvania Magazine of History and Biography* 18 (1894): 43–51.

———. "Stephen's *Life* Written for B. Rush in 1775." *Pennsylvania Magazine of History and Biography* 8 (1894): 43–50.

Stewart, Irene, ed. *Letters of General John Forbes Relating to the Expedition Against Fort Duquesne in 1758.* Pittsburgh: Allegheny County Committee, 1927.

Stromberg, Roland. *An Intellectual History of Modern Europe.* New York: Appleton-Century-Crofts, 1966.

Thomson, Charles. *An Enquiry into the Causes of the Alienation of the Delaware and Shawanese Indians from the British Interest, and into the Measures Taken for Recovering Their Friendship*. London: J. Wilkie, 1759.

Thwaites, Reuben Gold, ed. *Early Western Travels, 1748–1846*. New York: AMS Press, 1966.

Tilberg, Frederick. "Washington's Stockade at Fort Necessity." *Pennsylvania History* 20, no. 3 (July 1953): 240–57.

Trudel, Marcel. "The Jumonville Affair." Translated by Donald H. Kent. *Pennsylvania History* 21, no. 4 (October 1954): 351–81.

Volwiller, Albert T. *George Croghan and the Westward Movement, 1741–1782*. Cleveland, OH: Arthur H. Clark, 1926.

Wahll, Andrew J., ed. *Braddock Road Chronicles: Compiled from the Diaries and Records of Members of the Braddock Expedition and Others Arranged in a Day by Day Chronology*. Westminster, MD: Heritage Books, 2006.

Wallace, Nesbit Willoughby. *A Regimental Chronicle and List of Officers of the 60th, or King's Royal Rifle Corps, Formerly the 62nd, or the Royal American Regiment of Foot*. London: Harrison, 1879.

Wallace, Paul A. W. *Indian Paths of Pennsylvania*. Harrisburg, PA: Pennsylvania Historical and Museum Commission, 2005.

———. *Indians in Pennsylvania*. Harrisburg, PA: Pennsylvania Historical and Museum Commission, 1970.

Walne, Peter. "George Washington and the Fairfax Family: Some New Documents." *Virginia Magazine of History and Biography* 77, no. 4 (October 1969): 441–63.

Washington, George. *George Washington Remembers: Reflections on the French and Indian War*. Edited by Fred Anderson. Lanham, MD: Rowan & Littlefield, 2004.

———. [GW Journal of 1754 captured by French]. See Kent, Donald H., ed. "Contrecoeur's Copy of George Washington's Journal for 1754." *Pennsylvania History: A Journal of Mid-Atlantic Studies* 19, no. 1 (January 1952): 1–32.

———. *Journal of My Journey Over the Mountains; While Surveying for Lord Thomas Fairfax, Baron of Cameron, in the Northern Neck of Virginia, Beyond the Blue Ridge, in 1747–8*. Edited by J. M. Toner, M.D. Albany, NY: Joel Munsell's Sons, 1892.

———. "Journey to the French Commandant (1753)." [1753 journey; originally published 1754] *The Journal of Major George Washington*. Edited by Paul Royster. Electronic Texts in American Studies, 33. Lincoln, NE: University of Nebraska, [n.d.]. http://digitalcommons.unl.edu/etas.

———. *The Washington Papers*. Richmond, VA: University of Virginia. http://gwpapers.virginia.edu/, retrieved October 15, 2017.

Weiser, Conrad. "Journal of Conrad Weiser." *Minutes of the Provincial Council of Pennsylvania, from the Organization to the Termination of the Proprietary Government* 6 (April 1754 to January 1756): 151–152; 2nd. ed. Harrisburg, PA: Theo. Fenn, 1851.

Weslager, C.A. *The Delaware Indians: A History.* New Brunswick, NJ: Rutgers University Press, 1972.

Wright, John W. "Sieges and Customs of War at the Opening of the Eighteenth Century." *American Historical Review* 39, no. 4 (July 1934): 629–44.

CREDITS

Young Washington's Sketch Map of His Wilderness Journey. Library of Congress.

The Washington Family Runs Aground in British Virginia. Edwin Remsberg/ Alamy Stock Photo.

George's Revered Older Brother Lawrence. Courtesy of Mount Vernon Ladies' Association.

George Washington's Boyhood Home at Ferry Farm. Image courtesy of L. H. Barker © 2017. All rights reserved.

Washington's Youthful Writings on Self-Improvement. Library of Congress.

Land Surveyor Instead of Sailor. Courtesy of Mount Vernon Ladies' Association.

On Barbados, Washington Glimpses a Wider World. Yale Center for British Art via Wikimedia Commons.

Longhouses of Delaware Indians. Courtesy of the New York Public Library Digital Collections.

The Hardships of a Winter Journey into the Wilds. Photographers of the La Salle Expedition II and Elgin Public Museum, Elgin, Illinois.

A Tidewater Virginia Plantation in the Eighteenth Century. Wikimedia Commons.

Young Washington's First Commander, Governor Robert Dinwiddie. National Portrait Gallery.

Fort Necessity, Young Washington's Makeshift Redoubt. The Boston Public Library via Digital Commonwealth.

The Surrender of Fort Necessity and Admission of Assassination. Archives of Montreal.

Virginia's Fanciful Claims to North America. Library of Congress.

Sally Fairfax, Young Washington's Infatuation. Rendering of a painting, now lost, as reproduced in Wilson Miles Cary's biography, *Sally Cary.*

Apollo Room of Raleigh Tavern. 1850s sketch of the Apollo Room shortly before its destruction, in Benson Lossing's *Pictorial Field-Book of the Revolution,* via Wikimedia Commons.

Belvoir Manor, Estate of the Fairfaxes. Granger.

General Edward Braddock, Commander of the British Wilderness Assault. Library of Congress.

Sir John St. Clair, "A Mad Sort of Fool." Sir John St. Clair by Allan Ramsay, 1754, Fort Ligonier, Ligonier, Pennsylvania.

British Grenadiers, Elite Strike Force of the Royal Army. Royal Collection Trust/© Her Majesty Queen Elizabeth II, 2017.

General Braddock's Advance Column. Library of Congress.

The Might of the British Empire Marches into the Wilds. Library of Congress.

Indian Scouts Watch Braddock Crossing. *Indian Scouts Watch Braddock Crossing,* painting by Robert Griffing.

The Wounding of General Braddock. *The Wounding of General Braddock,* painting by Robert Griffing.

Indian War Club. National Museums of Scotland.

Indian Warrior Scalping a Victim. *Costumes de Different Pays* by Jacques Grasset de Saint-Saveur Labrousse "Guerrier Iroquois." Alamy Images.

General Braddock's Red Silk Sash. Courtesy of Mount Vernon Ladies' Association.

French Fort Duquesne. Courtesy of the Smithsonian Libraries.

Washington's Frontier Post at Winchester and Fort Loudoun. Winchester and Fort Loudoun artist's rendering by Eric Cherry.

A Journey to Philadelphia. "An East Prospect of the City of Philadelphia from the Jersey Shore" by George Heap, 1752. Library of Congress.

Disciplining the Troops. Hulton Archive/Getty Images.

Layout of British Wilderness Camp. Library of Congress.

Washington Confesses His Love for Sally Fairfax. Susan Dwight Bliss autograph collection, Houghton Library, Harvard University.

Martha Washington. Bequest of Mary Custis Lee, Washington and Lee University, Lexington, Virginia.

From Wilderness to Mount Vernon. From engraving by Alexander Robertson and Francis Jukes (1800), via Alamy.com.

The Middle-Aged Washington in His Old French and Indian War Uniform. From original by Charles Willson Peale hanging at Washington and Lee University via Wikimedia Commons.

INDEX

Note: Page references in *italics* refer to maps; pages followed by "n" refer to footnotes.